Submolecular Biology and Cancer

The Ciba Foundation for the promotion of international cooperation in medical and chemical research is a scientific and educational charity established by CIBA Limited—now CIBA-GEIGY Limited—of Basle. The Foundation operates independently in London under English trust law.
Ciba Foundation Symposia are published in collaboration with Excerpta Medica in Amsterdam.

Excerpta Medica, P.O. Box 211, Amsterdam

Submolecular Biology and Cancer

Ciba Foundation Symposium 67 (new series)
In honour of Albert Szent-Györgyi on the occasion of his 85th birthday

1979

Excerpta Medica

Amsterdam · Oxford · New York

ISBN Excerpta Medica 90 219 4073 6
ISBN Elsevier/North-Holland 0 444 90078 0

Published in August 1979 by Excerpta Medica, P.O. Box 211, Amsterdam and Elsevier/North-Holland, Inc., 52 Vanderbilt Avenue, New York, N.Y. 10017.

Suggested series entry for library catalogues: Ciba Foundation Symposia.
Suggested publisher's entry for library catalogues: Excerpta Medica.

Ciba Foundation Symposium 67 (new series)
359 pages, 87 figures, 43 tables

Library of Congress Cataloging in Publication Data

Symposium on Submolecular Biology and Cancer, London, 1978.
Submolecular biology and cancer.

(Ciba Foundation symposium; 67 (new ser.)
Proceedings of the symposium held at the Ciba Foundation, London, Sept. 25-27th, 1978.
Bibliography: p.
Includes indexes.
1. Cancer-Research-Congresses. 2. Electron mobility-Congresses. 3. Glyoxals-Congresses. 4. Molecular biology-Congresses. I. Wolstenholme, Gordon E. W. II. Fitzsimons, D. W. III. Whelan, Julie. IV. Title. V. Series: Ciba Foundation. Symposium; new ser., 67.
RC267.S94 1978 616.9'94'00157488 79-14324

ISBN Elsevier/North-Holland 0-444-90078-0

Printed in The Netherlands by Casparie, Heerhugowaard

Contents

R.J.P. WILLIAMS Introduction 1

A. SZENT-GYÖRGYI The living state and cancer 3

S. RESLOVA-VASILUKOVA and R.J.P. WILLIAMS A note on cancer and possible relationships to submolecular biology 19

K. LAKI, S. SUHAI and J.C. KERTESZ Energy bands and charge transfer in proteins 33
Discussion 42

J.J. LADIK Electronic structure of proteins and DNA: solid-state aspects 51
Discussion 58

T.J. LEWIS The mechanism of conduction in proteins 65
Discussion 78

S. BONE and R. PETHIG Electronic and dielectric properties of protein–methylglyoxal complexes 83
Discussion 98

H.M. SWARTZ Free radicals in cancer 107
Discussion 124

R. DAMADIAN, L. MINKOFF and M. GOLDSMITH Whole-body nuclear magnetic resonance scanning: n.m.r. studies of tumour cells 131
Discussion 138

C. THOMSON and J.R. BALL Quantum chemical investigations of charge-transfer interactions in relation to the electronic theory of cancer 143
Discussion 160

G. FODOR, R. MUJUMDAR and J. BUTTERICK The search for new cancerostatic agents 165
Discussion 169

S.W. FOX, R.M. SYREN and C.R. WINDSOR Thermal copoly(amino acids) as inhibitors of glyoxalase I 175
Discussion 187

S.F. ABDULNUR Interactions of methylglyoxal with methylamine 195
Discussion 205

R.B. BRANDT and S.A. SIEGEL Methylglyoxal production in human blood 211
Discussion 220

E. SCHAUENSTEIN and H. ESTERBAUER Formation and properties of reactive aldehydes 225
Discussion 241

M.U. DIANZANI Biological activity of methylglyoxal and related aldehydes 245
Discussion 265

P.J. CONROY Carcinostatic activity of methylglyoxal and related substances in tumour-bearing mice 271
Discussion 298

T.F. SLATER Biochemical studies of transient intermediates in relation to chemical carcinogenesis 301
Discussion 321

General discussion 329

Index contributors 343

Subject index 345

Participants

Symposium on Submolecular Biology and Cancer, held at the Ciba Foundation, London, 25th–27th September 1978

Chairman: R.J.P. WILLIAMS Inorganic Chemistry Laboratory, University of Oxford, South Parks Road, Oxford OXI 3UD

S.F. ABDULNUR Research Professor (NFCR) Laboratory at College of Arts and Sciences, Department of Chemistry, The American University, Massachusetts and Nebraska Avenues, N.W., Washington D.C. 20016, USA

P.K. BONDY Veterans Administration Hospital, West Spring Street, West Haven, Connecticut 06516, USA

R.B. BRANDT Medical College of Virginia, Department of Biochemistry, Virginia Commonwealth University, Box 727 MCV Station, Richmond, Virginia 23298, USA

T.A. CONNORS MRC Toxicology Unit, Medical Research Council Laboratories, Woodmansterne Road, Carshalton, Surrey SM5 4EF, UK

P.J. CONROY Cancer Center of the University of Rochester Medical Center, 601 Elmwood Avenue, Box 704, Rochester, New York 14642, USA

M.J. CRUMPTON National Institute for Medical Research, The Ridgeway, Mill Hill, London NW7 IAA, UK

R. DAMADIAN Regional Director (NFCR) Biophysical Laboratory, Department of Medicine, State University of New York, Downstate Medical Center, 450 Clarkson Avenue, Brooklyn, New York 11203, USA

M.U. DIANZANI Regional Director (NFCR) Laboratory at Faculty of Medicine and Surgery, University of Turin, Corso Raffaello 30, 10125 Turin, Italy

D.D. ELEY Department of Chemistry, University of Nottingham, Nottingham NG7 2RD, UK

H. ESTERBAUER Institut für Biochemie der Universität Graz, A-8010 Graz, Halbärthgasse 5/1, Austria

G. FODOR Regional Director (NFCR) Laboratory at College of Arts and Sciences, Department of Chemistry, West Virginia University, Morgantown, West Virginia 26506, USA

S.W. FOX* Regional Director (NFCR) Laboratory at The Institute for Molecular and Cellular Evolution, University of Miami, 521 Anastasia Avenue, Coral Gables, Florida 33134, USA

H.L. KORNBERG Department of Biochemistry, University of Cambridge, Tennis Court Road, Cambridge CB2 lQW, UK

J. J. LADIK Regional Director (NFCR) Laboratory at The Institut für Physikalische und Theoretische Chemie, Universität Erlangen-Nürnberg, D-8520 Erlangen—Egerlandstrasse 3, Federal Republic of Germany

K. LAKI Regional Director (NFCR) Laboratory at American University, Washington D.C., Section of Physical Biochemistry, National Institute on Arthritis, Metabolism and Digestive Diseases, National Institutes of Health (Bldg. 4, Bl-10), Bethesda, Maryland 20014, USA; and Regional Director (NFCR) Laboratory at the Medical University, Department of Clinical Chemistry, 4012 Debrecen, Hungary

T.J. LEWIS Regional Director (NFCR) Laboratory at The University College of North Wales, School of Electronic Engineering Science, Dean Street, Bangor, Gwynedd LL57 IUT, Wales, UK

N.H. MARCH Department of Theoretical Chemistry, University of Oxford, South Parks Road, Oxford OXl 3TG, UK

---

* Contributed *in absentia*.

A. NEUBERGER Department of Biochemistry, Charing Cross Hospital Medical School, Fulham Palace Road, London W6 8RF, UK

R. PETHIG Regional Director (NFCR) Laboratory at The University College of North Wales, School of Electronic Engineering Science, Dean Street, Gwynedd LL57 lUT, Wales, UK

D.C. PHILLIPS Department of Zoology, Laboratory of Molecular Biophysics, University of Oxford, South Parks Road, Oxford OXl 3PS, UK

T.F. SLATER Regional Director (NFCR) Laboratory at Brunel University Workshop, Department of Biochemistry, Uxbridge, Middlesex, UK

H.M. SWARTZ Department of Radiology and Biochemistry, The Medical College of Wisconsin, Milwaukee County Medical Complex, 8700 West Wisconsin Avenue, Milwaukee, Wisconsin 53226, USA

A. SZENT-GYÖRGYI Scientific Director, National Foundation for Cancer Research, PO Box 187, Woods Hole, Massachusetts 02453, USA

C. THOMSON Regional Director (NFCR) Laboratory at the University of St Andrews, Department of Chemistry, The Purdie Building, St Andrews KY16 9ST, Scotland, UK

Editors: G.E.W. WOLSTENHOLME *(Organizer),* D.W. FITZSIMONS and JULIE WHELAN

Editors' note

The editors are exceptionally grateful to Professor T.F. Slater for his expert and willing help in the preparation of this book for publication.

Introduction

R.J.P. WILLIAMS

Inorganic Chemistry Laboratory, University of Oxford

Any conference which has as its title the conjunction of two things, X and Y, must start with some attempt at a statement of what is X and what is Y and then go on to say why X and Y may have a connection. A book entitled Love and Sex has obvious sense in its title, but is there a similar relationship for Submolecular Biology and Cancer?

In this particular case no fruitful development of discussion could be made without looking at the man who first forced these words together. In the development of our thinking it is often a word or group of words without clear definition which catches the imagination and forces inspection. Words such as Liberty, Justice and Molecular Biology are clear examples. However, these words gather a peculiar strength only when spoken by very particular people. Here I must mention Professor Albert Szent-Györgyi, and not for the last time.

Throughout his working life, Professor Szent-Györgyi has loved the intellectual and imaginative approach to biological systems. His approach is not that of the strict analytical mind which gathers data painstakingly and then from the data produces summarizing theory. Rather, he stands back and says 'When I look at X and then look at Y, I sense that there might well be a connection. Now I shall look for that connection'. This is not strictly an intuitive approach since it has as a basis the accumulated knowledge of X and Y separately in the mind of the originator of the attack, and some coincidental factor has forced them into conjunction in his mind. Professor Szent-Györgyi forces us today, as he has at other times, to ask about the conjunction of a particular X and Y, here Submolecular Biology and Cancer.

At the beginning it will pay us to seek some clarification of terms. I take submolecular biology to refer to those chemical systems in biology at a lower level than whole molecules. The smallest species of interest is then the electron and, judging from many of the papers in this volume, it is the properties of unpaired electrons in biology which are to concern us. Thus, the first papers are given over to the possible modes in which the electron may be found in biology in some small molecules, in proteins and in DNA. Later papers relate the function of aldehydes to the generation of these species. I think, if I dare to be so bold, that there are other molecular fragments which we may have to consider within our title where the electrons are still paired. I mean relatively stable cations and anions such as NO^+, Cl^+, CH_3^+, H^-, or CH_3^-; and there are, of course, stable triplet states. Again certain elements, notoriously transition metal elements, are scant believers in molecular chemistry based on paired electron systems. Frequently, the chemistry of iron or nickel is effectively that of free-radical chemistry, but is this submolecular? It is therefore important that each contributor clarifies concepts as he uses them, defining the nature of the submolecular system he wishes to discuss.

I am in equal difficulty with the language of cancer. The Shorter Oxford English Dictionary gives by way of definition of a cancer 'a malignant growth or tumour that tends to spread and reproduce itself'. It gives for malignant the close synonym 'virulent' and for tumour an 'abnormal swelling'. Thus cancer is defined not at the molecular level but only as a virulent abnormal swelling. There is in fact no clear approach yet to the biology of cancer at the molecular level, never mind at the submolecular level. If we refer to a transformed stem cell as one which will lead to a clinically overt cancer we do not know why it will do so, for we do not understand the processes which initiate the transformation nor the processes of growth which are associated with malignancy. Neither do we know the nature of the transformation which caused the single cell to become malignant. May I plead with you to state clearly, when discussing cancer at the cellular level, what molecular and/or cellular changes you are describing. Only then can we relate them to submolecular events.

We are clearly in danger of falling into very deep water when we try to put together the gross overall disease or group of diseases called cancer and any submolecular events such as those which involve the free electron. There are no clear stepping stones to help us to link the two concepts together. As I have said, it is at present a mental leap by Professor Szent-Györgyi which forces us to look at them in juxtaposition. I look forward to his introduction and the subsequent discussion, expecting that as we develop the theme some stepping stones will be found which help the less adventurous mind to see the value of the conjunction.

The living state and cancer

ALBERT SZENT-GYÖRGYI

National Foundation for Cancer Research, Marine Biological Laboratory, Woods Hole, Massachusetts

Abstract The surrounding world can be divided into two parts: alive and inanimate. What makes the difference is the subtle reactivity of living systems. The difference is so great that it is reasonable to suppose that what underlies life is a specific physical state, 'the living state'.

Living systems are built mainly of nucleic acids and proteins. The former are the guardians of the basic blueprint while the business of life is carried on by proteins. Proteins thus have to share the subtle reactivity of living systems. A closed-shell protein molecule, however, has no electronic mobility, and has but a low chemical reactivity. Its orbitals are occupied by electron pairs which are held firmly. The situation can be changed by taking single electrons out of the system. This unpairs electrons, leaves half-occupied orbitals with positive electron holes, making the molecules into highly reactive paramagnetic free radicals. The reactivity of the system depends on the degree of its electronic desaturation. Electrons can be taken out of protein molecules by 'electron acceptors' in 'charge transfer'.

When life began, our globe was covered by dense water vapour. There was no light and no free oxygen. Electron acceptors could be made out of trioses by concentrating their carbon atoms as carbonyls at one end of the molecule. The resulting methylglyoxal is a weak acceptor which made a low level of development possible. When light appeared, free oxygen was generated by the energy of photons. Oxygen is a strong electron acceptor. Its appearance opened the way to the present level of development. The transfer of electrons from protein to oxygen is effected by a complex chemical mechanism which involves ascorbic acid.

What is life? This is the main problem of biology. Many have asked this question, but nobody has answered it. Science is based on the experience that nature answers intelligent questions intelligently, so if she is silent there may be something wrong about the question. The question is wrong because life, as such, does not exist. What we can see is material systems which have this wonderful quality of being alive. What is this quality? This is the problem. It must be a very fundamental quality because it allows us to divide the whole surrounding world into two parts: 'animate' and 'inanimate', alive and not-alive. The division is sharp and unequivocal, which suggests that the *living state* is a special physicochemical state, a state which can be described in terms of exact sciences and has to fit into the great order of the universe, having been created by the same forces as the universe itself. We must search for an understanding and an answer to our question with a wide natural philosophical outlook and fit life into the great scheme of creation.

What makes the difference between 'animate' and 'inanimate' is the wonderful subtle reactivity and flexibility of the animate. So our first question has to be: where to look for the physical basis of this reactivity, in what substance? The two main components of our body are nucleic acids and proteins. The nucleic acids are the guardians of the basic blueprints of structure, while the business of life is carried on by the proteins. So we can expect the proteins to share the subtle reactivity of life. Proteins are macromolecules, built from simpler constructional units: the amino acids. I could never believe that the wonderful subtlety of biological reactions should be brought about by clumsy, relatively unreactive macromolecules without the concurrence of much smaller and more mobile units which could hardly be anything else than electrons. But electrons are mobile only on a conductor and so, more than 30 years ago (1941), I proposed that proteins may be conductors. As far as any attention was given to it, my proposition was unanimously rejected. It was pointed out to me that already a large number of proteins have been isolated and thoroughly studied, none of which showed any signs of electrical conductivity.

This was a powerful argument and nothing could be said against it, but as time went by it became clear to me that science had overlooked here a very important circumstance: that proteins are the most versatile substances capable of performing the most different functions. On first approach, we have to distinguish between two kinds of functions: very simple ones which can be performed by single molecules in molecular dispersion, and more complex ones which can be performed only by integrated systems of molecules. These latter perform the great biological functions by which we know life, like motion and secretion or nervous activity. Such integrated systems have to be, by definition, insoluble. The simple primitive functions, like maintenance of osmotic pressure, or enzymic activity, which could be performed by single molecules in solution, in molecular dispersion, demanded no electronic mobility. But only these simple molecules were readily soluble, and for their analysis the protein chemists needed solutions. So what they did was to extract from tissues the soluble proteins, call the extracted tissue '*the residue*', and send it down the drain. With them they sent down the systems responsible for the higher biological functions, which had complex electronic structures.

If we were able to detach the integrated proteins that perform the vital functions, extract, precipitate, purify and crystallize them, I doubt whether they would still have the subtle qualities which characterize life and could tell us the difference between 'animate' and 'inanimate'.

Having decided to focus on proteins our next question is: in what dimension to search? Present-day biology is a *molecular biology,* which searches for answers mainly in the molecular dimension. Our body is built of molecules, so its reactions have to be molecular reactions, but molecules are built of atoms, and atoms are built of nuclei and electrons. So there is another dimension below the molecules which has been disregarded by biology.

The electrons surrounding the nuclei are in 'orbitals' which, in a way, can be looked on as boxes containing electrons in pairs. The two electrons of the pairs spin in opposite directions, compensating each other's magnetic moments, which makes them coupled. The electron pairs in their boxes are held firmly, have no mobility or high reactivity. In inanimate systems all these boxes in the ground state are occupied, making 'closed shell molecules'. An electron placed on such a molecule would find no place to go to. Shockley (1950) compared such a closed shell molecule to a completely filled parking lot to which no car could be added and in which the cars would have no mobility. It is difficult to see how such a clumsy closed shell molecule could produce the subtle reactivity of living systems. A molecular reaction between two of them would mean only sharing a superficially-lying electron pair. So the question arises whether there is a possibility of transforming such a clumsy unreactive macromolecule into a highly reactive unit with a measure of electronic mobility.

Returning to Shockley's parking lot, if we take out one single car we could make all other cars mobile, and having created an empty place we make shuffling possible. By taking out a single electron from the closed shell molecule we could create a 'positive hole' in it, opening the way to the shuffling of electrons. Taking out single electrons we also have to uncouple electron pairs and leave the earlier partner of the eliminated electrons with an uncompensated magnetic moment behind. A molecule containing such uncoupled electrons is a 'free radical', and radicals are known to be very reactive. We have upset the balance of the whole molecule. It seems natural that the more single electrons we take out, the more we upset the balance and make all electrons more mobile and reactive. By desaturating the molecule electronically we also do something very important, discovered lately by Laki & Ladik (1976): we greatly increase the interaction between the molecules. These forces which hold structures together are usually summed up as 'van der Waals attractions'. By desaturating the molecule, we thus strengthen the forces by which molecules can be linked together to give higher and more complex structures capable of increasingly complex and subtle reactions: we have opened the way to development and differentiation, opened the way to evolution. Without desaturation, these forces are very weak and could not hold structures together against normal molecular agitation. This leads us to the first rule of electrobiology: *the living state is the electronically desaturated state of molecules, and the degree of development and differentiation is a function of the degree of electronic desaturation.* Electronic desaturation is a central problem of biology, and so the next sections will be devoted to the chemical mechanism of this process.

CHARGE TRANSFER AND PERMITTIVITY

Electrons can be taken out of molecules by other molecules by means of 'charge transfer'. If two molecules are held close together so that their orbitals overlap, the two form a single electronic system in which the electrons can rearrange themselves. If an electron in molecule A can decrease its free energy and increase its entropy by going over to molecule B, it will tend to do so, leaving its own molecule behind with a positive charge. Molecule A becomes a donor, while the 'acceptor' molecule B acquires a negative charge. The transfer of a whole electron to another molecule, where it stays put, is a rare event which occurs only in 'strong' charge transfer. What mostly will happen is that the transferred electron or electrons oscillate between the two molecules. Depending on various factors the oscillating electrons may not divide their time equally between the two molecules but may spend, say, 1% more on A than on B. It is customary to say, in such a case, that only one hundredth of an electron has been transferred. Such a partial transfer of electrons may play a very important part in biology and contribute to the subtle adjustment of biological reactions. It may have also a major importance for the mechanism of evolution.*

Here we meet a difficulty: by transferring an electron we create two electrically charged free radicals (Fig. 1: A and B). Electrically charged free radicals are exceedingly reactive and it is doubtful whether this reactivity is compatible with life. There is a way out. Let us suppose that before transferring the electron from B to A we incorporate B into A, as shown in Fig. 1C. In this case the transfer of electrons could take place as before and transform both molecules into reactive free radicals, but no net charge would be

*It is believable that by sending out 'fractions of electrons', molecules could explore simultaneously a number of situations and would stay only where they can decrease their free energy, increase their entropy and do something useful.

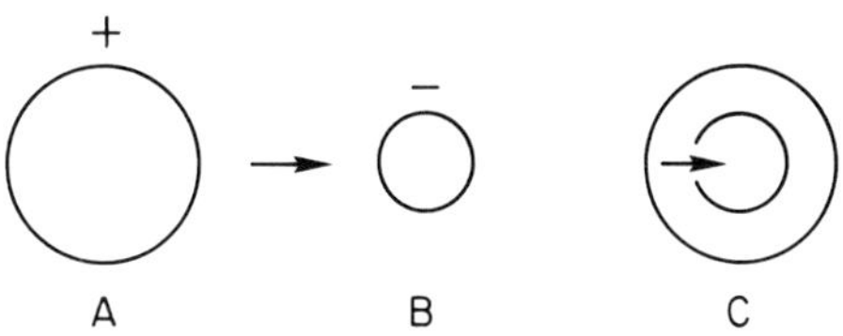

FIG. 1. A and B symbolize charge transfer; C stands for doping.

generated, the transfer having taken place inside the complex. Such incorporation with intramolecular charge transfer is called 'doping', which is one of the most important reactions on which the electronics industry is built, in which poor semiconductors are made into strong ones by doping them with electron donors or acceptors.

The creation of life demanded 'donors' and 'acceptors'. How do we find them? The universe has been transformed into one coherent system by the periodic chart of atoms of Mendeleev, the top rows of which are reproduced in Fig. 2. Where do donors and acceptors fit into this system? As we all know, this chart, which contains all the elements, consists of horizontal and vertical rows. Each horizontal row begins and ends with a noble gas. The noble gases are the most stable ones, and all physical systems tend to acquire stability. So all elements tend to resemble a noble gas by having the same number of electrons in their outermost shell. The elements on the right side of the chart have less, those on the left side have more electrons than the nearest noble gas; and so the former tend to take up electrons and the latter tend to give off electrons. Thus the former become electron acceptors, the latter electron donors. According to the table the best acceptors are fluorine and the other halogens. In fact, they are too strong as acceptors to be used by life, so for a good biological acceptor we have to turn to the next column, headed by oxygen, the universal biological acceptor. The energy driving life is derived from the transfer of an electron from hydrogen to oxygen.

Charge transfer depends to a great extent on the dielectric constant, ϵ, of the solvent, which decides its permittivity. A high ϵ corresponds to a high permittivity, and a low ϵ to a low permittivity. A given ϵ may both promote and impede charge transfer in which positive and negative charges have to be separated. Such separation will be promoted by a high ϵ which depolarizes the charges and so facilitates separation. At the same time, for charge transfer to take place, the two interacting molecules have to be held together in close proximity by conventional forces which are electropolar. So a high ϵ

	H							
He	Li	Be	B	C	N	O	F	Ne
Ne	Na	Mg	Al	Si	P	S	Cl	Ar

FIG. 2. Top lines of Mendeleev's periodic chart of atoms. To make the figure correct it would have to be cut out and rolled into a spiral in which the two neons overlap.

will tend to depolarize and disrupt such complexes, interfering with charge transfer. One special case of complex formation is dimerization (or polymerization in general), which can greatly promote charge transfer, as shown in Fig. 3. In Fig. 3, A represents a charge transfer complex in which the charges have been transferred to the ends of the particle. A dimerization, as shown in B, will greatly depolarize and so facilitate the transfer. These opposing influences will allow some charge transfer to take place only at a certain ϵ where the opposing forces balance each other most favourably. This is illustrated in Fig. 4, in an experiment in which a series of test tubes were filled with different solvents containing an increasing amount of water, the concentration of water increasing from tube to tube by 10%. The far left tube contained the pure organic solvent and the last one on the right, pure water. Then in all tubes a charge transfer reaction was performed which could be recognized by its colour. The + sign indicates the intensity of colour and charge transfer. The organic solvents had a relatively low ϵ (45 for dimethyl sulphoxide and 37 for dimethylformamide). As can be seen in Fig. 4, there was no reaction in the pure organic solvents; the reaction gradually became stronger as the ϵ increased, then weaker again in pure water.

As will be discussed later, I believe that cancer, essentially, is a failure to desaturate proteins and so substances which interfere with charge transfer have to be carcinogenic. According to Fig. 4, water is a carcinogen. This car-

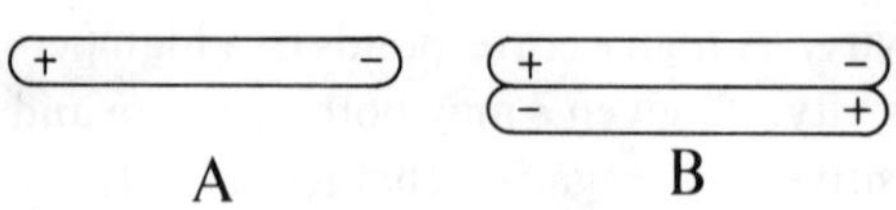

FIG. 3. A. A fibrous molecule with separated charges. B. The same dimerized. The figure is intended to show that the separation of charges promotes dimerization and dimerization promotes the separation of charges.

DIMETHYLSULFOXIDE	0	±	(+)	+	++	+++	++	+	+	±	±
DIMETHYLFORMAMIDE	0	(±)	±	+	(++)	++	++	++	+	(+)	±
GLYCEROL	+++	++++	++++	++++	++++	+++	+++	++	++	+	±

FIG. 4. Intensity of the charge transfer reaction between methylamine and methylglyoxal in a mixture of dimethyl sulphoxide or dimethylformamide and water. The first sample on the left contains the pure organic solvent, the last on the right, pure water. The maximum of the reaction is around $\epsilon = 60$. Glycerol (bottom line) has no such maximum.

cinogenic activity of water around the living protein structure is eliminated by the water structures the proteins build around themselves, which decrease ϵ. The ϵ of water is 81.5; that of ice is 2, similar to that of paraffin wax.

The third line in Fig. 4 shows glycerol. I showed many years ago (1949) that the motility of muscle can be preserved for years in 50% glycerol. Up to now this activity has been mystery. It seems to be the result of the favourable permittivity of the solvent, which allows the conservation of electronic relations.

METHYLGLYOXAL

When life originated, about three and a half billion years ago, it was probably pitch dark on our globe and there was no oxygen, the earth being covered by a dense layer of water vapour. There was no free oxygen. The oxygen was present in bound form as water, carbonate, phosphate, etc. which is of no use as an acceptor for protein. So if our rules of electronic biology are valid and there can be no life without desaturation, and oxygen is the universal acceptor, then we have to suppose that life found ways of using the bound oxygen as acceptor. Nature achieved this by taking a molecule of water out of triose and crowding all the oxygens as carbonyl, C=O, at one end of the molecule, the hydrogens at the other. The substance formed is methylglyoxal (Fig. 5).

Nature is simple but subtle (P. Ehrenfest, personal communication). Methylglyoxal is a unique substance. There is no other substance with its special characteristics. In spite of its small size it contains a reactive aldehydic group, and a ketonic C=O, which has a low lying triplet orbital (Abdulnur 1976), making it an acceptor. Triose is the smallest molecule which could be transformed in this way.

That I have not lost myself in meaningless speculation and am still close to the central problems of biology is indicated by the fact that more than sixty years ago a most reactive and apparently ubiquitous biological enzymic system was discovered, *'glyoxalase'*, which catalyses a two-step reaction using glutathione as coenzyme. It can transform, at an extreme speed, methylglyoxal into D-lactic acid. Nature does not indulge in luxuries, and if there is such a widely spread and active enzymic system, it must have something very important to do. But up to the present nobody has been able to find any use for it, D-lactic acid and methylglyoxal not being known to have any major biological function.

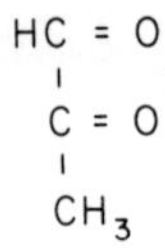

FIG. 5. Structure of methylglyoxal.

Methylglyoxal can attack protein by means of its aldehydic C=O interacting with an amino group of protein. Protein has but one typical NH_2 group which is in its lysine residues.* Before discussing this interaction I should like to consider briefly a simpler model in which the protein's place is taken by the simplest aliphatic amine, methylamine.

If aqueous solutions of methylamine and methylglyoxal are mixed, the appearance of a yellow colour indicates the formation of a Schiff base which contains the chromogenic N=C link:

$$R^1-C=O + H_2N-R^2 \rightarrow R^1-C=N-R^2 + H_2O$$

If the same reaction is performed in a solvent of lower permittivity, such as methanol or acetone, instead of water, a purple colour appears. If 0.5M-acetone solutions of two reagents are mixed, then the purple product precipitates and can be isolated. Pohl *et al.* (1977) measured its molecular weight, which indicated a polymerization to a tetramer. In the electron spin resonance (e.s.r.) spectroscope the purple precipitate gives a strong signal with a rich hyperfine structure, showing that polymerization and charge transfer have occurred. In the first instance the precipitate is colourless; it turns purple in a fraction of a second, showing that the primary product of the reaction is a colourless substance. The same can be shown also for the Schiff bases by using butylamine instead of methylamine. The colourless primary product is in

*The NH_2 in arginine residues is part of a guanidine group and has specific qualities and reactions.

all probability a hemiacetal.* At the high permittivity of water the hemiacetal turns into the yellow Schiff base, while at a lower permittivity, such as that of acetone or methanol, it polymerizes and turns into the purple charge transfer complex.

METHYLGLYOXAL AND PROTEIN

If we represent the peptide chain of protein by a straight line, as has been done in Fig. 6, then the lysine side-chain can be drawn as a straight line which hangs out to one side, consisting of four carbon atoms with an NH_2 at its end. This side-chain can be looked upon as a fishing rod, fishing for methylglyoxal. Methylglyoxal molecules can be expected to attach themselves to the amino group forming a Schiff base which is but slightly weaker as an acceptor than methylglyoxal (J. Ladik, unpublished calculations).

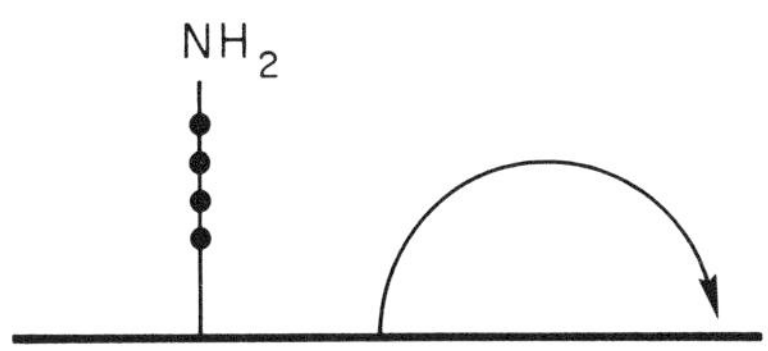

FIG. 6.

While the lysine side-chain is stretched out it can take with its Schiff base no electrons from the peptide chain from which it is separated by four saturated carbon atoms, through which no electrons can be transmitted. However, the side-chain is pliable and can fold. Otto *et al.* (1978) made a thorough study of this folding and found that it will bring the Schiff base into touch with the second-neighbour peptide bond of the peptide chain of the protein, enabling it to enter a charge transfer reaction with it and take electrons from it. It will form an n-p junction which will charge the Schiff base negatively, the peptide chain positively, making it into a p-type conductor.† In many proteins, about

---

$$*CH_3-NH_2 + O=\underset{\displaystyle R}{\underset{|}{C}}-H \rightarrow CH_3-\underset{\displaystyle H}{\underset{|}{N}}-\overset{\displaystyle H}{\overset{|}{\underset{\displaystyle R}{\underset{|}{C}}}}OH$$

†It deserves consideration whether anaerobic life may not be based on an n-type conductivity. The very first forms of life had to be anaerobic.

each eighth amino acid residue is a lysine (Fig. 7), and so the peptide chain has a boosting station at every eighth amino acid. This lysine side-chain with the attached Schiff base is the kernel of the mechanism proteins carry with them for their desaturation.

If a solution of lysine is treated with methylglyoxal it turns yellow because of the formation of a Schiff base. It can give no purple complex, not even in alcohol or acetone, because the bulky molecule interferes with polymerization. If a protein, such as casein, is treated with aqueous methylglyoxal, its granules being suspended in methylglyoxal solution, it turns yellow, but if the amino groups of the lysine residues are methylated, the protein remains colourless. If treated with a methanol solution of methylglyoxal, the colour of the protein becomes darker, brown. Since the lysine side-chains cannot polymerize this darker colour has to be due to a more intimate complexing of the Schiff base with the peptide chain, due to the lower permittivity of the solvent.

NH_2
|
CH_2
|
CH_2
|
CH_2
|
CH_2
|
CNH_2
|
COOH

FIG. 7. Lysine.

The brown colour of the methylglyoxal-treated casein is very similar to that of the liver, which makes it seem likely that the liver actually owes at least part of its colour to its electronic desaturation by methylglyoxal. Fodor *et al.* (1978) have in fact isolated and identified the methylglyoxal linked to the liver protein.

In the first period, life had to be based on a desaturation by methylglyoxal, which is a weak acceptor, and so life, in this period, could have developed but only the simplest forms which have left no traces behind.

Fig. 6 suggests that the Schiff base, being at the end of the lysine side-chain, at a distance from the peptide chain, is formed in random water which promotes the formation of Schiff bases. The association of the Schiff base with the peptide link in the peptide chain may be promoted by the lowered dielectric constant, lowered by the protein, which builds water structures around itself.

According to the theory presented, the desaturation of the peptide chain should lead to an increase in its electronic conductivity. This conductivity has been measured by Pethig & Szent-Györgyi (1977) as well as by Bone *et al.* (1978) and found to be increased considerably.

The brown colour of the liver indicates unpaired electrons and suggests that protein radicals are an important part of the structure of liver cells. Pohl *et al.* (1977) isolated these coloured proteins and found them to be paramagnetic, giving a strong e.s.r. signal, similar to the signal given by casein treated with methylglyoxal. This suggests that the e.s.r. signals given by living material are actually due to the protein radicals composing the cellular structures, and are not entirely due to undefined free radicals in solution, as hitherto supposed. The structural proteins in Pohl's experiments were isolated by homogenizing the tissue in ice-cooled 50% ammonium sulphate and subsequent centrifugation. The soluble proteins were isolated similarly in the centrifuge by 100% saturation, after the structural proteins had been eliminated; the soluble proteins gave but a very weak signal.

THE $\alpha \longrightarrow \beta$ TRANSFORMATION

The history of life was divided into two parts by the appearance of light. In the first, dark period the globe was covered by dense water vapour, and the atmosphere had to be reducing. There was no free oxygen, and there could be no stable electron acceptors. Accordingly, life could reach only a very low degree of development, which has left practically no traces behind. The protein in this period had to be desaturated by methylglyoxal, which is a weak acceptor and could develop only the simplest forms of life, the main function of which had to be the proliferation which made life perennial. This unbridled proliferation had to be favoured by the low level of cohesive forces and the poverty and simplicity of structures. I have called this first dark proliferative part 'the α period'.

This situation changed when, because of cooling, the water vapour condensed and light could reach the surface of the globe. What life did with this light was to use its energy for the separation of the elements of water, producing oxygen, which is a strong acceptor, and so could start up the development and differentiation, the end result of which is us. I have called the second, light oxidative part of life's history the β period. The cohesive forces generated by the electronic desaturation led to the building of increasingly complex structures with increasingly complex and subtle reactions.

Cohesion and structures interfere with proliferation, so when the cell divides it has to lower its cohesive forces and dismount part of its structure.

These changes can be summed up as a partial return to the α state, which is the ground state of life. It has to be the more stable state, having the lower free energy and higher entropy. So the $\alpha \longleftrightarrow \beta$ transformation had to be kept reversible. After it had completed its division, the cell had to build up its β state again. So in the β period the unbridled proliferation was replaced by regulated growth. Should the cell, after completing cell division, find the way of return to the β state perturbed, it has to persist in the proliferative α state and continue to divide when no division is needed, leading to a tumour. This also explains why very rapidly dividing cells resemble one another, be they embryonic, cancerous, or simply very rapidly dividing normal cells.

The more complete the return to the α state, the faster the proliferation and the lower the cohesive forces will be, and the more malignant the tumour produced.

ASCORBIC ACID

Free oxygen, O_2, is a strong electron acceptor and so its appearance in the β period opened the way to a higher degree of electronic desaturation, and to the corresponding higher degree of differentiation and development. If nature develops a new method, as a rule, she does not throw the old one out, but simply adds the new one to it, improving it. So, in the β period, methylglyoxal was not replaced by O_2, but the O_2 was added to it, boosting up its acceptor strength.

However, oxygen does not interact with protein, nor does it interact with methylglyoxal. So in order for it to be used as an acceptor for protein, a link had to be developed by which the oxygen could be linked to methylglyoxal, transferring to it part of its acceptor power. This link had to have very specific qualities: it had to be able to make a bond simultaneously both with oxygen and methylglyoxal and had to have an electronic mobility by which it could transfer the acceptor strength of O_2 to the ketoaldehyde. This new substance, which nature developed when light and oxygen appeared, is ascorbic acid, discovered by me in the early thirties. Later it was found to have an antiscorbutic activity, and to be identical with the then unknown antiscorbutic component of fresh vegetables called 'Vitamin C'.

In a remarkably short time the chemical nature of ascorbic acid was cleared up and the substance became available at low cost in crystalline condition in an unlimited quantity. But while we have learned everything worth knowing about its chemistry, its biological function remained unknown, preventing medicine from making full use of its remarkable reactivity. We can control only what we understand. Even a simple question, such as the magnitude of the recommended daily dose, has remained unsettled, oscillating between megadoses and a few milligrams.

The development of ascorbic acid is one of the landmarks of evolution, as was the appearance of light and oxygen. That ascorbic acid readily interacts with oxygen can be shown by placing a sample of its Na salt into the e.s.r. spectrometer in the absence of oxygen. On admission of air the typical free radical signal appears, consisting of a doublet with a splitting of 1.75 gauss. Each part of the doublet has a further triplet splitting of 0.175 gauss.

That ascorbic acid can readily interact with methylglyoxal has been shown by Fodor, who will describe his observations later in this symposium (Fodor *et al.*, pp. 165-169).

The great mobility of the unpaired electron inside the ascorbate radical can be demonstrated by e.s.r. which shows that this electron can interact with many of the protons of ascorbate.

Also, the spectrometer gives valuable information about the interaction of oxygen, ascorbate and methylglyoxal, which information has hitherto been utilized only to a small extent. If methylamine and methylglyoxal solutions are mixed a very strong signal appears with a rich hyperfine structure. In the presence of ascorbate the signal is much stronger, showing that the ascorbate catalyses charge transfer, without changing the character. If casein is treated with methylglyoxal in the presence of ascorbate, a new signal appears (P.C.R. Gascoyne, personal communication), indicating that the ascorbate is built into the charge transfer complex—an observation which may have far-reaching biological consequences and introduces new viewpoints into the medical use of the acid.

Most fascinating and complex spectroscopic reactions have been found by my colleague Jane McLaughlin (unpublished results). If both cuvettes of the spectrophotometer contained a mixture of a 0.0256M solution of methylglyoxal and methylamine (0.077M), and 0.0077M-ascorbic acid was added, a very strong absorption appeared around 400 nm, which disappeared slowly (Fig. 8). A second peak appeared around 500 nm. In the absence of oxygen the addition of ascorbic acid made no difference, showing that it catalysed the charge transfer between O_2 and methylglyoxal. If no ascorbate was present, admission of oxygen to the mixture of methylglyoxal and methylamine inhibited the appearance of the absorption spectrum.

Oxygen, essentially, brings the life-giving light into the living system. Ascorbic acid catalyses this reaction. It is involved in bringing matter to life. This opens up new aspects of the medical application of ascorbic acid.

The actual role played by oxygen in these reactions is yet unknown and demands more study on the basic level. It is possible that the O_2 molecule

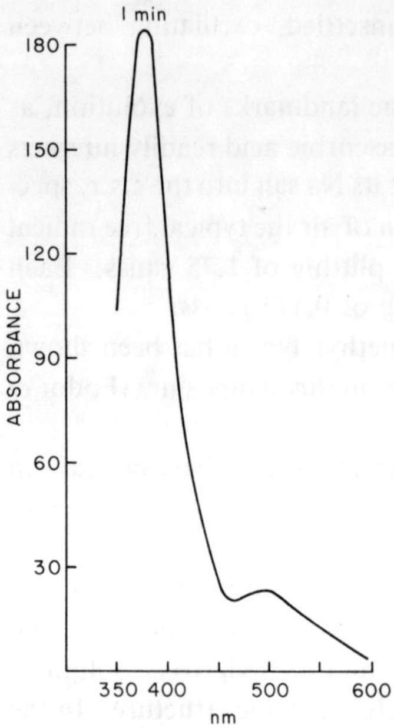

FIG. 8. See text for explantation (p. 15).

simply takes an electron over and dissociates off, but it seems more likely to me that it takes over only parts of an electron and remains attached to the charge transfer complex. If, in this case, the oxygen pressure is decreased, the oxygen may dissociate off, upsetting the electron balance in the whole charge transfer mechanism. This may explain why consciousness instantaneously fades out if the oxygen supply to the brain is cut off. This suggests also that in the comatose state after a heart attack the infusion of ascorbic acid and artificial respiration might be indicated.

A THEORY OF CANCER

Two cardinal symptoms of a cancerous transformation are a low paramagnetic susceptibility and a low cohesion. Both are the consequences of a defective desaturation of the structural proteins, due to the loosening up and disorganization of the chemical machinery, a shift towards the α state.

A cell is a strongly integrated system and all its reactions are coupled to one another. In the normal cell the reactions are coupled in such a way that activity improves activity. Desaturation promotes further desaturation, and exercise makes us stronger, but coupling is a two-way street, and factors which inhibit normal interactions may be coupled in such a way that they cause further inhibition. An inadequate desaturation may inhibit desaturation. This can push the cell into a *vicious circle* which the cell is unable to break, and leads it into a state of disorganization, landing it in the proliferative α state. If the situation is not corrected this situation may become constitutive, irreversible.

To be able to correct a deficiency we need a still better and more detailed knowledge of the chemical machinery of electronic desaturation, which can be achieved by basic research. The blindfold search for a cure for cancer seems a hopeless waste. Until correction becomes possible the best defence against cancer is keeping the machinery in perfect working order. More than sixty years of research on living systems has convinced me that our body is much more nearly perfect than the endless list of ailments suggests. Its shortcomings are due less to its inborn imperfections than to our abusing it.

One factor which deserves special attention is ascorbic acid, for the supply of which we depend on food, the expensive nature of which is not a guarantee of its quality. Some experiments suggest that this acid is not merely a catalyst but that it is built into the machinery—and is part and parcel of it. Since we are building and rebuilding this machinery all the time, a continuous supply of ascorbic acid is important. A machinery built without ascorbic acid cannot be corrected by suddenly administering megadoses. Correction of defects may take the better part of a year.

The ideal of medicine is the curing of all diseases. The ideal should be full health, which leaves no room for any shortcoming. In the U.S.A. we are still losing the battle against cancer: every day there are eight hundred casualties. I strongly believe that cancer is accessible to a complete analysis on the basic level, and that understanding it means also the ability to control it.

References

ABDULNUR, S.F. (1976) The interactions of glyoxals with proteins and DNA in relation to cancer. *Int. J. Quantum Chem. Quantum Biol. Symp. 3,* 59-64

BONE, S., LEWIS, T.J., PETHIG, R. & SZENT-GYÖRGYI, A. (1978) Electronic properties of some protein-methylglyoxal complexes, *Proc. Natl. Acad. Sci. U.S.A. 75,* 315-318

FODOR, G., MUJUMDAR, R. & SZENT-GYÖRGYI, A. (1978) Isolation of methylglyoxal from liver. *Proc. Natl. Acad. Sci. U.S.A. 75,* 4317-4319

LAKI, K. & LADIK, J. (1976) A note on the 'electronic theory' of cancer. *Int. J. Quantum Chem. Quantum Biol. Symp. 3,* 51-57

OTTO, P., LADIK, J., LAKI, K. & SZENT-GYÖRGYI, A. (1978) Internal charge transfer in proteins to the Schiff base of their lysine side chains. *Proc. Natl. Acad. Sci. U.S.A. 75,* 3548-3550

PETHIG, R. & SZENT-GYÖRGYI, A. (1977) Electronic properties of the casein methylglyoxal complex. *Proc. Natl. Acad. Sci. U.S.A. 74,* 226-228

POHL, H.A., GASCOYNE, P.R.C. & SZENT-GYÖRGYI, A. (1977) Electron spin resonance absorption of tissue constituents. *Proc. Natl. Acad. Sci. U.S.A. 74,* 1558-1560

SHOCKLEY, W. (1950) *Electrons and Holes in Semiconductors,* Van Nostrand, Princeton, New Jersey

SZENT-GYÖRGYI, A. (1941) The study of energy levels in biochemistry. *Nature (Lond.) 148,* 157-159

SZENT-GYÖRGYI, A. (1949) Free energy relations and contraction of actomyosin. *Biol. Bull. (Woods Hole) 96,* 140-161

A note on cancer and possible relationships to submolecular biology

S. RESLOVA-VASILUKOVA and R.J.P. WILLIAMS

Inorganic Chemistry Laboratory, University of Oxford

Abstract A general discussion of the nature of cancer is given in order to set the scene for the symposium. Tentatively the possible relationship with submolecular biology is indicated.

Before we come to the main papers of the symposium it might be useful if I, with the help of a colleague, give an outline of the complexity of cancer, so that those more unfamiliar with the disease who might wish to help in our understanding of it can realize the nature of the problem of linking submolecular biology and cancer. We give this outline not as experts in the details of the cancer problem but in an effort to highlight major areas of knowledge which must be appreciated from the outset.

DISEASE IN GENERAL

Cancer is already known to be a disease quite different from other diseases, as we indicate in Table 1, in which distinguishing features of cancer are enumerated. The simplest diseases are really due to Darwinian competition between higher and lower organisms—for example, an invasion of the higher organisms by bacteria, fungi, or lower animals or plants. Such diseases are usually infectious or contagious for obvious reasons. Cancer is not in general infectious or contagious and cannot have an origin similar to most other diseases. A second class of disease arises from a *germ cell* mutation which is inherited on Mendelian lines and is due to a sustained mutation which is generally weakening to the whole organism and affects all cells. As cancer is not inherited in most cases, although there are well-known inherited forms of the disease, this excludes mutation of the *germ* cell as the general origin of

TABLE 1

Some peculiarities of cancer

1. Extremely widespread among species
2. All types of cells of one organism are susceptible
3. Not generally infectious or contagious
4. Susceptibility to disease may be inherited
5. No known cure except death of the affected cells—i.e. no known detransformation
6. Sustained, not necessarily rapid, growth rate of aberrant cells
7. Tumours do not have a growth form or shape
8. Chromosome damage in some cases
9. Abnormal enzyme patterns and isoenzyme compositions
10. Unusual antigenic character
11. Associated with wide variety of chemicals in the environment
12. Associated with some viruses
13. Cancer cells are autonomous and may be transplantable

cancer but does not exclude mutations of somatic cells, to which we shall return. Neither does it exclude embryonic-cell or stem-cell mutation, but there is no known general connection between teratogenically damaged organisms and cancer incidence. We conclude that certain types of mutation are excluded as the general source of cancer, but not others.

Chemical diseases such as poisonings are different from cancers, too, in that they cause cell injury but, in general, do not increase cell growth rate. However, some chemical contaminations do lead to cancers. Environmental chemicals of one kind or another act as one major factor in the generation of 80% of cancers. As biological systems are chemical systems in balance we must look, too, at the variation of internal cell chemical composition and its effect on cancer incidence. We do not know yet how sensitive cells are to changes in the steady-state level of metabolites, although they are obviously very sensitive to hormone levels.

There are also viral diseases. Viruses lead an independent stable existence outside higher organisms but do not multiply there. Viruses can become a stable part of the replicon and be reproduced in all cells (proviruses). Finally, viruses can multiply in cells. Diseases such as influenza and measles are caused by viruses and are infectious. There is known to be a relationship between some viruses and cancer, but this has not been demonstrated in humans in any general sense. There is a strong suspicion that cancer is a product of contamination of genetic material by viruses or viral fragments, possibly incorporated in DNA, and some activating events caused by changes in the internal or external environment. The very incorporation of a virus or a viral fragment into DNA is not too dissimilar to the incorporation of other genetic material to give a mutant.

Finally, there are the self-generated diseases in which the host's own defence mechanisms (autoimmune diseases), or its protective devices of other kinds (detoxification systems), or its digestive activity, or even its ability to lay down skeletal material, undertake misplaced action so that the host is attacked or damaged. Cancer has some connection with this type of disease, as we shall show under the heading of protection (p. 26). We now turn to cancer itself.

If we are absolutely honest we have to say that we do not know what cancer is in any precise molecular scientific terms (Table 1). It is obviously something to do with aberrant excessive growth and it is clearly of cellular origin. Note that the extracellular matrix may have a key role in controlling the spread of the aberrant cells. After that there is a confusion of factors including the environment, viruses, and even the disposition of individuals which affect the disease. Even so, it is agreed that the origin of cancer is in all probability an initial fault of cell regulation—though whether this fault is always related to mutations (changes in DNA) or to wrong steps in the interpretation of unaltered DNA is not agreed. Thus our molecular or submolecular examination must start from a discussion of regulation which, although simplified, will ask about every factor which could cause changes in regulation, and must aim to uncover control features at the cell level. However, cancer is associated with several other peculiarities (Table 1), upon which we want to make brief comment at both a general and a molecular level.

In order to discuss the relationship between regulation and a cancerous condition at a molecular level we need to distinguish cancer from other cell regulations such as general differentiation, for cancer is only recognized by a *sustained* derestriction of cell multiplication. We note immediately that 'cancerous' cannot refer to a condition in bacteria or unicellular organisms. Unfortunately, these are the organisms about which most is known and in which adaptation (and regulation) are well studied. However, the cancerous condition would appear even by definition, and to some extent from experience, to be common—though not equally common—to all animals and plants having multicellular organization. In a simple sense this must be the case, since once an organism has developed a multicellular *form*, where cells are differentiated within a centralized body, there must be a finite probability that the controls which generate the form will fail while for some cell multiplication continues. It is also of the essence of modern genetic thinking that each cell contains all the information for the complete generation of a new organism, no matter to what degree the information is used in a specific organ of a whole system. Thus somatic cells of any kind always have a finite possibility of multiplying too rapidly, which is one failure of control, or of going over into an incorrectly defined cell type and multiplying. Both these conditions are abnormal growth patterns. If sustained they kill the victim. As we see it, then, the occurrence of cancer is an *inevitable risk* in the development of multicellular form within the whole of biology, just as mutation is a risk at the level of unicellular organisms. We shall ask what factors can act on the genetic material and its controls so as to cause aberrant growth. Quite obviously, to answer this implies a knowledge of the factors controlling normal growth.

GROWTH CONTROLS

Growth itself does not have a fixed pattern in lower life forms or higher life forms. It is easy to see that the controls on differentiation are much stronger in some fully grown organisms than in others. Thus, growth control in plants is much less than the control in advanced animals, as expressed in the variations in *form* of the adult. Moreover, controls are exercised much more strongly in some cells of one organism than in others and the controls develop as the organism grows.

A given multicellular organism has many different types of cells and these have different regulation. Two obvious additional factors of importance will be the incidence of cancer in a particular organ rather than another of the same multicellular organism and the time of appearance of each growth. Thus in humans tumours of the brain and urinary tract, and leukaemia, are more prevalent in childhood, whereas tumours of the digestive tract and respiratory system are more common in old age. Some tumours appear more frequently in late middle life, such as breast cancer. Muscle cell cancers are relatively rare at all ages and epithelial cells are the common sources of cancer. Thus the site and age at which cancer is observed must be explained. The growth patterns and controls of different somatic cells are very different.

It is now clearly seen that in the development of an organism there are differentiation steps, and that after these steps there are maintained states of cells which multiply. The typical pattern is

$$\text{cell A} \xrightarrow{(1)} \text{cell A}' \xrightarrow{(2)} n \text{ cell A}' \xrightarrow{(3)} \text{cell death}$$

There are special controls on step (1) which are strongly exerted in the fetal stage and general controls on stage (2) which give an organ within the organism its form. Step (3) may well have controls, too. Even later in life than the fetal stage, some cells of the type A give rise to more than one differentiated cell type A′ , for example the bone marrow cells. The rates of steps

(1) and (2) are controlled but varied throughout life, are altered after damage, and are very different for different classes of cells. The complexity of control is very great, but in all cases the controls must be chemicals which are, in some sense, gene products. Obviously the above scheme is the simplest one and more complicated patterns could be drawn and clearly exist.

Cancer can now be treated as some failure of this growth pattern. The simplest pattern of causation requires the initiation of a fault *and* a failure of subsequent growth control, for cancer has two parts: an initiating transformed cell and a failure of control over the growth of this cell. Once again, this is the simplest description. The next step is to look at the factors known to initiate cancer and to control cell multiplication.

INITIATION OF CANCER

It is commonly accepted that cancer is in some way related to a mistake in or a persistent misreading of genetic information. Thus it has obvious relationships with a mutation (though this is not a necessary relationship). The pattern of events in all mutations and in some cancers can be written:

host cell → daughter cell → type of growth (progeny)
↘ physical or ↗
chemical change
to DNA

Chemical change includes incorporation of pieces of DNA of viral origin. The first step may be chemically similar (or the same) for mutation, transformation to a cancerous cell, or even for several other transformations, both good and bad, which include death (see below). This circumstance arises because DNA (RNA), the genetic material, is open to a variety of different reactions with the same reagents, since DNA is a diversely organized linear polymer of only four monomers. (In fact this empirical relationship between mutation and cancer is the basis of the Ames' test for carcinogens.) There may be agents which can act on DNA at many places at the same time and, even though the actual local site of attack is the same, for example, the N-7 atom of guanosine, it can be at 10^6 different guanosine residues in a given DNA molecule. There can, therefore, be 10^6 effects of a given chemical reaction. We may say that some of these, perhaps in combination, lead to cancer, but we stress that the attack is then on a cell which can go on to reproduce itself while retained in an organism. There may not be chemical or physical effectors *per se* which are lethal, *or* teratogenic, *or* carcinogenic, since all could be involved simultaneously but our inspection methods may be devised so that we can only observe one at a time. It could even be that different times of attack on a given base of DNA during the development of an organism could give different effects. One such attack would be by the radicals which interest Dr Szent-Györgyi so much.

We have stated the chemical case so far in terms of attack on DNA itself, because this is a basic genetic material and cancer is usually a fixed condition of cells derived from one errant cell, as shown by cloning, strongly indicating a 'genetic' origin. Thus a change in chromosomal DNA seems the most obvious reaction which could cause cancer. A similar effect is possible through damage to extranuclear DNA, for example plasmids. However, it is possible to imagine ways of inheriting *instruction* to DNA without direct chemical change of the DNA and we should not exclude this possibility. The multiplication of differentiated somatic cells of given organs, such as liver cells, provides clear examples of the inheritance of controlled instruction without mutation. There is the possibility that some permanently expressed *change* in DNA *may* have occurred at a critical stage in embryogenesis that reflects itself in the stability of the differentiated state seen for example in liver cells. This possible explanation can be ruled out for plant cells. Clearly, a differentiated state can be maintained during cell multiplication. In theory, at least, a new differentiated state could arise which was not an ingredient of normal development and was caused by an external perturbation of the inherited instructions to DNA. Once formed the new differentiated cell could multiply and become cancerous. In what follows, 'chemical attack' will include direct chemical attack on DNA or any other modification which can be passed on directly to the daughter cells. We must examine what information is transmitted, for it is not just DNA. In all such circumstances we need to look carefully, too, at the *environmental* conditions which lead to chemical alteration in instructions. This includes external and internal cell chemicals as well as events in a cell at the level of binding to DNA itself. Note that only a part of this environment is transmitted during division. Environment must be treated in a very broad sense in describing cancer incidence (see below). (We note too that spontaneous changes in DNA are not excluded, for DNA is a polymerized energized state of the monomer bases.)

If the attack is not on DNA, it could be on a component control protein or on RNA. However, we do not know how such a change would be inherited and we have to invent a feedback scheme with the following characteristics. Damage produces an inherited material which reproduces the damage. Plausible schemes are: (1) damage to an RNA which is related to DNA reversibly, for example viral RNA: (2) damage to a protein Y which normally binds to gene B; removal of Y allows production of a derepressor X which acts on other genes, A and C. The derepressor X is maintained in sufficient quantity in the daughter cells after cell division to keep them derepressed. X also stimulates gene B which produces the derepressor by a positive feedback loop through, say, genes A and C, even if Y is made. A steady state in X can be seen to be inheritable after a control has been switched.

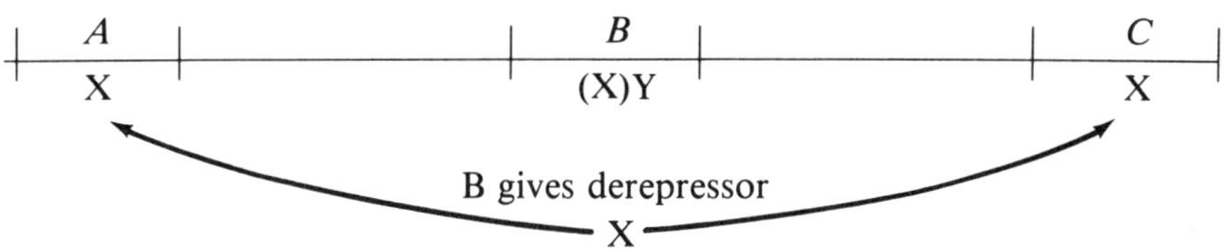

The damage is then inherited through failure of a control by Y due to overproduction of X and not through changes in DNA. This type of system must be related to the production of all simple differentiated cells. Such a differentiated cell which failed to recognize its neighbours would be a cancer if its production was maintained—that is, if it became autonomous.

The relationship between the incidence of some animal tumours and the length of time of exposure (t) to a given environment, for instance an initiating chemical carcinogen, can be expressed in the form (1),

$$\begin{aligned} \text{Incidence rate} &= k\,(t-a)^n \\ \text{i.e. } \log(\text{incidence rate}) &= \log(k) + n\log(t-a) \end{aligned} \qquad (1)$$

where a is a constant and n is a power term of time $(t-a)$. The rate constant, k, is related roughly linearly to the dose of an 'initiator'; that is to say, it is a measure of the environmental stress to which the organ is exposed. Not all cancers obey this incidence law and we may separate different types of cancer on this basis; for example, leukaemias do not seem to fit into this pattern. However, the general nature of this incidence equation for some groups of tumours needs an explanation.

The equation depends on the assumption that the living system which is open to cancer is in a constant environment. The rate constant, k, is then fixed. However, in life there are various changes in environment and of internal chemistry. Thus the equation could not be expected to apply to tissues which change markedly with time. The breast is an example of an organ where the levels of hormone have a particular influence over a restricted period of the life of the whole organism. The incidence of breast cancer is related to changes in the activity of this organ—that is, to changes of k.

PROTECTIVE DEVICES AND CANCER INITIATION

The development of protection in organisms as a response to attack seems to have followed different lines, such as the evolution of:

(1) external shells or coats,

(2) pigmented layers,

(3) enzymes for the oxidative removal of chemicals by, for example, the cytochrome *P*-450 system,

(4) organelles within cells for oxidative attack, e.g. in myelocytic cells and peroxyzomes,

(5) organelles for hydrolytic attack, after detection by antibodies, by, for example, lymphocyte lysosomes.

Unfortunately, all these defence mechanisms can damage the host by accident or in the course of their aggressive activity. There then arise the autoimmune diseases from (5), and especially (2), (3), (4) the possibility of attack on DNA followed by repair, which then raises the spectre of cancer. In fact, much of cancer may arise accidentally from the protective devices. We must never forget that cancer is a disease which may only manifest itself clinically some 10 or 20 years after initial attack (see equation 1). For example, the initial damage to DNA will surely occur very rapidly after exposure to a chemical (e.g. $OH^{\bullet}$). The clinical development will depend on very many factors that usually do not allow clinical recognition until many years have passed. An infection develops seriously within days. Cancer is, therefore, very largely a risk which is worth the organism taking while preventing infection or removing an immediately damaging chemical. Cancer is mainly destructive in old age.

In the context of cancer we note particularly the protective devices which use oxidation. These are the *P*-450 cytochromes of the detoxification system, the myelocytes producing H_2O_2 and perhaps $OH^{\bullet}$, and the cells producing pigments. In all these cases there will be side-products of the chemical reactions which are electrophiles capable of attacking DNA. (Some of the stimulation which has forced us to look at oxidation, especially radical chemistry, in the context of cancer, comes from Szent-Györgyi's views on the role of electron-deficient molecules.) The problem is a simple one in the following sense. The aggressive chemistry which protects by killing also causes disease, and one disease which can be so caused is cancer. Cancer itself is poorly protected against by the protective systems since it is a modified form of 'self'.

THE SECOND STEP: GROWTH OF A TUMOUR

Continuing with the features noted in Table 1 (p. 20), we stress not only the condition of the initial loss of control of regulation but also that this loss of control must lead to unusually *sustained* growth. An unusual growth which stops will be treated as cancer only while it grows. The growth rate itself is then of interest and has two phases, cell multiplication and death. Cells which have a longer life than other cells and the same rate of division, as well as those which have a greater rate of division but the same life expectation, grow to a greater mass. If death is regulated and is a controlling feature of organ growth, cancer could arise also from failure of the regulation of death, and we cannot exclude this possibility. We note that a cancer cell is usually committed as are other somatic cells, though it can sometimes adapt (differentiate) as in a bacterial cell growth pattern or an evolving stem cell pattern.

The sustained growth must be seen together with the lack of *form* of cancer (Table 1). The cancer cell must have lost the control common to all normal cells in a body which allows them to multiply only to the stage of giving a fixed form—a form controlled though varied throughout life. It is this striking feature of life which is so central to Szent-Györgyi's thinking. *Living cells must communicate. Cancer cells are autonomous.* What is the difference?

This way of thinking about the over-all relationship of control of growth and its connection with cancer is stressed again and again in Szent-Györgyi's papers. The question is: 'Are we missing an essential principle which is based on complex chemistry but may well be simple in fundamentals?' While others look closely at the membrane surfaces for this organization control, Szent-Györgyi points to the interior of the cell. Undoubtedly there are some things peculiar about both the external surface and the internal systems of tumour cells (see, for example, Damadian's chapter in this volume, pp. 131-138). We do not know yet whether any of these peculiarities are related to the concepts of submolecular biology.

Finally, we must be aware that the complexity of growth patterns associated with cancer can be very great. In addition to the simplest systems in which there is excessive continuous growth of a given cell type there are: *(i)* pleomorphic tumours in which cells of many types grow together in a disorganized manner; *(ii)* benign tumours where growth is stopped by an overgrowth of fibrous tissue—this situation is known in both animals and plants; *(iii)* apparently reversible situations in which the tumour is reabsorbed by some response of the host. Szent-Györgyi has stressed especially this last observation—the reversibility of tumour growth. Lastly, the growth of a tumour is not the same as excessive growth of all or some host cells stimulated by, for example, hormones—but growth which remains controlled and organized. The latter is referred to as stimulated hyperplasia.

TREATMENT OF CANCER

As well as the stresses which cause cancer, the types of treatment which cure it should be evidence of the characteristics of the disease. This is a different aspect of cancer chemistry, namely therapy. We note again that the agents which cure cancer are often multipotent, like the agents which cause it. In fact, the very same agents which cure cancer can very frequently cause it (Haddow's Paradox), and, as stated above, they also can cause a range of other effects related to changes in inherited material—mutations, teratogenic effects, and cell death. Killing, in fact, is often the very essence of cancer therapy. It will be important to notice any differences in chemistry which distinguish cause and cure of cancer. (Note again the paradox that for a cancer patient, killing cells with a carcinogen can be worth the risk since killing is fast and new tumour development slow. The patient may gain 10 years of life.)

Once again it must be stressed that there could be quite another way of treating cancer. This would be to cause the cancer cell to regain 'communication' with other cells—that is, to normalize its regulation. Here again, Szent-Györgyi is almost the only worker who has made specific proposals, for his theory of the origin of cancer in proteins or DNA is very different from the theories of others. The cancer cell for him is not irredeemable.

Lastly, in a comparative study such as this, we must look for biological parallels with cancer, not only with mutation and with differentiation (in particular), but also with events such as lysogeny in bacteria that are known to be related to changes in the genetic equipment of cells. The connection between viral infections and cancer is a major topic.

SUMMARY OF SOME KNOWN FACTORS

We have mentioned many general features of cancer: (1) the incidence patterns within an organism, (2) its occurrence in a population, and age-related effects, (3) the growth rate, (4) its generality in living things, (5) the vertical line of propagation from a stem cell and its possible relationship to DNA, (6) the environmental influences so apparent in the initiation of much of cancer, (7) the relation between cancer cures and cancer causes, (8) the somewhat parallel biological phenomena of mutation, differentiation, lysogeny and perhaps death, (9) the contrast with other diseases, and (10) the absence in general of inheritance and of infectious or contagious elements in the disease. We have not yet referred to individual disposition to the disease.

DISPOSITION TO CANCER

Although a major cause of cancer is an attack on a single cell by environmental agents, especially chemicals, we must be careful not to miss the possibility that there can be a disposition or susceptibility to cancer which varies widely from individual to individual and with the general well-being (health) of that individual. The first factor to consider here is the vertical inheritance of information in replicons. There is good reason to suppose that transformation in cells could be related to such factors as:

(1) the levels of hormones in individuals,

(2) the strength of the immune system,

(3) the strength of the DNA repair apparatus, and

(4) the lability of the cell in other ways, such as the strength of its regulation.

All these factors are inheritable. Since we are describing cancer as an initial attack on genetic material and its regulation, there must be a wide range of individual cellular conditions which could predispose the DNA to attack or protect it from attack. These factors will be expected to depend on age and on changes of internal conditions at particular stages in life, such as puberty. Some cancers, for example breast cancer, have an incidence pattern very different from that defined by equation (1) (p. 25). Other cancers that we may suspect arise from predisposition in individuals and that do not follow this equation of incidence rate against age are the cancers of childhood, especially the leukaemias. There is also a strong predisposition to skin cancer as a result of lack of pigmentation, as in albino horses, and in patients with malignant melanoma, where the sufferers are now known to lack a DNA repair mechanism.

Alongside these vertically inherited predisposition factors, we must also remember changes during the life of an organism which can affect the steady states of cells and may then alter the ease of attack on the cellular genetic equipment, so altering the ease of transformation. Such changes can be thought of as internal 'horizontal' changes of disposition. An example would be the 'accidental' effect of normal internal metabolism which caused a labilization of genetic information or which damaged a repair mechanism.

There can also be an invasion of a living system—that is, an environmental effect (a diseased condition)—that is not itself carcinogenic but may leave the cell system in a condition more readily disposed to transformation at a later stage. For example, some viruses deposit genetic material in the genome. Likewise, stress may permanently alter the internal chemical balance of an organism much as age does. We must remember the lability of organisms, for there can be factors, both environmental and internal, which alter the relative number of cells in different cell compartments. A fully normal steady state may not be readily regained once a system is perturbed. Diet is now recognized as one major factor affecting susceptibility to cancer.

We can now draw up, at least in principle, a scheme which shows the interrelationship between cancer and factors arising both from disposition and environment. Underlying this scheme is the concept that in each person there is an idiosyncratic cellular balance of the kind:

$$\begin{matrix}\text{Totally 'normal'} & \underset{(?)}{\rightleftarrows} & \text{More labile} & \underset{(?)}{\rightleftarrows} & \text{Transformed} & \underset{(?)}{\rightleftarrows} & \text{Cancer} \\ \text{cells} & & \text{cells} & & \text{cells} & & \text{cells}\end{matrix}$$

The question marks are placed on the back reactions which Szent-Györgyi postulates as a possible way of curing cancer. The populations of normal, labile, transformed and carcinogenic cells then vary from individual to individual and the rates of forward and back reactions also vary. It is also true that the strengths of the various protections against cancer amongst other diseases vary from individual to individual (see above, p. 26).

RELATION TO SUBMOLECULAR BIOLOGY

We hope that this set of notes on cancer allows the reader to sense the complexity of the cancer problem. The *very simplest* model of cancer has to explain both carcinogenesis (normal cell → transformed cell) *and* the maintained state of this transformation in a long series of cell divisions.

In this symposium we shall be looking at the possible relationship between submolecular states and both these steps. Dr Szent-Györgyi himself has concentrated on the way in which cancer cells, as distinct differentiated cells, might differ from other types of cells. He has the idea that this difference could lie in the amounts of certain chemicals which themselves would alter the amounts of conducting proteins, and therefore of communication between proteins, in a cell. Thus his theory involves concepts about the properties of proteins generally (in particular that they could act as electronic conductors), and about the relationship of the loss of these properties to cancer. (Szent-Györgyi has always looked for some essential communication principle of all living cellular systems.) Extension of these ideas has led research groups not only to seek conductivity itself in proteins, by bulk electrical measurements, but also to search for the charge carriers, by looking for radicals. This symposium includes papers on both these points. The effect of three substrates as sources of radicals has always been stressed: oxygen, methylglyoxal, and ascorbic acid. The chemistry of these last two compounds will therefore be considered in detail.

Following on these ideas, it has been proposed that free radicals could be the cause of cancer (that is, they could act in the first step above); it has also been suggested that compounds derived from methylglyoxal could be carcinostatic. By this is meant that cells would no longer multiply abnormally as a result of the action of chemicals taken as drugs. If such drugs could be found, this step would be of immense significance, since the drugs used now in cancer therapy are mostly killers of cancer cells (and other cells). The symposium will look into these matters.

I hope it will be seen that Szent-Györgyi's ideas are of the greatest importance, since they focus on real problems in cancer studies. We shall examine their validity carefully in this symposium. Even if the central idea proves to be incorrect the fall-out from the analysis of the hypothesis by experiment may well be considerable, and it must force upon us the need to have some general theory relating cellular information content and the form of organisms.

Energy bands and charge transfer in proteins

K. LAKI, S. SUHAI* and J.C. KERTESZ

*National Institute of Arthritis, Metabolism and Digestive Diseases, National Institutes of Health, Bethesda, Maryland and *Institute of Theoretical Chemistry, University of Erlangen-Nürnberg, Erlangen, West Germany*

Abstract The effects of salts on protein—the causing of a shift in isoelectric point and the altering of the melting temperature—are proposed to be the result of binding to the protein peptide chain, which is considered as a one-dimensional solid. The interaction of methylglyoxal with protein and polylysine to give charge-transfer complexes and allow electrical conductivity are viewed as further support for the band structure of proteins. Calculations on protein chains resembling real proteins show that conductivity should be much less than expected for homopolypeptides.

Toward the end of the 1930s, the idea was advanced that proteins may be looked on as pieces of solid matter: as insulators or conductors (Jordan 1938; Möglich & Schön 1938; Riehl 1940). In the early 1940s, one of us (Laki 1942) proposed that these ideas be used to explain the effect of salts on proteins.

Salts have pronounced effects on proteins. For example, the isoelectric point may be shifted according to the nature of the salt; and neutral salts may alter the melting temperature of gelatin (Fig. 1). As a starting point, Laki took the data collected in Szent-Györgyi's laboratory on the effect of salts in shifting the isoelectric point of casein (Erdös 1942). He assumed that the electrons of the regularly repeating elements (CNOH) of the peptide chain form an energy continuum. As a theoretical approach, he proposed adding atoms of the salt to the protein which, depending on the properties of the protein energy band, the ionization energy and electron affinities of the atoms, would extract or add portions of an electron to the band. In this picture, chlorine would extract electron, thus producing a hole in the continuum, and potassium would add electron. The result would be that ions are attached to the protein which would gain a charge, depending on the nature of the salts.

On the basis of these considerations, the isoelectric point of casein was plot-

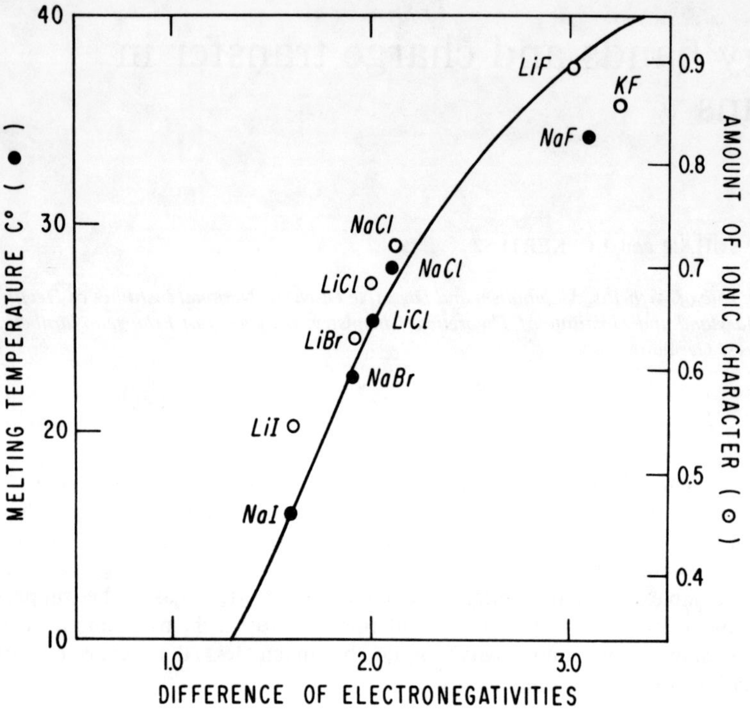

FIG. 1. The open circles and curve depict the amount of ionic character of the salts plotted against electronegativity differences as given by Pauling. Filled circles give the melting temperature of gelatin in the presence of 1M solutions of salts, also plotted against the electronegativity differences. (Data from von Hippel & Wong 1962).

ted against the electronegativity difference of the atoms composing the given salt: a smooth curve resulted (Fig. 2). This was a satisfying result because there was no need to list the effect of the salt separately according to its cations and anions.

No experiment with sodium or chlorine as visualized here has been done with proteins but, since the proposition was made, electron transfer between naphthalene and sodium producing an anion-radical has been observed. Similarly, naphthalene dissolved in sulphuric acid is known to give a cation-radical (Rozantsev 1970). Ion-radicals of this kind give e.s.r. signals.

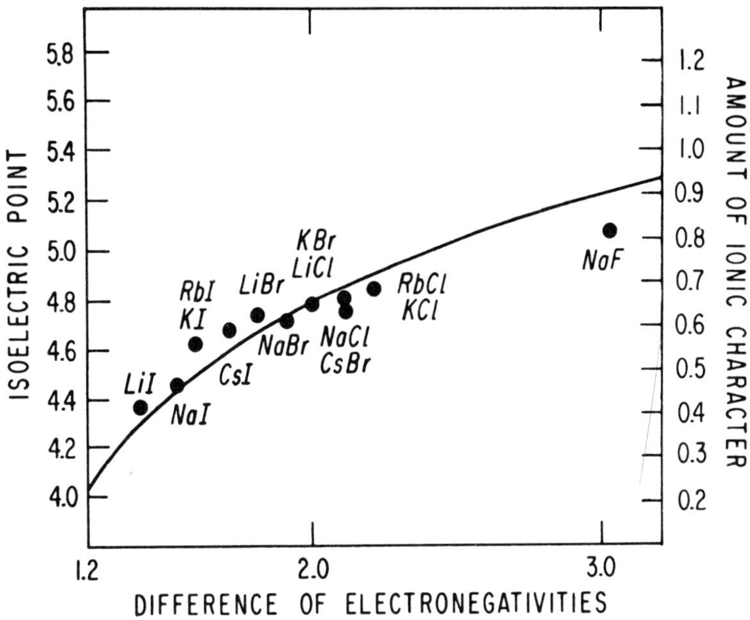

FIG. 2. The isoelectric point of casein (•) in the presence of various salts is plotted against the electronegativity differences of various salts. The curve is the ionic character of the bonds plotted against the electronegativity differences as given by Pauling.

The solid-state approach implied that proteins, along the hydrogen-bonded chain, as a one-dimensional continuum, should in certain conditions (e.g. salts and pH) become electronic conductors, perhaps feeble ones, but nevertheless conductors. To most biochemists at that time the proposition that an organic macromolecule could become a conductor was uninteresting (except for Szent-Györgyi who, himself, had similar ideas). Today we know that compounds like $(SN)_x$, a linear polymer of S and N, act in certain conditions like metallic conductors (Friend *et al.* 1977).

Theoretical calculations, especially those of Suhai & Ladik (1972) and Suhai (1974), showed that simplified models of proteins could become fairly good conductors along the hydrogen-bonded chain and also along the peptide chain, on account of the broad width of the electron energy bands.

Several attempts have been made to measure protein conductivity, but those results are not generally accepted as proofs of electronic conductivity, chiefly on account of the inherent difficulties of the measurements (Eley 1968).

Recently, Laki & Ladik (1976) have reported that protein molecules with partially filled valence or conduction bands should attract each other much more strongly than the same proteins with completely filled bands do. With partially filled energy bands, the proteins would acquire net charges, thereby increasing the electrostatic attraction between them. This, and the increased polarization and dispersion forces, most probably overcome the increased repulsions between the charged units of the chains (interchain interactions). Thus, on the basis of what was said above, one should expect salts to influence the aggregating properties of proteins. For gelatin it has been convincingly demonstrated that neutral salts influence the transition temperature of the sol–gel transformation (von Hippel & Wong 1962).

An interesting aspect of the studies on gelatin is the observation that the salts were effective even when ionizable groups of the gelatin were made inoperative. From this, Bello *et al.* (1962) concluded that the elements of the salts were bound to the peptide chain.

One of the propositions made by Bello & Bello (1962) and Bello *et al.* (1966) about the binding of LiCl to model peptides is worth considering:

$$-\underset{|}{C}=O \ldots Li \qquad \underset{\underset{H}{|}}{O}-H \ldots Cl \ldots H-N-$$

This proposition is based on X-ray diffraction studies on crystals formed when *N*-methylacetamide is mixed with lithium bromide. The ... Li ...O–H ... Cl ... unit inserted into the hydrogen-bonded peptide chain of a protein could act as an 'impurity' centre.

On model amides, it has actually been demonstrated that cations bind to the carbonyl oxygen and, in so doing, alter the spectral properties of the amide group (Balasubramanian & Shaikh 1973). This result provides further experimental support for the original proposition that salts affect the properties of proteins by 'binding' to the peptide chain (see Fig. 2, p. 35).

The solid-state model of proteins implies that, in an analogous manner, 'doping' with compounds that interact with the proteins should lead to charge transfer. Szent-Györgyi proposed that methylglyoxal would be such a compound which, by interacting with protein, should make it conductile; the charge transfer should be detectable by e.s.r. spectroscopy. Apparently this is so: a complex of casein with methylglyoxal exhibited an e.s.r. signal and weak electronic conductivity (Pethig & Szent-Györgyi 1977).

According to calculations, proteins should be fairly good conductors once the energy band becomes partially filled. All experimental measurements indicate rather feeble conduction. The explanation could be that the calculations have been done for simplified protein models, but the measurements have been made on real proteins. Our recent calculations suggest that real proteins should be poor conductors.

The basic common feature of earlier calculations was their use of simple strictly periodic models instead of aperiodic proteins. The translational elementary cell of these models consisted either of the four atoms of the planar peptide unit, with only the π-electrons of the resulting hydrogen-bonded chain being considered, or of a glycine residue forming a polyglycine chain in which all valence electrons have been taken into account explicitly (Evans & Gergely 1949; Suard-Sender *et al.* 1961; Ladik 1964; Suard-Sender 1965). However, real proteins are by their very nature always aperiodic, owing to the presence of different side-groups bound to the β-carbon atoms of the polypeptide backbone. As a first attempt to investigate the disorder effect of these side-groups on the electronic structure of a protein, we have calculated the energy bands and the densities of the electronic states in poly(Gly-Ala-Ser) by the same method.

The structural data of the polymers were taken from the work of Pauling & Corey (1953). The most interesting new feature of the band structure so obtained is that in about the same energy region where the conduction and valence bands of the three homopolypeptides were originally situated, we have obtained six new bands. Three lie between 0.3 and 1.2 eV and can be traced back to the conduction bands of the pure systems but, owing to the aperiodicity within the new elementary cell, instead of the broad original band $\Delta E_{cond} = 0.8—1.0$ eV, we now have three much narrower bands with $\Delta E = 0.135$, 0.339 and 0.175 eV. They are separated by two smaller gaps of 0.138 and 0.471 eV. In the same manner, the valence bands of the pure systems (lying in the region of -9.8 to -9.2 eV) have split into three new filled bands having $\Delta E = 0.718$, 0.171 and 0.002 eV with two small gaps of 0.134 and 0.093 eV between them. Fig. 3 shows the calculated densities of the electronic states for the three new bands in the valence band region. (The plots have been scaled to make the highest peak for each band of equal height.)

Recalling the examples of various organic semiconductors (Glarum 1963), we know that in materials of this kind the carrier mobilities are usually much smaller than 1 $cm^2 V^{-1} s^{-1}$ and consequently the mechanism of the charge transport may differ from that predicted earlier for homopolypeptides. Recent experimental results (Bone *et al.* 1978) have led to the same conclusion, namely, that in some proteins the mobility value is compatible only with an activated-type transport between localized states.

Conductivity calculations based on the electronic properties reported here and using a localized hopping-type transport theory are in progress.

In recent publications Szent-Györgyi has proposed that ketoaldehydes that

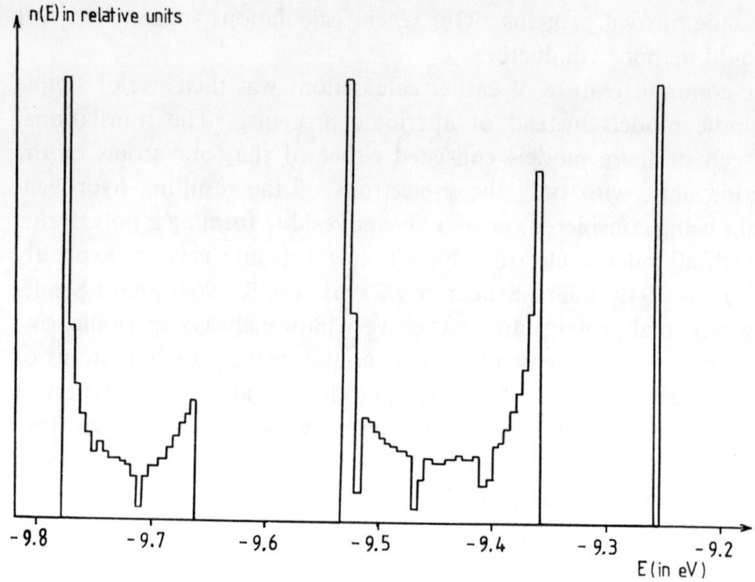

FIG. 3. The density of electronic states n(E) plotted for the three new bands of poly(Gly-Ala-Ser) in the original valence band region of the pure homopolypeptides.

can bind to proteins should act as electron acceptors and that the formation of a charge-transfer complex would create a hole in the energy band of the protein. Several reports indicate that with some proteins such a charge-transfer complex is formed with methylglyoxal and, as a result, the complex conducts (Bone *et al.* 1978). The exact mechanism of how methylglyoxal reacts with the protein is not yet known. Perhaps methylglyoxal interacts directly with the protein or a Schiff base is formed with lysine residues.

In our approach to this question, we used polylysine as a protein analogue to interact with methylglyoxal. From the work of Kertesz-Crisba (1977) we know that polylysine interacts with keto- and aldo-hexoses. When, similarly, polylysine interacts with methylglyoxal in methanol, a yellow-brown compound is formed. By ultracentrifugal analysis we found that this compound had twice the molecular weight of the polylysine; this indicates that enough methylglyoxal was bound, supposedly to the ϵ-NH_2 groups of the lysine residues.

However, titration with fluorescamine revealed a considerable number of free lysine residues and elementary analysis gave three oxygen atoms for every nine carbon atoms (carbon atoms of methylglyoxal and polylysine). This finding makes it doubtful that we are dealing with a Schiff base type reaction.

The yellowish-brown colour alone cannot be taken as evidence for Schiff base formation, since methylglyoxal in alkaline conditions forms a similarly coloured polymer. Lysine catalyses the formation of methylglyoxal polymers (Bonsignore *et al.* 1973). We have to assume, then, that in our polylysine–methylglyoxal complex, a polymer of methylglyoxal combines with the polylysine. ^{13}C n.m.r. studies of the complex not showing the characteristic line for the CH_3 group of methylglyoxal indicate polymer formation in which the CH_3 group of methylglyoxal participates in a manner proposed by Bonsignore *et al.* (1977).

^{13}C n.m.r. studies further revealed that the ϵ-carbon atom of polylysine is hardly affected by the reaction with aldehyde but, as Fig. 4 shows, the α and β-carbon atoms are drastically affected. If the polymeric methylglyoxal with its carbonyl groups really formed a charge-transfer chain, we could expect n.m.r.

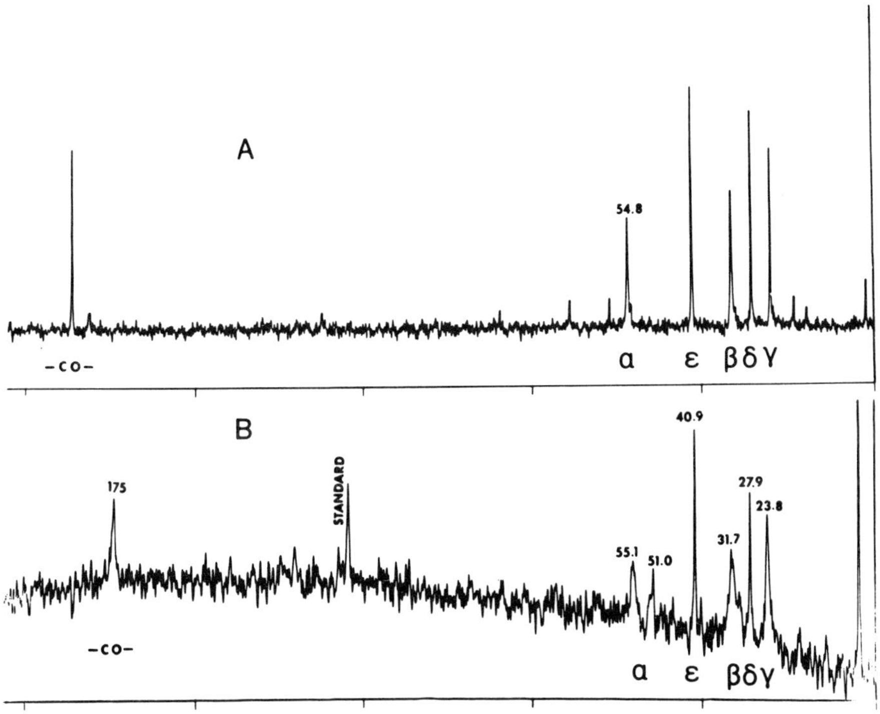

FIG. 4. ^{13}C n.m.r. spectra of (A) polylysine and (B) polylysine-methylglyoxal complex. In the complex the α-and β-carbon atoms of polylysine are influenced the most. In these experiments commercial (Sigma) polylysine (mol. wt 13 000) and methylglyoxal (ICN Pharmaceutical, Inc.) were used. For the reaction, the mixture was heated under reflux in methanol for five hours. After evaporation of the methanol the yellowish-brown solution was extensively dialysed against distilled water and lyophilized.

spectroscopy to show an effect on the α- and β-carbon atoms of polylysine.

One can imagine that a reaction between the carbonyl groups and the peptide nitrogen causes changes in the n.m.r. spectrum. We prefer the idea of the charge-transfer-type interaction since the methylglyoxal complex in the solid state gives a pronounced e.s.r. signal (Fig. 5). We do not dismiss the significance of this finding on the ground that lyophilized proteins often give e.s.r. signals. G. Szabo, Jr in our laboratory finds that the poly-lysine–methylglyoxal complex also gives such a signal in aqueous solutions.

Polylysine is a homopolymer; therefore, we could expect, in contrast to real proteins, a wide energy-band structure to exist with large mobilities for the charge carriers. Thus, this complex appears to be more suitable material for

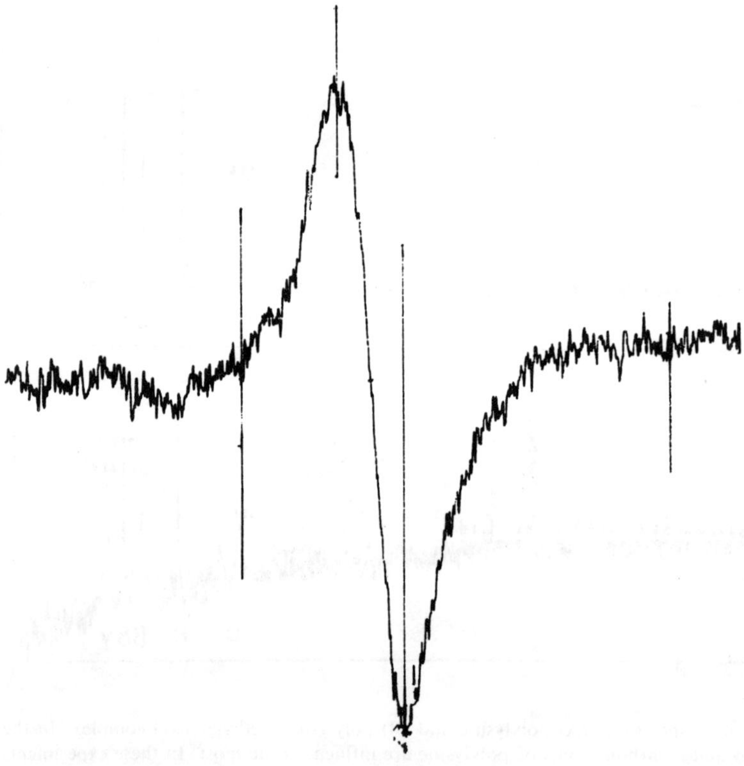

FIG. 5. E.s.r. spectrum of the polylysine–methylglyoxal complex measured at room temperature in a powdered form. Signal has the g value of 2.0058, and the peak-to-peak line width of about 11 G. The signal has no hyperfine structure resolved.

conductivity measurements than proteins. Such measurements could further support the contention that we are dealing with a charge-transfer complex.

At this stage, it seems worth considering that not only methylglyoxal but also its linear polymer interact in the cell with a particular protein. In such a case one could expect a catalyst, either an enzyme or simply a polyamine, to bring about the polymerization.

In summary, it appears that the solid-state approach to proteins is a sound one and that, as Szent-Györgyi's electronic theory of cancer indicates, it gives a new insight into basic biological phenomena.

ACKNOWLEDGEMENTS

We thank H. Kon, National Institutes of Health, Bethesda, Maryland, USA, and G. Szabo Jr, National Foundation for Cancer Research, Bethesda, Maryland, USA for the e.s.r. measurements, Dr J. Kaspar for his help in the preparation of the structural data and Dr R.D. Singh, Erlangen, West Germany, for the implementation of the MINDO/3 parametrization into our crystal-orbital program, as well as for the use of his program for calculating the densities of states. We also thank Professor J. Ladik for his interest and advice.

References

BALASUBRAMANIAN, D. & SHAIKH, R. (1973) On the interaction of lithium salts with model amides. *Biopolymers 12,* 1639-1656

BELLO, J. & BELLO, H.R. (1962) Evidence from model peptides relating to the denaturation of proteins by lithium salts. *Nature (Lond.) 194,* 681-682

BELLO, J., BELLO, H.R. & VINOGRAD, J.R. (1962) The functional groups in the gelation of gelatin. *Biochim. Biophys. Acta 57,* 222-229

BELLO, J., HAAS, D. & BELLO, H.R. (1966) Interaction of protein-denaturing salts with model amides. *Biochemistry 5,* 2539-2548

BONE, S., LEWIS, T.J., PETHIG, R. & SZENT-GYÖRGYI, A. (1978) Electronic properties of some protein–methylglyoxal complexes. *Proc. Natl. Acad. Sci. U.S.A. 75,* 315-318

BONSIGNORE, A., LEONCINI, G., RICCI, D., SIRI, A. & BIGNARD, G. (1973) Properties of the polymer formed from methylglyoxal in the presence of lysine. *Ital. J. Biochem. 22,* 105-116

BONSIGNORE, A., LEONCINI, G., ANDISIO, G., ZETTA, L. & FERRUTI, P. (1977) Characterization of the polymer formed from methylglyoxal in the presence of L(+)-lysine. *Ital. J. Biochem. 26,* 162-168

ELEY, D.D. (1968) Semiconducting biological polymers, in *Organic Semiconducting Polymers* (Katon, J.E., ed.) pp. 259-294, Dekker, New York

ERDŐS, T. (1942) The effect of salts on the isoelectric point of casein. *Stud. Inst. Med. Chem. Univ. Szeged 2,* 27-41

EVANS, M.G. & GERGELY, J. (1949) A discussion of the possibility of bands of energy levels in proteins—electronic interaction in non-bonded systems. *Biochim. Biophys. Acta 3,* 188-197

FRIEND, R.H., JEROME, D., REHMATULLAH, S. & YOFFE, A.D. (1977) Electrical conductivity in polymeric sulphur nitride at high pressures. *J. Phys. C. Solid-State Phys. 10,* 1001-1012

GLARUM, S.H. (1963) Electron mobilities in organic semiconductors. *J. Phys. Chem. Solids 24,* 1577-1583

JORDAN, P. (1938) Über die physikalische Struktur organischer Riesenmoleküle. *Naturwissenschaften 26,* 693-694

KERTESZ-CRISBA, J. (1977 *Etude des Corrélations entre Structure et Activité Biologique des Dérivés Glucidiques de la Lysine et de la Poly-L-Lysine,* Thèse de Doctorat D-Etat, L'Université Pierre et Marie Curie, Paris

LADIK, J. (1964) Energy band structure of proteins. *Nature (Lond.) 202,* 1208-1209

LAKI, K. (1942) On the electronegativity of atoms and their influence on the isoelectric point of casein. *Stud. Inst. Med. Chem. Univ. Szeged 2,* 43-48

LAKI, K. & LADIK, J. (1976) A note on the 'electronic theory' of cancer. *Int. J. Quantum Chem. Quantum Biol. Symp. 3,* 51-57

MÖGLICH, P. & SCHÖN, M. (1938) Zur Frage der Energiewanderung in Kristallen und Molekülkomplexen. *Naturwissenschaften 26,* 199

PAULING, L. & COREY, R.B. (1953) Two rippled-sheet configurations of polypeptide chains, and a note about the pleated sheets. *Proc. Natl. Acad. Sci. U.S.A. 39,* 253-256

PETHIG, R. & SZENT-GYÖRGYI, A. (1977) Electronic properties of the casein–methylglyoxal complex. *Proc. Natl. Acad. Sci. U.S.A. 74,* 226-228

RIEHL, R. (1940) Die 'Energiewanderung' in Kristallen und Molekülkomplexen. *Naturwissenschaften 28,* 601-607

ROZANTSEV, E.G. (1970) *Free Nitroxyl Radicals,* p. 6, Plenum Press, New York & London

SUARD-SENDER, M. (1965) Etude théorique de la délocalisation électronique dans les systèmes polypeptidiques. *J. Chim. Phys. 62,* 79-98

SUARD-SENDER, M., BERTHIER, G. & PULLMAN, G. (1961) Sur les états electroniques des protéines. *Biochim. Biophys. Acta 52,* 254-265

SUHAI, S. (1974) Theoretical investigations of semiconductive properties in protein. *Biopolymers 13,* 1731-1737

SUHAI, S. & LADIK, J. (1972) The electronic structure of periodic protein models. *Theor. Chim. Acta (Berl.) 28,* 27-35

VON HIPPEL, P.H. & WONG, K.-Y. (1962) The effect of ions on the kinetics of formation and the stability of the collagen-fold. *Biochemistry 1,* 664-674

Discussion

Williams: You are suggesting that the interaction of ions with proteins is one way in which the properties of a protein can be controlled by a special mechanism. Salts, in particular iodide, you say, create charges in the protein (excesses or deficiencies in electrons) and this changed state, the charged protein, is possibly of interest to us. To my mind, there are two ways in which the observations could be explained: (1) your way — a deficiency or excess of electrons created in the protein, and (2) ionic binding. You further pointed out that the effect of salt was also seen on the melting points of proteins. Consequently you are referring to a new state of a protein — a rather degraded state — which corresponds to melting. You imply that because salts will alter the equilibrium between the native protein and the 'charged' protein they will, therefore, affect the way in which both these species are related to the state that corresponds to a melt.

To get our ideas about proteins straight let me ask a question in a slightly aggressive way: who really believes that salts convert proteins into electron-excessive or electron-deficient molecules, rather than simply binding as ions? Once we have answered that we may be able to look at how one can induce charges in proteins.

Eley: Salts are unlikely to induce electron deficiency in proteins. The relative effectiveness of ions in altering macromolecular configuration is the classical lyotropic or Hofmeister series, which is observed in aqueous solution. Hey *et al.* (1976) studied the Hofmeister series in relation to the cloud point of certain micellar systems and suggested that the effect of ions was to change the free volume available in the solvent (water). So I conclude that the interaction between the salt and the water is important in Professor Laki's system. It may be a general explanation of the Hofmeister series.

Neuberger: Am I right in saying that the Debye-Hückel theory does not explain the Hofmeister series? It seems to me that Professor Laki has tried to explain the Hofmeister series in terms of electron deficiency. But can you explain it qualitatively?

Laki: Rather than considering the effect of adding salts to protein, in my paper I started out with an imaginary situation that I described in 1942. I assumed that proteins have band structure, like a piece of solid, and when an atom, say, chlorine, is added to it, an electron is attracted from the valence band creating a hole and resulting in an ion bound to the protein. Similarly, if we add, say, sodium, it will send an electron into the system, thus creating a positive ion and eliminating the holes totally or partially. A consequence of this picture is that, in certain conditions, depending on the salts, the protein would become a conductor, or would give an electron spin signal. I do not know how water would affect this picture. I would assume that in an experimental situation one would reduce the water content of the protein to a low level where we could still talk about bound ions.

A direct consequence of this picture was that by considering the ionization potential and electronegativities of the elements composing the salt, the isoelectric point changes of casein could be correlated to the electronegativity differences of the atoms composing the salt. This was a pleasing feature of this proposition, since the usual approach to salt effects on proteins is to talk separately about anions and cations.

In terms of chemistry, my approach implied that ions affecting proteins are bound in some manner to the peptide bond. In early 1940 the proteins were looked upon as giant ions and the effect of salts was explained as influencing the ionizable groups of the side-chains. Since then, it has been experimentally shown that in the case of the sol–gel change of gelatin the salts have their effect even when the side-chain ionizing groups are eliminated, compelling one to conclude that the ions are bound somewhere to the peptide bond. Experiments with peptide models show directly that ions do bind to peptide bonds.

What I have presented here does not prove that proteins have a band structure but that, if they do, salts will have such an effect. The question will not be decided on who believes in it, but on what further experiments will tell.

Williams: So you reject the idea of changes in the ionization of the protein groups which could be related to the Debye-Hückel theory. That leaves two alternatives: binding of the ions, and some induction of charges which appear in the unsaturated part of the protein system.

Eley: The binding of the ions is important but, in order to be bound to proteins, ions have to lose solvent water. This may be the connection between the solvent water idea that I favour and your concept of ion binding.

Ladik: If the protein has some kind of energy-band structure, binding of an ion allows the possibility of electron mobility. There will then always be a partial charge shift that will generate holes.

Williams: That brings me to the next point that Professor Laki raised. Obviously we feel that there is something still left unexplained about the lyotropic series and the effect of ions on proteins. But that in itself doesn't introduce the notion of electron *mobility,* which was part of Professor Szent-Györgyi's idea. For the naive chemist your conduction ideas may be interpreted by considering the ground and excited states of a system. The ground states are usually one of paired electrons while the excited states contain no electrons. Unsaturation simply makes the gap between the ground and excited states smaller. When one constructs a large unit from many molecules, many orbitals of the same type are brought together. When each molecule is sufficiently far away from every other, each molecule has the same energy level. But when the orbitals coalesce into what is called a band structure, each level is split into several close energy levels (see Fig. 1, p. 45). This behaviour applies to any structured solid—diamond, metal or whatever. The distinction between conducting systems and non-conducting systems is then the relative energy difference E_a between the ground state and the first excited state. As the highest ground state level comes nearer to the lowest excited state level it becomes easier to transfer an electron from the former to the latter. In the ground state (of proteins, DNA, or whatever) the ground states are paired states—they contain no unpaired electron density: e.s.r. spectroscopy shows

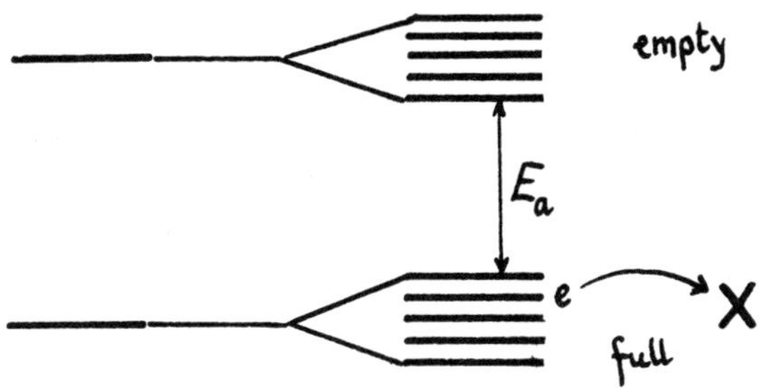

FIG. 1. (Williams). Ground (full) and excited (empty) states of a molecule (left) and of a large unit of many molecules (right), showing diagrammatically the band structure. E_a is the difference between the ground state and the first excited state. X is an impurity.

nothing. However, as E_a decreases the Boltzman factor means that some electrons will appear in the excited state. They may be unpaired and thus identifiable. The questions relevant to a theoretical discussion are: how big is the energy gap, E_a? And what affects this gap? (Professor Ladik will discuss some of these points in his paper.) Professor Laki, you believe that a chemical, specifically methylglyoxal, will affect this transfer of electrons. Is that correct?

Laki: Yes, but not in the way you propose. Removal of an electron from a filled band will create a hole. In other words, it is not necessary for the electron to be raised to a higher energy level.

Williams: That is, addition of an external component, an impurity X (see Fig. 1), allows 'sideways' transfer of an electron and creates a mobility path. Of course I accept that possibility.

Eley: In our original note on hole injection into bovine plasma albumin (Davis *et al.* 1960) we shook an aqueous solution of bovine plasma albumin (BPA) with *p*-chloranil (i.e. *p*-tetrachlorobenzoquinone), filtered off excess chloranil, and evaporated the water to give an albumin–chloranil complex film (0.5 wt % chloranil) which showed a weak e.s.r. signal ($g = 1.998$) and an enhanced electrical conductivity, by a factor 3×10^5, over the dry albumin film alone. In experiments done ten years later (Beales *et al.* 1979) it was found that chloranil alone reacted with water to give a narrow single e.s.r. line ($g = 2.005$, linewidth 1.0 G), characteristic of the semiquinone radical anion. On addition of bovine plasma albumin the line broadened to 8.1 G in solution, the solid complex formed by evaporating off the water showing a similar signal of width 7.3 G. Trichlorobenzoquinone similarly reacted with water to give two narrow lines, separated by 1.7 G, the splitting presumably due to hyperfine interactions with the single proton on the quinone. This splitting was lost and the line broadened on addition of BPA to the solution. Tetrafluorobenzoquinone on the other hand retains some hyperfine structure on interacting with dissolved BPA. These results show that the previously observed e.s.r. signals in chloranil–BPA complexes were due to the chloranilsemiquinone formed by previous reaction with water, followed by a subsequent weak association of this free radical with a site on the protein molecule which causes the line broadening. Whether the increased conductivity of the protein arises from a chemical reaction with the semiquinone ion, or by simple charge transfer to uncharged chloranil molecules which may also be present, cannot be decided at present. The e.s.r. signal for BPA/H_2O/ trichlorobenzoquinone was enhanced by illumination, on a time scale of minutes, rather like the effects reported earlier (Ilten *et al.* 1968). For a fuller understanding of the system, more work is necessary.

Pethig: In collaboration with Peter Gascoyne at Woods Hole we have found that the coloured protein-methylglyoxal complexes mentioned by Professor Szent-Györgyi do exhibit e.s.r. signals with hyperfine structure. One of the two main components of the e.s.r. signal can be interpreted in terms of proton interactions and these interactions appear to be directly related to the degree of hydration of the coloured protein. Indirect evidence leads us to believe that the proton splitting of the e.s.r. spectrum is produced at the same site at which methylglyoxal interacts with the protein and also where water preferentially adsorbs on the coloured protein–methylglyoxal complex. Although the coloured proteins tend to be insoluble, Joyce Eden in our laboratory has found from hydration isotherm studies of casein that at low partial pressures of water the casein–methylglyoxal complex takes up more water than normal casein, and only at higher partial pressures of water does the complex exhibit an isotherm of the form expected for insoluble material. This observation meant that either the number of primary adsorption sites had increased (maybe owing to denaturing of the protein) or, as we believe, the adsorption of water on the coloured protein had initially increased because the local charge-transfer interactions at lysine groups (see pp. 83-98) had produced regions of larger polarizability where water molecules could be held more strongly. In this way the proton splitting of the e.s.r. signal and the charge-transfer interaction that produced the unpaired electrons can possibly be related directly to each other.

Laki: We cannot yet say whether the e.s.r. signal from the complex of methylglyoxal with polylysine is related to the possible conductivity. Polymerization of methylglyoxal by heat or addition of amines gives a yellowish-brown solution. The polymer gives an e.s.r. signal. On cooling the polymer to 77 K we found that the signal disappears but that from the polylysine–methylglyoxal complex does not. For that reason we believe that the interaction is one of charge transfer.

Fodor: With regard to the way methylglyoxal interacts with amines and with proteins, I see no need to invoke a polymer of methylglyoxal because there is no evidence for the formation of such a polymer. We have studied the chemistry of methylglyoxal (see pp. 165-174). We detected no appreciable decrease in the intensity of either the ^{1}H or ^{13}C n.m.r. signals of the methyl groups. However, in aqueous solution the ^{1}H signal at the carbonyl of the aldehyde disappears. This disappearance is incompatible with an aldol-type structure for the polymer. The type of polymer proposed by Bonsignore *et al.* (1977) does not explain this. Professor Ladik's idea about the folding of the methylglyoxal–lysine Schiff base towards the peptide bond to form a charge-transfer complex is much easier to accept. Following Professor Szent-Györgyi's qualitative experiments, we found that there is free-radical formation between an excess of a primary amine (RNH_2) and methylglyoxal. So, although the aldehyde group is blocked, the keto carbonyl is capable of such an interaction. Stereochemically this is possible. Also, the sequence of reactions proposed by Bonsignore *et al.* for the reaction with amines to form pyrroles is unlikely.

Laki: I see no reason why the polymer proposed by Bonsignore *et al.* (1977) shouldn't form a Schiff base, but the Schiff base is not formed to the extent one would expect with the monomeric methylglyoxal.

Connors: Professor Laki, you stated that in your experiments the reaction of methylglyoxal and polylysine gave a product which had twice the molecular weight of polylysine (p. 38). A large amount of methylglyoxal or its polymer has therefore reacted with the polypeptide. Is such a large amount of methylglyoxal needed to remove an electron from the valence band? If so, I can hardly see how this could be an important physiological event.

Laki: Instead of a protein with a few lysine groups we are using polylysine. Not all the lysine groups in polylysine bind methylglyoxal: some free amino groups can be titrated.

Neuberger: Does polylysine dimerize? After all, methylglyoxal has two reactive carbonyl groups that could cross-link two polylysine molecules.

Laki: That is possible, but elemental analysis of the complex suggests that it does not. There was almost one methylglyoxal (three extra carbon atoms) for every lysine (N:C ratio, 1:4). The cross-linked polylysine dimer could give a very different analysis (N:C ratio, 1:2.5).

Williams: One could react 1 mol of protein with 1 mol, 2 mol etc. of methylglyoxal.

Abdulnur: The lack of reactivity of methylglyoxal with the ϵ-amino group of lysine that Professor Laki described might reflect a peculiarity of the polylysine. The work of Bonsignore *et al.* (1977) shows that on treating lysine itself with methylglyoxal the maximum shift in the ^{13}C n.m.r. spectrum occurred in the ϵ-carbon atom of the lysine, indicating the involvement of the nearby NH_2 group in the reaction.

Ladik: What is the evidence for the formation of the polymer of methylglyoxal?

Laki: The experimental evidence so far is the formation of the yellowish-brown compound from methylglyoxal.

Pethig: Joy Behi and Michael Arnold in our laboratory found that when polylysine was treated with randomly ^{14}C-labelled methylglyoxal (kindly supplied by Professor Slater) at pH 4.5 a yellow product was formed. After vigorously washing to remove all the remaining unreacted methylglyoxal, according to the ^{14}C measurements on average only one in two of the lysine groups had a methylglyoxal molecule attached to it.

Williams: Was the methylglyoxal itself polymerized?

Pethig: At pH 4.5 methylglyoxal does not readily polymerize and so we believe that the initial reaction with protein essentially involves the monomeric form of methylglyoxal. I agree with Dr Connors' point about the small number of methylglyoxal molecules required to interact with a protein. Our experience with more conventional electronic materials suggests that a methylglyoxal concentration of only one part in 10^6 is sufficient to alter significantly the electronic properties of a macromolecule.

Laki: I agree; very few methylglyoxal molecules (whether monomeric or polymeric) need react with the ϵ-amino group of lysine, yet the fact is that a lot of methylglyoxal was bound to polylysine. This implies the formation of a polymer.

Lewis: I should like to make a comparison with solid-state physics to emphasize Dr Pethig's point. Professor Williams has also mentioned the concept of an 'impurity' capable of taking electrons from the filled valence band. In most solid-state electronic devices the matrix in which the different electronic processes occur is crystalline silicon, a completely covalently bonded and insulating structure. However, the addition of appropriate acceptor material to this in concentrations no greater than 1 part in 10^9 results in a highly active electronic conductor. It could be, therefore, that the degree of reaction or the amount of charge transfer required to effect a significant change in the electronic properties of the macromolecules we are considering here is well below that traditionally considered to establish a biochemical reaction.

Kornberg: I am very much intrigued by the concept of polymer formation from methylglyoxal in an aldol-type condensation, but one has then to assume that this reaction is proportional to some power of the concentration of methylglyoxal. Professor Szent-Györgyi reminded us of a powerful enzyme present in most tissues which rapidly removes methylglyoxal. Therefore, the concentration must be extremely low. I don't want to anticipate Dr Brandt's paper by asking where methylglyoxal comes from in the first place but I wonder what is known about the concentration of methylglyoxal in living matter and what is the likelihood of a polymer being formed in those circumstances?

Laki: Amines catalyse the formation of the polymer, and so the question should be: is the concentration of methylglyoxal high and is the concentration of amines that would catalyse it high enough?

Fodor: The low concentration of methylglyoxal in the tissues underlines Professor Neuberger's point (p. 47); a doubling of the molecular weight of polylysine is much more likely to be due to inter-chain linking than to polymerization. The aldehyde will form a Schiff base while the carbonyl group will, for stereoelectronic reasons, form a charge-transfer complex with another chain of polylysine. Frankly I don't believe that elemental analysis of such a polymer would distinguish between a dimer and a cross-linked product.

Williams: So far, several points have been raised. First, we don't seem to be too sure about the very simple chemistry of proteins. We should clear that up later in our discussions. Second, we are not sure what is the reaction of methylglyoxal. I should like to hear a clear account of what must be straightforward chemistry. (That topic, too, is bound to be raised later.) These two topics may be nothing but molecular chemistry of the kind that organic chemists can understand. Third, we do not know the concentrations of the reactants *in vivo*. So, even if we know what the reaction is we do not know the extent to which it will occur in biology. This must be established before we discuss relationships to cancer, which is a biological phenomenon, not just a chemical one. Fourth, the question arose of whether the special chemistry of the aldehydes occurs through the production of free radicals. So far we have not established that; e.s.r. signals have been mentioned but nobody has said what they are. After all, practically any organic crystal will give an e.s.r. signal. It is also very important to show that these free radicals have something to do with the 'submolecular biology' part and then the 'cancer' part of the title of this symposium. Finally, we have to relate them to an energy-level diagram in order to appreciate the mobility that Dr Szent-Györgyi discussed.

References

BEALES, K.J., COOPER, W.D. & ELEY, D.D. (1979) Protein quinone complexes. Bovine plasma albumin and halogenated *p*-quinones. *J. Bioenerg. Biomembr.*, in press

BONSIGNORE, A., LEONCINI, G., ANDISIO, G., ZETTA, L. & FERRUTI, P. (1977) Characterization of the polymer formed from methylglyoxal in the presence of L(+)-lysine. *Ital. J. Biochem. 26,* 162-168

DAVIS, K.M.C., ELEY, D.D. & SNART, R.S. (1960) Semiconductivity in proteins and haemoglobin. Enhanced semiconductivity in protein complexes. *Nature (Lond.) 188,* 724-725

HEY, M.J., CLOUGH, J.M. & TAYLOR, D.J. (1976) Ion effects on macromolecules in aqueous solution. *Nature (Lond.) 262,* 807-809

ILTEN, D.F., KRONENBERG, M.E. & CALVIN, M. (1968) Reversible EPR photosignals in charge-transfer complexes. *Photochem. Photobiol. 7,* 331-340

Electronic structure of proteins and DNA: solid-state aspects

JÁNOS J. LADIK

Laboratory of the National Foundation for Cancer Research at the Institute for Theoretical Chemistry, Friedrich-Alexander-University Erlangen-Nürnberg, Federal Republic of Germany

Abstract The generalization of the Hartree–Fock method to periodic systems (polymers or crystals) using a linear combination of atomic orbitals (LCAO) makes it possible to calculate *ab initio* self-consistent-field LCAO band structures of periodic protein and DNA models. The results obtained for polyglycine, polyalanine, and for the mixed poly(Gly-Ala) periodic chain, as well as for the four homopolynucleotides, are presented. The correction of these band structures for excitonic effects (in the excited state) and for long-range correlation effects is shown also. Furthermore, it is outlined how the short-range correlation in insulator biopolymers and correlation effects in proteins with a partially filled valence band (for instance, due to charge transfer) can be calculated.

The Coherent Potential Approximation is outlined and its possible application to aperiodic proteins and DNA is pointed out. Finally the effect of an electron acceptor on proteins or of a chemical carcinogen bound to DNA is discussed as a local perturbation of the band structures of these systems on the basis of the self-consistent resolvent method.

BAND STRUCTURE CALCULATION METHODS FOR BIOPOLYMERS

Biopolymers like proteins and DNA play a fundamental role in life processes. To understand the different physical and chemical mechanisms which underlie their biological functions, including the detailed mechanism of tumour development, one has to have a fair knowledge of their electronic structure.

The proteins and DNA are built up from molecules linked together in a polypeptide chain or stacked in a polynucleotide. This means that for their quantum mechanical treatment those methods are needed which take into account properly both the molecular and macromolecular (solid-state physical) aspects. The self-consistent-field (SCF) linear combination of atomic orbitals (LCAO) crystal orbital (CO) method formulated by Del Re *et al.* (1967) was found to satisfy these requirements. This method builds up the orbitals of the electrons in a crystal (or polymer) by linear combination of atomic orbitals (AOs) centred on each atom of the polymer. For the determination of the coefficients of the linear combinations one can derive a matrix equation in which the matrices have the order of infinity, if the number of AOs goes to infinity. If, however, the polymer (or crystal) possesses translational symmetry, the problem can be reduced easily to a set of matrix equations in which each matrix has only the order of the number of orbitals in the unit cell. One can also generalize the theory (using elementary group-theoretical arguments) to the case of those periodic systems in which the periodicity is due not to a simple translation but to a combined symmetry operation (for instance: helix operation = translation + rotation) (Merkel 1977).

To obtain a more accurate description of the COs one has to take into account the fact that the potential which acts on an electron in a given CO depends on the distribution of all the other electrons. Since the distribution of the electrons depends on their COs, we arrive at a non-linear system of equations (or, in a more condensed form, a matrix equation in which the matrix elements depend on the eigenvector components) for the determination of the orbitals in the polymer or crystal. To solve this problem one assumes a set of starting COs, and, with the aid of them, forms the matrix elements and solves the matrix equation. This iterative procedure is repeated until one obtains consistent (SCF) results for the COs.

The programs which have been applied to calculate the band structures of different periodic protein and DNA models (homopolypeptides and homopolynucleotides, respectively) take into account explicitly all the electrons, all the interaction integrals between them, and also determine the COs $[\psi_i(k)]$ and their energies $[\epsilon_i(k)]$ in an SCF way (k is the crystal momentum which specifies the $\psi_i(k)$ and $\epsilon_i(k)$, respectively, in the ith energy band). Therefore we can say, in accordance with the convention used in quantum chemistry, that the crystal-orbital outlined methods provide *ab initio* Hartree–Fock band structures. Finally, it should be pointed out that in the calculations the periodicity due to the combined symmetry operations (and not only due to the simple translation) has been taken into account.

APPLICATION TO PERIODIC PROTEIN AND DNA MODELS

We have applied the *ab initio* SCF LCAO CO method outlined above to the two simplest homopolypeptides, polyglycine and polyalanine and to the mixed but periodic Gly-Ala-Gly-Ala-Gly-Ala ... chain (Suhai *et al.* 1978; Suhai & Ladik, unpublished results 1978) as well as to the four homopoly(nucleotide bases), polycytosine, polythymine, polyadenine and polyguanine using a minimal basis (Suhai *et al.* 1977; Ladik *et al.* 1977; Suhai & Ladik, unpublished results 1978). In the polypeptides and the homopoly(nucleotide bases), the parallel chain β-pleated chain geometry and the configuration which corresponds to the Watson–Crick model of DNA B, respectively, were applied.

The results indicate that poly(Gly) has a valence band with a width of 2.10 eV and a conduction-band width of 1.38 eV. The equivalent values for poly(Ala) are 1.82 eV and 1.17 eV, respectively. The gap between the valence and conduction bands is 12.97 eV for poly(Gly) and 13.54 eV for poly(Ala) (see Table 1). We can see from these figures (as well as from the more detailed data) that poly(Gly) and poly(Ala) have a similar band structure and Table 1 shows that the positions of these bands are also similar. I should mention that since 12 of the 20 essential amino acids have side-chains which start with a CH_2 group one can expect that the homopolypeptides built up from these amino acids would also have band structures similar to the poly(Ala) band structure.

In the poly(Gly-Ala) mixed chain the valence and conduction bands become essentially narrower because the original bands of poly(Gly) and poly(Ala), respectively, split into two bands with small gaps between them (a gap of 0.22 eV for the valence band and of 0.11 eV for the conduction band). The occurrence of these gaps is the consequence of the aperiodicity in the poly(Gly-Ala) mixed system. From this first result we can conclude that in proteins the aperiodicity of the amino acid sequence most probably strongly affects the electronic structure.

Turning now to the four homopoly(nucleotide bases), in Table 2 we give their valence and conduction bands. As the table shows, the widths of the valence and conduction bands of the four systems are again similar (owing to the same overall stacked arrangement) but, in contrast to poly(Gly) and

TABLE 1

The valence and conduction bands (in eV) of polyglycine, polyalanine and of the poly(Gly-Ala) mixed chain

Polypeptide	*Valence band* ϵ_{min}^{CO}	*Valence band* ϵ_{max}^{CO}	*Band width* *($\delta\epsilon$)*	*Conduction band* ϵ_{min}^{CO}	*Conduction band* ϵ_{max}^{CO}	
Poly(Gly)	−11.252	− 9.154	2.098	3.817	5.195	1.378
Poly(Ala)	−11.383	− 9.567	1.816	3.975	5.146	1.171
Poly(Gly-Ala)[a]	−10.559	− 9.342	1.217	3.973	4.369	0.396
	−11.093	−10.782	0.311	4.481	5.127	0.646

[a] For the mixed poly(Gly-Ala) chain we have given the data for the highest filled (valence) band and for the next band below it, as well as for the conduction band and for the next unfilled band.

TABLE 2

The valence and conduction bands (in eV) of the homopoly(nucleotide bases)

Homopolynucleotide	*Valence band*		*Band width*	*Conduction band*		
	ϵ_{min}^{CO}	ϵ_{max}^{CO}	*(δϵ)*	ϵ_{min}^{CO}	ϵ_{max}^{CO}	
Poly(C)	−9.665	−9.133	0.552	1.535	2.775	1.240
Poly(T)	−9.579	−9.153	0.426	3.003	3.928	0.925
Poly(A)	−9.179	−8.767	0.412	2.815	3.778	0.963
Poly(G)	−8.033	−7.398	0.635	3.679	4.731	1.052

poly(Ala), the positions of the bands show larger deviations (due to the larger differences of the electronic structures of the single molecules). The Hartree–Fock gaps obtained, similarly to the homopolypeptides, are again too large: between 10.5 eV for poly(C) and 12.4 eV for poly(T).

EXCITONIC EFFECT

If one compares the experimental u.v. spectra of the homopolypeptides and homopolynucleotides with the Hartree–Fock gaps, one finds in accordance with general experience in solid-state physics that the latter are far too large. The physical reason for this is that the Hartree–Fock method treats the excited electron as if it sees the potential of all N electrons (V^N potential), while in reality the potential of the other $N-1$ electrons (V^{N-1}) acts only on it. There is a comparatively simple way to correct this failure by introducing instead of the original Fock operator a so-called excitation Hamiltonian that takes into account this effect in a proper way in the case of the excited states and leaves the filled levels unchanged (for the description of this so-called ÔÂÔ method see Collins & Kunz 1974). By correcting the energy levels of the single units and shifting the conduction bands of the chains built up from these units one obtains gaps between 6.9 eV [poly(C)] and 4.7 eV [poly(G)] for the homopolynucleotides (for details see Ladik *et al.* 1977) and a gap of 6.6 eV for poly(Gly); these values agree tolerably well with the experimental data. For a more accurate description of the excited states of polymers (which would provide also the width and the dispersion of the exciton bands) one has to apply more sophisticated methods (such as the intermediate exciton theory; for a description see, for instance, Ladik 1978). Such calculations are in progress.

CORRELATION CORRECTIONS

The *ab initio* crystal orbital method already described, of ordering two electrons with opposite spins to a CO, does not take account of an essential physical effect, namely the fact that electrons in reality try to avoid each other because of their mutual repulsion (around each electron there is a so-called Coulomb hole). To take account of this effect one has to describe the distribution of the electrons in a correlated way. For the treatment of this so-called electron correlation well-established methods are available for atoms and smaller molecules but their application to extended systems and solids presents formidable problems.

In the case of polymers with completely filled valence bands and empty conduction bands the problem is comparatively easier. We can subdivide the correlation into a long-range and a short-range part. The latter can be treated with the methods used for molecules after transforming the COs delocalized over the whole polymer into localized so-called Wannier functions. Such an investigation, applying a method due to Čižek (1966), is in progress for insulator biopolymers with not-strongly-interacting units (such as a stacked homopolynucleotide).

For the treatment of the long-range correlation one can use the electronic polaron model. According to this model an electron moving in a polymer (or in a crystal) polarizes the distribution of the other electrons in such a way that the next electron entering the same region sees a more attractive (less repulsive) potential field. This phenomenon has the result that the motion of electrons far apart from each other will also not be independent but becomes correlated (long-range correlation). Mathematically this polarization of the electronic distribution can be described by a proper mixing of the states belonging to the valence and to the conduction bands (see, for instance, Collins *et al.* 1973). This method has been applied to the *ab initio* band structures of the polypeptides and homopolynucleotides under consideration. In contrast to the cases of simple metals and ionic crystals (Collins *et al.* 1973) where this correction has large effects on the band structures, we have found that the widths of the valence and conduction bands and the gap between them decreased only by about 10% in the case of poly(Gly) (Suhai *et al.* 1978) and by 5–10% in the cases of the four homopolynucleotides (Ladik *et al.* 1977; Suhai & Ladik, unpublished results 1978). From these results we conclude that most probably the long-range correlation effects are not essential in the organic polymers.

The treatment of correlation for polymers with a partially filled valence band (semiconductor polymers such as proteins which are interacting with electron acceptors) presents a much more difficult problem, because (1) in this case the correlation cannot be divided into short- and long-range effects, and (2) one cannot localize the filled delocalized COs without mixing with them unfilled COs belonging to the same band. One could obtain a reasonable estimate of the correlation by applying the inhomogeneous electron gas model (see, for instance, Stoddard *et al.* 1974) using the electronic density obtained from the Hartree–Fock band structure calculation (N.H. March, personal communication 1978). Another, more tedious, way would be to try to localize states also in a partially filled band by introducing a suitable complex Hamiltonian and to perform a configuration interaction between those states that fall into regions of high density of states (T.C. Collins, C. Nicolaides, J. Ladik & P. Winkler, unpublished results 1978). Theoretical investigations of these possibilities are in progress.

THE TREATMENT OF APERIODICITY AND LOCAL PERTURBATIONS IN BIOPOLYMERS

In reality both proteins and DNA are aperiodic, having 20 and four components, respectively. To treat these substitutionally disordered polymers one can start from the band structures (density of states) of the periodic components and apply the Coherent Potential Approximation (CPA) method. This method in the simplest two-component, one-band case is based on the assumption that one replaces the average of the two- (or more) component system by an effective medium determined so that the average fluctuation (scattering) through the medium is zero. This requirement in its mathematical formulation gives equations which serve for the determination of this coherent potential, or better for the effective Green's function of the medium (for a recent review of the CPA method see Martino 1978).

When the density-of-states curves of the two- (or more) component chains differ considerably from each other (which seems always to be the case for multicomponent polymers), one has to apply a more refined version of the CPA method than in the case of geometrically disordered systems. [More technically, one cannot apply a constant self-energy, but has to work with a k- and energy-dependent self-energy, $\Sigma(k,E)$.] This newly developed and refined version of the CPA method has been applied to the $(SN)_x$–$\underset{\displaystyle H}{(SN)_x}$ mixed system (Seel *et al.* 1978), because at IBM San Jose 5–10% hydrogen has been found in $(SN)_x$ (S. Faulkner & B. Györffy, personal communication 1977). Even at 3 molar % hydrogen spikes and dips have been found in the density of states of the mixed system. The occurrence of small gaps in the mixed system seems to be a genuine disorder effect which until now has not been found in simple (single-site) CPA calculations. The nature of these gaps is now under study.

Using the CPA method outlined (by generalizing it to more than two component systems and to the multiband case) we can start to investigate aperiodic DNA and proteins also. In the latter case to decrease the number of components we intend to perform the calculations on randomly mixed protein chains with four or five essentially different amino acids. These calculations are also in progress. In this way we shall obtain a fairly good description of the density-of-states curves and energy dispersions [$\epsilon_i(k)$ curves] of aperiodic proteins and DNA.

If an electron acceptor binds to a protein chain (see Szent-Györgyi, this volume) or if a chemical carcinogen is bound to DNA (see, for instance, Ladik 1978), one has to treat a local perturbation acting on the band structure of an extended system. To discover the extra levels of the perturbing molecule embedded in the extended system and the change in the band structure of the (periodic or aperiodic) extended system due to this local perturbation one can apply a generalization of the resolvent method originally introduced by Koster & Slater (1954) to the case of an *ab initio* SCF procedure with an arbitrary number of orbitals in the impurity molecule (Ladik & Seel 1976). When the necessary computer programs are completed we shall be able to treat the effect of local perturbations in one-dimensional protein and DNA chains and also in two-dimensional protein networks (the programming of the *ab initio* CO method for two- and three-dimensional periodic systems is already nearly completed).

CONCLUSION

This brief review has perhaps shown that with the appropriate adaptation of the methods of theoretical solid-state physics complex biopolymers can and will be treated in the near future in a sufficiently accurate way. By going one step further we can investigate the interactions between biopolymers on the one hand and their transport properties (determining first their phonon spectra) on the other hand. In this way the theory can provide essential contributions for the understanding of the mechanisms of conduction in proteins and of the detailed effects of chemical carcinogens bound to DNA. The solution of both problems should make an important contribution to the understanding of the microscopical mechanism of tumour development. This understanding—as Szent-Györgyi has pointed out many times—is necessary for the control of cancer.

ACKNOWLEDGEMENTS

I express my deep gratitude to Professor Albert Szent-Györgyi for his continuous encouragement and for many stimulating discussions with him. I am further indebted to Professors T.C. Collins, F. Martino and J. Čižek and to Drs S. Suhai and M. Seel for their cooperation and many fruitful discussions.

References

ČIŽEK, J. (1966) On the correlation problem in atomic and molecular systems. Calculation of wave function components in Ursell-type expansions using quantum-field theoretical methods. *J. Chem. Phys. 45,* 4256-4265

COLLINS, T.C. & KUNZ, A.B. (1974) Excitation Hamiltonian of electronic systems using ÔÂÔ. *Int. J. Quant. Chem. 58,* 437-446

COLLINS, T.C., KUNZ, A.B. & DEVREESE, J. (1973) An excited state of alkali halides. *Int. J. Quant. Chem. 57,* 551-557

DEL RE, G., LADIK, J. & BICZÓ, G. (1967) Self-consistent-field tight binding treatment of polymers. I. Infinite three-dimensional case. *Phys. Rev. 155,* 997-1003

KOSTER, G.F. & SLATER, J.C. (1954) Simplified impurity calculations. *Phys. Rev. 96,* 1208-1223

LADIK, J. (1978) Electronic structure of biopolymers, in *Quantum Theory of Polymers* (André, J.-M., Delhalle, J. & Ladik, J., eds.), pp. 257-278, Reidel, Dordrecht & Boston

LADIK, J. & SEEL, M. (1976) Electronic structure of aperiodic polymers. I. The average-matrix method and the effect of a cluster of impurities on the band structure of a periodic system. *Phys. Rev. B 13,* 5338-5346

LADIK, J., SUHAI, S., OTTO, P. & COLLINS, T.C. (1977) Estimation of the band structure of DNA on the basis of an *ab initio* SCF LCAO band structure of polycytosine. *Int. J. Quantum Chem. Quantum Biol. Symp. 4,* 55-63

MARTINO, F. (1978) Methods for the treatment of disordered systems with possible applications to aperiodic polymers, in *Quantum Theory of Polymers* (André J.-M., Delhalle, J. & Ladik, J., eds.), pp. 169-184, Reidel, Dordrecht & Boston

MERKEL, C. (1977) *Elektronische Eigenschaften von Molekülkristallen* [Electronic Properties of Molecular Crystals], Thesis, Technical University Munich

SEEL, M., COLLINS, T.C., MARTINO, F., RAI, D.K. & LADIK, J. (1978) $(SN)_x$ with hydrogen impurities in the coherent potential approximation. *Phys. Rev. B,* in press

STODDARD, J.C., STOREY, P., MARCH, N.H. & ORTENBURGER, I.B. (1974) Exchange and correlation contribution to one-body potential in metallic Be. *Nuovo Cimento 23B,* 15-26

SUHAI, S., MERKEL, C. & LADIK, J. (1977) *Ab initio* energy band structure of polycytosine. *Phys. Lett. 61A,* 487-489

SUHAI, S., COLLINS, T.C. & LADIK, J. (1978) Theoretical investigation of semiconductive properties, II. On the possibility of charge transfer between proteins and different acceptor molecules. *Biopolymers,* in press

Discussion

Williams: You claimed that there was no real thought in your mind, from theory, that proteins or DNA would show strict thermal conductivity without the introduction of impurities with band gaps that were smaller than 10 eV. Is that correct?

Ladik: Of course not!

Eley: The d.c. dark conductivity found in dry proteins and DNA (and indeed cellulose also) has 'energy-gaps' $\Delta\epsilon$ in conductivity $\sigma = \sigma_0 \exp(-\Delta\epsilon/2kT)$, in the range 2.4 – 3.0 eV (cf Eley 1968). On the other hand, light of wavelength corresponding to this energy range only appears to stimulate a 'bolometric' effect, i.e. thermal increase in dark conduction, for DNA (Eley *et al.* 1975) and also proteins (the same authors, unpublished). Such dark conduction experiments as have been done on bovine plasma albumin indicate that the effects are independent of the metal constituting the electrodes, so the charge carriers do not arise from electrode injection (Eley & Thomas 1968). The general reproducibility of the small currents found in a wide range of dry proteins ($\sigma \sim 10^{-17}\ \Omega^{-1}\ cm^{-1}$ at 298K) tends to rule out a dominant role of adventitious impurities (Eley & Spivey 1960).

Over the years, quantum mechanical calculations with improved assumptions and approximations have yielded increasing values for the energy gap $\Delta\epsilon$ between valence and conduction bands; for example, for β-polyglycine using MINDO/2, $\Delta\epsilon = 4.7$ eV (Suhai 1974), and recently by the *ab initio* method $\Delta\epsilon = 19.2$ eV (Kertész *et al.* 1977). We have recently suggested (Eley *et al.* 1978) that the charge carriers in a protein crystal, holes and electrons, might arise from a thermally activated charge transfer from a donor group such as SH or CO, to a fairly distant acceptor group such as NH, possibly on two different molecules, or perhaps even two separated sites on the same molecule. Using Lyons' (1957) equation $\Delta\epsilon = I_g - A_g - (P_+ + P_-)$ with an ionization potential of the donor $I_g = 8.4$ eV, electron affinity of the acceptor $A_g = 1.5$ eV, then an experimental value of $\Delta\epsilon = 3.0$ eV would require a value for $P = \frac{1}{2}(P_+ + P_-) = 2$ eV for the average polarization energy of the hole and electron. Rosenberg & Postow (1969) using an independent value of $R = 0.36$ nm and a dielectric constant for the 'medium' $\epsilon = 3$ have calculated

$$P = \left(\frac{e^2}{2R}\right)\left\{1 - \frac{1}{\epsilon}\right\} = 1.35 \text{ eV}.$$

Therefore to reconcile the experimental results we clearly require an enhancement of the expected P value, which would need to arise from dipolar interactions, perhaps with residual H_2O (it is difficult to remove the last 1%), or at special sites, or defects in the protein crystal.

If this enhancement occurred only for P_- the electron would tend to be localized and the hole left mobile in the crystal, as found for dry haemoglobin (Eley *et al.* 1979). This charge-transfer mechanism was originally suggested for DNA (Burnel *et al.* 1969).

Ladik: The large energy gaps I quoted referred to a single chain. With two interacting chains there is the possibility of charge transfer and everything is changed. Then, of course, you can have conductivity, but I wouldn't call it intrinsic conductivity for a single chain. Consider crystals of tetracyano-quinone–tetrathiafulvalene (TCNQ–TTF). Crystals of TCNQ alone are insulators but the two-component TCNQ–TTF crystals are conductors, because of charge transfer.

March: I would like to make a comment which Professor Ladik will be aware of but which relates to the energy-band diagram that Professor Williams mentioned (p. 45). We can correctly describe by band theory the electron distribution in the materials that Professor Ladik has been discussing, if we choose the self-consistent field properly. But the question of whether we can derive excitation energies, like an energy gap, correctly is much more difficult to answer. The problem there depends on weighing kinetic energies against electronic Coulomb repulsion energies. It seems to me that in proteins the band widths are not large and it may be that we are near the limits where one can apply the energy-band theory, for obtaining excitation energies. I should stress that, although I don't know how to do it, it is important for people to think more about experiments to check some of the aspects of the band-theory calculations.

Ladik: Any kind of Hartree–Fock theory gives the wrong excitation energy in solids because the excited electron sees a V^N potential instead of a V^{N-1} potential; therefore it is too high. There are simple remedies such as the ÔÂÔ method (Collins & Kunz 1974; see my paper, p. 54); this excitation Hamiltonian corrects at least the major part of the error. For a better result one can start from the Hartree–Fock Bloch functions and formulate an intermediate exciton theory. This has also been worked out but no calculations have been done, at least in the *ab initio* form. In semi-empirical form, M. Kertész did it, and that worked out well. So excitation energy has not been a major problem. To put your remark more provocatively, one might say that the bands in biopolymers are so narrow that the Bloch (delocalized) description breaks down. Until one has a strictly periodic system, the band widths are not small—about 1-2 eV—and so in those cases it does not break down. Disorder, however, changes the situation.

March: As you say, disorder is a problem, but it might help, in a way, because, if the band theory is right, the effect of disorder is to blur out the sharp band edges that one would get in a periodic material. Is it possible to test that experimentally? It certainly has been done for covalently bonded silicon (cf. p. 48), for which one can compare the periodic case with the amorphous case. Experiment shows that the edges are not blurred in the amorphous material, which is readily understood in terms of chemical bonds but presents difficulties for energy-band theory.

Ladik: Disorder certainly splits the bands and narrower bands occur. Then, of course, it is questionable whether a band description makes sense, especially from the point of view of conduction mechanisms. The sort of disorder depends of course critically on the system, too. More calculations need to be done before we can answer this question.

Eley: Hopping of electrons between disordered states at band edges of amorphous inorganic materials has been detected by the frequency (ω) dependencies of a.c. conductivity, $\sigma_{ac}\alpha\omega^{0.8}$ (Mott & Davis 1971). Similar studies have been done on proteins, where, however, a problem arises because of the operation of the other two classical loss mechanisms, dipolar rotation and interfacial (Maxwell–Wagner) polarization. Sorting out the three loss mechanisms is proving to be a difficult matter.

Phillips: What did you mean by assuming a two-dimensional structure for the protein in these calculations?

Ladik: With a polypeptide chain, besides the bonds in the chain there may be interactions between chains or between different segments of the same chain through the hydrogen bonds perpendicular to the primary structure. In this way one ends up with a two-dimensional periodic system.

Phillips: In other words, the structure was a β-pleated sheet constrained to be planar?

Ladik: That is one instance; but one can also think of the α-helix (outside on the cylinder, perpendicular to the chain).

Phillips: How important do you think the detailed structure is likely to be in comparing your theory with experimental results?

Ladik: The detailed structure composition, sequence and so on of the proteins is certainly important: this brings us to the disorder effect. However, disorder has been studied only very little. We have started it, but it is extremely hard.

Phillips: The models of periodic proteins you used, polyglycine and polyalanine, have different structures: one is triple helix and the other is an α-helix. Is that taken into account?

Ladik: Both calculations were done on a single chain (linear model). Previous 2-D semi-empirical calculations have been done on the parallel chain β-pleated sheet.

Laki: The suphur–nitrogen polymer you mentioned (p. 56) is a most interesting material; it is as good a conductor as copper. That finding pleased me because people *a priori* rejected the whole idea that proteins in any condition could conduct, as they are made up of carbon, oxygen, nitrogen, hydrogen and sulphur. Such non-metallic conductors are thus not impossible. As Professor Eley pointed out, given the right circumstances, conduction along the protein backbone might be possible.

Williams: I should add that chemically the sulphur–nitrogen polymers are very different from proteins in conjugated bonding. They also contain a standing concentration of free radicals as end-groups and, therefore, they have holes ready-made.

Ladik: In the $(SN)_x$ system there is an odd number of electrons per unit cell, so one obtains a half-filled valence band (as in a metal). This is the reason for its high conductivity, and not the fact that it is a conjugated system. Further, in the ground state of $(SN)_x$ there are no holes (the end effects of long chains are not very important).

Eley: The DNA double helix contains two parallel stacks of purine and pyrimidine bases. The TTF-TCNQ crystal contains parallel segregated stacks of TTF-TCNQ molecules. In each case we are concerned with electrons and holes delocalized along the stack and perpendicular to the planes of the rings. To this extent TTF-TCNQ is a closer model to the DNA situation than the $(SN)_x$ chain polymer where electrons and holes are delocalized along the chain.

Ladik: TCNQ contains only carbon, hydrogen and nitrogen (I shall ignore TTF because it contains sulphur) and has a conduction band about 0.6 eV broad. When one injects charges into it there is charge transfer to the TCNQ acceptor—it is really a good conductor. It is a stacked system like DNA.

Williams: I do not accept TCNQ systems strictly as a chemical example for DNA. I accept this as a general model that may be used in the study of conductivity in certain circumstances. The chemicals in the TCNQ systems have been deliberately chosen to give a certain type of interaction of low energy.

Pethig: With regard to the effect of long-range correlation, would introduction of water into the model protein structures (so increasing the effective dielectric constant) in your calculations give an energy-band scheme which consists of a decreasing energy-band gap, with the valence band and conduction band becoming narrower, as the hydration content increases?

Ladik: Yes, it would affect the gap but, so far as I know, the gap is not extremely sensitive to the water content and the position of the water molecules.

Pethig: I ask because many enzymes seem to have a crevice in which the active site is located and where water could interact with the protein structure more than it could in the bulk hydrophobic region of the protein. As a result of the long-range correlation effects, the valence band of the protein molecule would tend to tilt upwards in energy in the direction of the water-filled crevice. Some of the charge-transfer interactions that we are discussing involve perhaps the removal of only a fraction of an electronic charge from the valence band, and we could imagine a collection of such partial charge deficiencies tending to migrate towards the active site in the crevice until there would be a probability of having an electron 'hole' of unit charge located there which could participate in a classical chemical reaction. An analogy would be air bubbles in a tank of water; tilting the tank would cause the air bubbles to migrate in the same direction.

Ladik: We have looked at that (A. Karpfen & J. Ladik, unpublished). There is no tremendously large dependence of the band structure on the water.

Williams: I should stress for the theoreticians that biology does use electronic-type movements of charge. That is not in question. What is in question is whether biology uses the backbone of proteins (or of DNA) for electron transport. When we have heard Professor Lewis's paper I shall comment on the way in which biology is known to use movements of electrons in protein systems, but not using these backbones (see p. 81).

References

BURNEL, M.E., ELEY, D.D. & SUBRAMANYAN, V. (1969) Semiconduction in nucleic acid and its compounds. *Ann. N.Y. Acad. Sci. 158,* 191-209

COLLINS, T.C. & KUNZ, A.B. (1974) Excitation Hamiltonian of electronic systems using ÔÂÔ. *Int. J. Quant. Chem. 58,* 437-446

ELEY, D.D. (1968) Semiconducting biological polymers, *in Organic Semiconducting Polymers* (Katon, J.E., ed.), pp. 259-294, Dekker, New York

ELEY, D.D. & SPIVEY, D.I. (1960) Semiconductivity of organic substances. Part 6. A range of proteins. *Trans. Faraday Soc. 56,* 1432-1442

ELEY, D.D. & THOMAS, P.W. (1968) Semiconductivity of organic substances. Part 14. Electrode effects in proteins. *Trans. Faraday Soc. 64,* 2459-2462

ELEY, D.D., METCALFE, E. & WHITE, M.P. (1975) Semiconductivity of organic substances. Part 17. Effects of ultraviolet and visible light on the conductivity of the sodium salt of deoxyribonucleic acid. *J. Chem. Soc. Faraday Trans. I 71,* 955-960

ELEY, D.D., LOCKHART, N.C. & RICHARDSON, C.N. (1979) Thermoelectric effects and dielectric polarisation in biopolymers. *J.Chem. Soc. Faraday Trans. I75*, 323.

KERTÉSZ, M., KOLLER, J. & AZMAN, A. (1977) Calculated forbidden band gap in periodic protein models indicating them to be insulators. *Nature (Lond.) 266,* 278

LYONS, L.E. (1957) Photo- and semiconductance in organic crystals. Part V. Ionised states in molecular crystals. *J. Chem. Soc.* 5001-5007

MOTT, N.F. & DAVIS, E.A. (1971) *Electronic Processes in Non-Crystalline Solids,* Clarendon Press, Oxford

ROSENBERG, B. & POSTOW, E. (1969) Semiconduction in proteins and lipids — its possible biological importance. *Ann. N.Y. Acad. Sci. 158,* 161-190

SUHAI, S. (1974) The electronic structure of periodic protein models. II. Energy band structure and anisotropy of electrical conduction in the parallel-chain β conformation of polyglycine. *Theor. Chim. Acta (Berl.) 34,* 157-163

The mechanisms of conduction in proteins

T.J. LEWIS

Laboratory of the National Foundation for Cancer Research, University College of North Wales, Bangor, Gwynedd

Abstract Theoretical calculations have already shown that the electronic properties of biopolymers may be described in terms of conduction and valence energy bands in which electrons are delocalized to a significant degree. Because these bands are separated by a large energy gap, it is concluded that electronic conduction in such systems must be by 'holes' in the valence band. The mechanism of this conduction is discussed with special reference to dry proteins where the charge-transfer role of appropriate acceptor molecules and the modulating influence of amino acid residues will encourage localized and delocalized hole production at the valence band edge. It is suggested that there are close similarities, which have shown up in measurements, in the electrical behaviour of proteins and other non-biological semi-crystalline solids.

The significance of such long-range conduction and attendant polarization is considered briefly in the light of evidence for the existence of proteins in cell membranes and in the skeleton of the cytoplasm.

Several papers describe the electrical behaviour of covalently bonded organic solids in terms of delocalized electron energy-band schemes. Eley and his collaborators did much pioneering work to establish these ideas and outlined the possible mechanisms of electron or hole transfer. A useful discussion of the models proposed by Eley & Spivey (1960) and by McCubbin (1963) has been given by Fleming (1970). He concludes that charge transfer is best described using energy bands in which electrons move as in more conventional inorganic semiconductors. It is significant that these conclusions are applied to a long-chain covalently bonded polymer such as polyethylene. The importance of such bands for polymers is that they allow facile transfer of electrons in the conduction band or holes in the valence band along pathways extending along the backbone of the polymer. This property of long-range electron transfer is a corner-stone of Szent-Györgyi's idea of submolecular processes in proteins (Szent-Györgyi 1941*a,b.*). Ladik and others (see for example Laki & Ladik

1976; Ladik 1976; Ladik *et al.* 1978) have established theoretically the energy-band scheme for polyglycine, a molecule having all the essential features of the polypeptide backbone of a natural protein. The determinations of the energies involved depend on the sophistication of the initial assumptions about atomic properties, and the latest estimates are that polyglycine has valence and conduction band widths of about 1.9 and 1.25 eV, respectively, with a band gap of more than 11 eV. In nature the polypeptide chains themselves may be ordered by hydrogen-bonding and such three-dimensional arrays will also give rise to energy-band structures where the valence and conduction bands are much less wide, although the calculated band-gap remains large (Ladik *et al.* 1978).

Electron and hole *mobilities* in the bands depend on band widths and have been estimated to be 59 and 45 $cm^2\ V^{-1}\ s^{-1}$, respectively, along the main polypeptide chain and perhaps an order of magnitude less perpendicular to it along the hydrogen-bonded network. In spite of these mobilities, the large calculated energy gap makes the probability of thermal excitation of electrons from valence to conduction band insignificant and thus the polypeptide structure is intrinsically a good insulator. This conclusion is, at first sight, a fundamental stumbling block to the acceptance of a theory of biology based on electronic conduction. It should be noted, however, that many other materials, the glasses, oxides and amorphous semiconductors, also have energy gaps sufficiently large to make them poor conductors but this has not prevented consideration of them in electronic terms and the establishment of a considerable body of experimental and theoretical evidence for long-range electron transfer in them (Mott & Davis 1971).

Examination of the electrical properties of these latter materials and of the dry proteins in fact indicates many similarities. For example, the so-called a.c. conductivity $\sigma(\omega)$ of bovine serum albumin, casein, lysozyme and collagen measured in an alternating electric field of frequency ω follows the law (1) (see

$$\sigma(\omega) = A\omega^n \tag{1}$$

Fig. 1), where A and n are weakly temperature-dependent parameters and n is slightly less than unity over a wide range of ω (Bone *et al.* 1978). This law is also found for such diverse non-biological materials as oxide and chalcogenide glasses, synthetic polymers, anthracene and silicon containing donor and acceptor concentrations arranged so as to partially compensate each other.

At a frequency below about 10^{-2} Hz, n tends towards zero and in this range the steady-state conductivity σ_0 overshadows $\sigma(\omega)$. Complexing the protein with methylglyoxal (Pethig & Szent-Györgyi 1977; Bone *et al.* 1978) increases σ_0

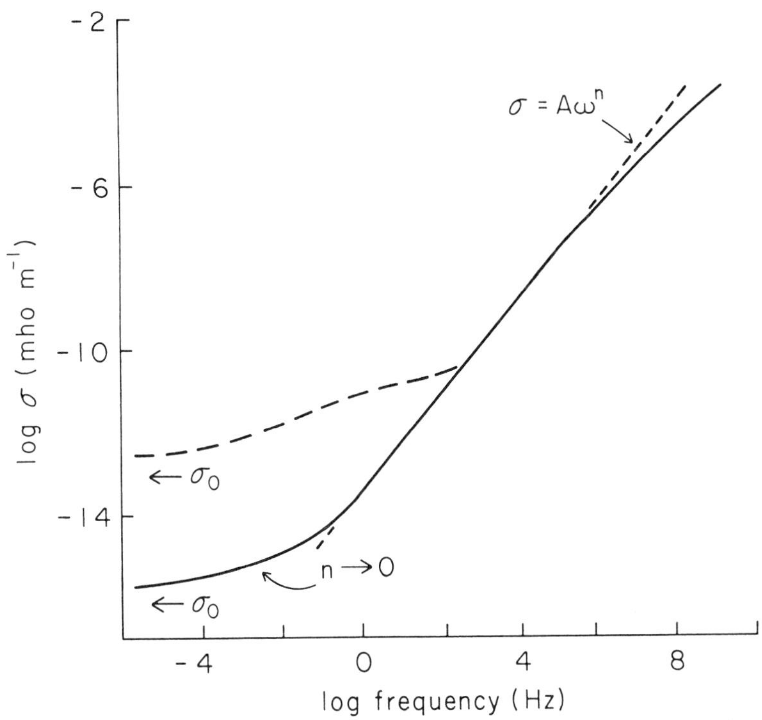

FIG. 1. Typical conductivity–frequency plot for dry protein samples. The conductivity follows closely an ω^n law over a wide range of frequencies. Steady-state conductivity σ_0 occurs at the limit $\omega \to 0$. The enhanced conductivity resulting from complexing proteins is shown by a broken line (Bone *et al.* 1978).

enormously and also changes $\sigma(\omega)$ in the low frequency range (Fig. 1). The conductivity σ_0 of the complexed proteins also depends exponentially on the square-root of the applied field (a result also found for other insulating solids, among which are the synthetic organic polymers), and is attributable to electronic mechanisms based on the Poole–Frenkel effect (Adamec & Calderwood 1975). Although little work has been done on the temperature dependence of $\sigma(\omega)$, it appears to be similar to that for inorganic insulating solids such as the glasses (Sayer *et al.* 1971) and even to partially compensated silicon (Pollak & Geballe 1961), following not a simple activation law with a constant activation energy but one in which the activation energy increases with increasing temperature.

In physics it has become acceptable (see for example Mott & Davis 1971) to describe these conductive processes in terms of electron or hole transfer between localized states arising from the degree of non-crystallinity in the solid

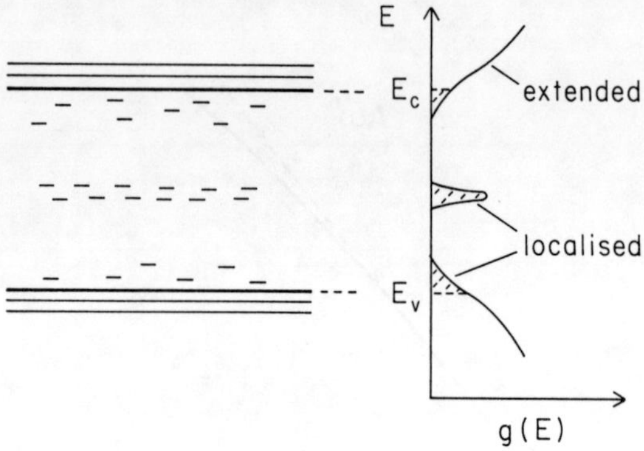

FIG. 2. Typical energy-band and density-of-states schemes for a non-crystalline solid showing localized states at the band edges and also in the mid-gap region (Davis & Mott 1970).

and superimposed on an energy-band scheme characteristic of the underlying periodic lattice structure. For such solids, the electron energy scheme adopted is like that shown in Fig. 2, where, because of loss of local order, the edges of the bands of delocalized states become ill-defined with tails extending, with increased localization, into the band-gap. When impurities, charged or uncharged, form a more gross disruption of the lattice, localized states may appear well into the band-gap, even forming impurity bands. The ideas embodied in Fig. 2 are now well established in physics for describing conduction in non-crystalline systems (Davis & Mott 1970) and they seem to be applicable and particularly valuable in biology.

ELECTRONIC ENERGY STATES IN PROTEINS

Examination of the protein structure suggests that the *periodic* potential created by the polypeptide backbone on which the theoretical calculations of band structure by Ladik and his co-workers were based must, in reality, be somewhat disturbed. First, the various side-chains and especially those which are charged in the physiological situation will modulate the periodic potential set up by the backbone (Brillouin 1962; Kertész *et al.* 1977) and so tend to induce localized states. It may prove to be significant that each of the 20 amino acid residues will produce a characteristic modulation of the potential and a characteristic localized state. A sequence of residues in a particular protein then produces a corresponding sequence of localized states which characterize the protein as much as the residues do. Appropriate sequences might build-in electric-field gradients in the manner suggested by Kroemer (1957).

Reactions of the proteins with other molecules are likely to introduce localized states into the band-gap away from the band edges. In the case of the reaction with methylglyoxal (Szent-Györgyi 1977), the localized states are likely to be acceptor-like and to be located on the valence band side of the band-gap where they can have a profound effect on conduction. The tertiary structure involving specific folding of the peptide chain, the degree of hydration, and the presence or otherwise of ions must also contribute to the formation of localized states. As mentioned already, hydrogen-bonding to form pleated structures will give the possibility of conduction normal to the chains. Thus Fig. 2 could justifiably be used to describe the likely electron states of proteins.

The bonding of suitable donor or acceptor molecules to the protein with attendant localized band-gap states will, as a result of ionization, introduce the possibility of electron or hole transfers. For example, the establishment of an ionized acceptor site A^- (Fig. 3) by electron transfer from a suitable donor site D at the valence band edge to give D^+ allows subsequent transfers of a 'hole' from the D^+ site to neighbouring D sites at the valence band edge. The Coulombic attraction between the hole and the parent ionized acceptor site will tend to confine the transitions to a relatively few localized D sites at the band edge in the neighbourhood of A^-, although long-range transfers are always possible through the delocalized states of the valence band.

Both transfer processes represent electronic conduction in the protein but the former limited transport will result in polarization. For this process the possible transfers within a set of D sites about an A^- site can usually be represented with little loss of accuracy by a simple two-site system, as in Fig. 4*a*, where the two donor sites *i* and *j* at a distance r_{ij} apart are separated by a potential barrier. One hole is associated with the pair element. The transition from the state *i* to *j*, which may involve tunnelling as well as activation, will have a rate given by equation (2) where ν is a transition-attempt frequency and

$$\nu \exp(-q_{ij}) = \nu \exp[-(\phi + W_{ij}/kT)] \quad (2)$$

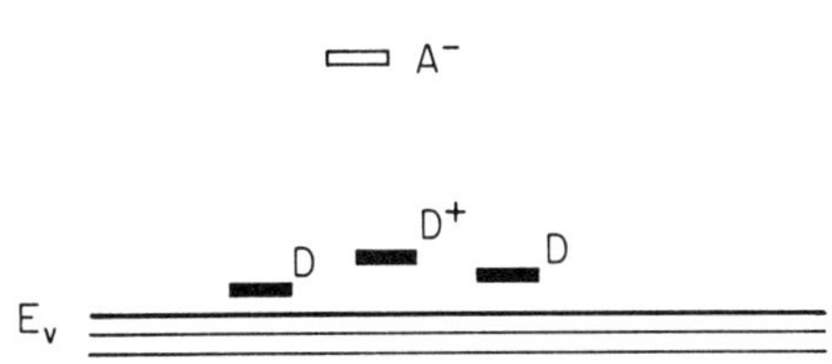

FIG. 3. Ionized acceptor A^-, associated donor D^+ and adjacent accessible donor states D arising from localized states at the valence band edge.

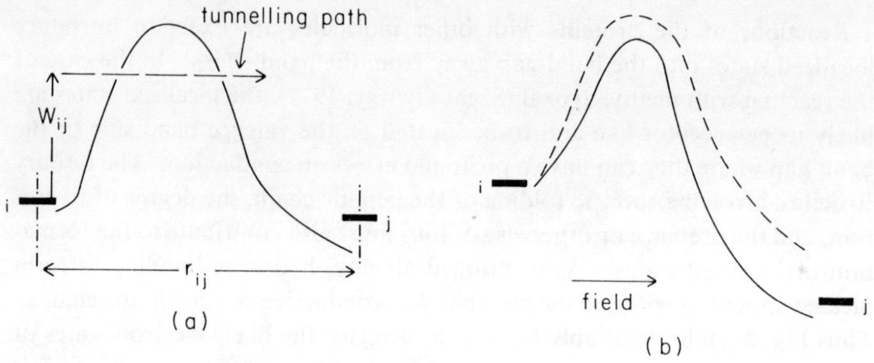

FIG. 4 (*a*) A pair element consisting of two donor sites *i* and *j* and a single associated hole. A transition from *i* to *j* is shown, requiring an activation energy W_{ij} followed by tunnelling. (*b*) Changes in the relative energies of the two sites and in the energy barrier induced by an applied field.

the parameter q_{ij} has a tunnelling component ϕ determined by wave-function overlap between *i* and *j* over a distance r_{ij} (Mott 1967) and an activation component W_{ij}/kT, as shown in Fig. 4*a*. A similar expression may be written for transitions from *j* to *i* in terms of W_{ji}. Although in many treatments (see Mott & Davis 1971) tunnelling is considered to be dominant, it may not always be so and, for the protein–methylglyoxal complexes discussed below, it appears to be much less important than activation. Transitions between *i* and *j* produce local changes in polarization and the conductive response to an applied electric field will result from the induced changes in W_{ij} and W_{ji} (Fig. 4*b*) and hence in the transition rates. The response to an alternating field is the a.c. conductivity $\sigma(\omega)$. In practice there will be some probability distribution function $P(q_{ij}, q_{ji})$ for q_{ij}, q_{ji} determined by the pair-element characteristics of the protein and also some distribution of r_{ij}. It is then possible to write equations (3) and (4) (Lewis 1977) where $\epsilon''(\omega)$ is the dielectric dispersion loss factor, *N*

$$\sigma(\omega) = \omega\epsilon''(\omega) \approx \pi\omega e^2 r^2 NP(q,q)/6kT \quad (3)$$

$$q = \phi + W/kT = \ln(2\nu/\omega) \quad (4)$$

is the density of pair elements and *r* is an appropriate mean value of r_{ij}. Equations (3) and (4) indicate that the major contribution to $\sigma(\omega)$ or ϵ'' at a frequency ω will come from those pair elements that are symmetrical and have a parameter *q* given by equation (4).

In a range of frequencies in which equation (1) holds, it is possible to express $P(q, q)$ by equation (5), where C is a normalizing constant. If $P(q)$ is the pro-

$$P(q,q) = \mathrm{C}\exp[(1-n)\mathrm{q}] \quad (5)$$

bability density function for a single state, then we can write equation (6)

$$P(q,q) = 2P^2(q)F(q)[1 - F(q)] \tag{6}$$

where $F(q)$ is the appropriate Fermi function for the state of activation energy W. For the dry uncomplexed protein it seems likely that these states are associated with the valence band edge but it is difficult to proceed without more knowledge of the balance between tunnelling and activation processes. However, since n is found to be nearer unity than 0.8, the value usually associated with purely single-phonon* tunnelling (Austin & Mott 1969), some degree of multi-phonon activation is probable. If activation is dominant and the tunnelling parameter ϕ is zero, then for the uncomplexed protein with ν being chosen to be the upper phonon frequency (about $10^{13}\ s^{-1}$) the range of W for a frequency range $1-10^6$ Hz becomes 0.75–0.40 eV (equation 4). If tunnelling is also involved or if there is an entropy change during the transition, the range will change. As discussed below, a lower value of ν (about $10^{10}\ s^{-1}$) might then be appropriate, in which case W has the range 0.58–0.23 eV. Such energies are easily accommodated in the wide band-gap of the protein. If the transitions are essentially through the extended states of the valence band, as in Fig. 5*a*, then $W = (E - E_v)$, where E_v marks the mobility edge and $P(q)$ is approximately proportional to $\exp[-(E-E_v)/2kT]$.

As measurements were made on compressed samples, however, inter- rather than intra-protein transition may be involved. If so, at least a part of W may be associated with the interprotein barrier (see Fig. 5*b*), but it is not possible to separate the two processes at this stage.

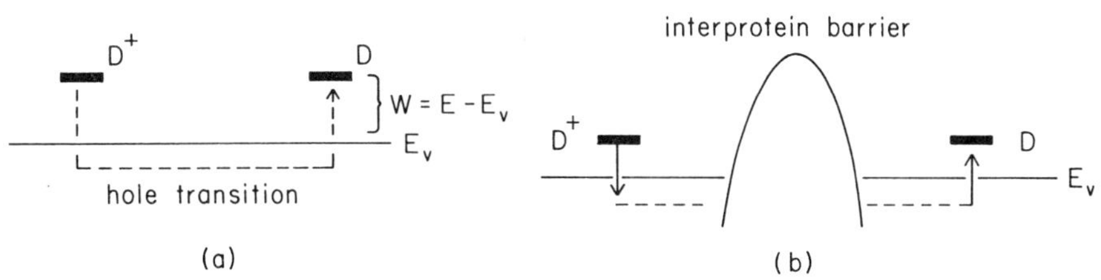

FIG. 5. (*a*) $D^+ \rightarrow D$ hole transitions through the valence band states below the mobility edge of a single protein molecule. (*b*) $D^+ \rightarrow D$ hole transitions involving activation across an interprotein barrier.

*The phonon is a quantum of energy associated with thermally induced vibrations of the solid lattice structure.

PROTEIN–METHYLGLYOXAL COMPLEXES

With regard to the protein–methylgloxal complexes, in which the glyoxal appears to act as an acceptor through the formation of a Schiff base (Bone *et al.* 1978), much enhanced values of $\sigma(\omega)$ at low frequencies and of σ_0 are found (Fig. 1, p. 67). When the dispersion loss ϵ'' rather than σ is plotted against frequency ω, equations (3) and (4) reveal that a much-enhanced density of pair elements has been introduced in a frequency range which varies somewhat with the type of protein (Pethig 1978) but is between 10^{-5} and 10 Hz. For example, the lysozyme–methylglyoxal complex showed a dispersion spread over the full range of frequency with a maximum given by equation (7), where

$$f_m = f_o \exp(-W_m/kT) \tag{7}$$

$W_n = 0.67$ eV and $f_0 = 5.3 \times 10^9$ Hz (Bone *et al.* 1978). Interpretation of this result according to equations (2) and (4) suggests that complexing has introduced new pair elements with W in the range 0.80–0.52 eV, with a maximum at 0.67 eV. The fact that the attempt-frequency found experimentally is some three orders of magnitude smaller than the theoretical limit suggests that some degree of tunnelling or structural reorganization associated with entropy change accompanies the transitions (Bone *et al.* 1978).

The area under the plot of $\epsilon''(\omega)$ against log ω is given exactly by the expression $\pi e^2 r^2 N/13.8kT$ (Lewis 1977) and thus it is possible to estimate the r^2N product. Bone *et al.* (1978) and Pethig (1978) have advanced arguments to show that the dispersion is unlikely to arise from conventional dipoles of the system and can be associated with hole transfer between localized states associated with the complexes. Assuming that N can be equated to half the enhanced electron-spin density found for these complexes since, according to the $A^- D^+$ model described above, only the hole on the D^+ site is mobile and allowing for a Lorentz correction for local field enhancement, r is found to lie between 3 and 6 nm for the complexes studied. Thus r is some 20–40 atomic spacings; this value suggests that the transition involves two protein molecules or, possibly, widely separated parts of the same molecule, and in both cases will include a protein–protein interface.

Assuming that transfer occurs across inter-protein boundaries, the following model may be proposed. Let the valence band mobility edges of the two molecules be at E_v and separated by the interface barrier E_i. This is shown in Fig. 6 where, although hole transfer is envisaged, *electron* states are shown. The localized glyoxal acceptor state (if we assume one per molecule) is shown at E_a, above E_v. As the estimated charge transfer to E_a is about 0.005 e (Pethig 1978; Otto *et al.* 1977) the Fermi energy of the system, E_f, will lie below E_a by about 0.13 eV and the density of spins will be $2N_a\exp[-(E_a-E_f)/kT]$ where N_a is the density of glyoxal complex sites.

The donor states arise from the valence band edge (Fig. 3) but, as the complex is formed, local distortion of the potential will move these states into the band gap. This interaction will arise partly from Coulombic attraction between the ionized acceptor and donor site, but it would be unjustifiable to consider this alone. In fact the donor states are likely to be found in a range of energies about some mean value E_d, as shown in Fig. 6.

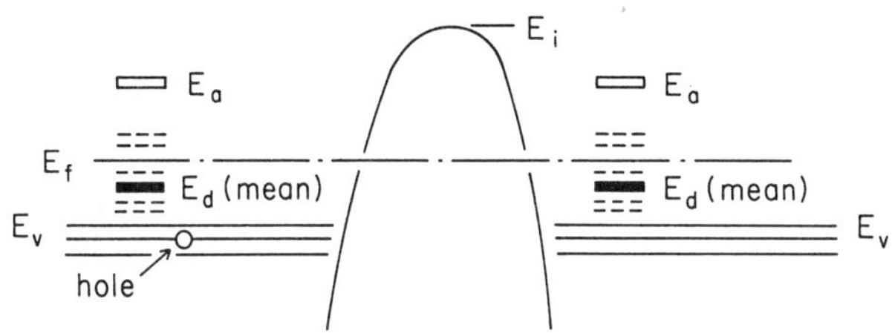

FIG. 6. The relevant acceptor and donor states involved in hole transfer between protein molecules and giving rise to not only the a.c. dispersion, $\epsilon''(\omega)$, but also the steady conduction σ_0.

The activation energy W for hole transitions between the donor states on either side of the interface is thus given by equation (8). It must be emphasized that a spread ($\pm\Delta E$) in E_a, E_i and E_d is to be expected; this would lead to a corresponding spread $\pm\Delta W$, as found from the ϵ'' dispersion measurements.

$$W = E_d + E_i - 2E_v \tag{8}$$

If the density of donor sites at E_d is $N(E_d)$, the effective number of symmetrical elements, N, is given by equation (9) where $F(E_d)$, the Fermi function evaluated at E_d, is given by equation (10). The density of acceptor states, N_a, is then equal to $\Sigma N(E_d)$. The function $F^1(E_d)$ has a sharp maximum equal to ¼ when $E_d = E_f$ but decreases rapidly as E_d moves away on either side of E_f (Fig. 7). The combined evidence from a.c. and d.c. conductivity (see below) suggests that E_f normally lies in the upper part of the range of E_d. Thus not only would N be less than N_a but also probably less than the effective number

$$N = \Sigma N(E_d)F^1(E_d) \tag{9}$$

$$F^1(E_d) = F(E_d)[1-F(E_d)] \tag{10}$$

of complex sites, $N_a\exp[-(E_a-E_f)/kT]$, determined from spin measurements. In that case the estimates of r would need to be increased.

The model illustrated in Fig. 6 should also explain the steady-state conductivity σ_0 for which the activation energy (about 0.72 eV) lies on the higher side of the range of energies determined from the ϵ'' dispersion (Bone *et al.* 1978). The steady-state observations may not truly be so but may instead represent the low-freqency tail of the dispersion arising from pair-element transitions, since all measurements below 10^{-5} Hz require time intervals well in excess of 36 hours. However, Bone & Pethig (see pp. 83-105) have demonstrated that the current is strongly field-dependent and follows the Poole–Frenkel law accurately (Frenkel 1938). This law holds for conduction in many insulating solids, notably, in the context of this paper, for organic polymers (Adamec & Calderwood 1975). The law suggests that conduction is by free or delocalized carriers thermally activated over energy barriers progressively lowered as the applied field is increased. Attempts to fit the results presented by Bone & Pethig to a law derived from a high-field version of the pair-element model of Fig. 4 have not resulted in such satisfactory agreement.

Now, free holes will exist in states in the valence band below E_v; their number density is given by $N_v\exp[-(E_f-E_v)/kT]$ where N_v is an effective

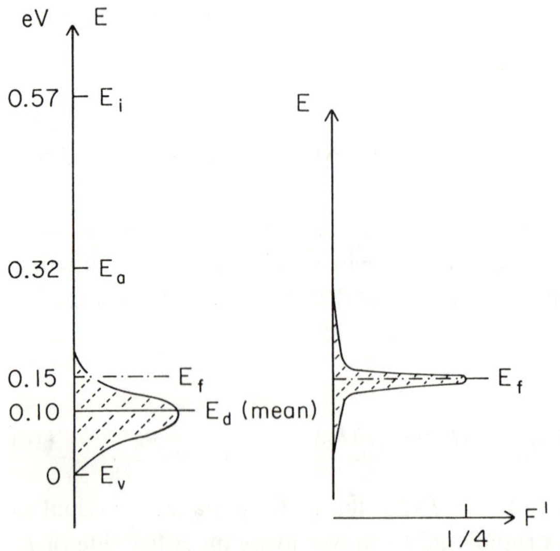

FIG. 7. Numerical estimates of the important energy states involved in hole transitions in complexed proteins. Note the band of donor states with a mean value below the Fermi energy and the form of the $F^1(E)$ function which peaks strongly at $E = E_f$.

density of hole states in the valence band. These free holes will be restricted in motion by the interprotein barrier E_i (Fig. 6) and will have a hopping-type mobility, μ, in the solid (equation 11; Mott 1969), where ϕ' is a tunnelling

$$\mu = e\nu' a^2 \phi' (kT)^{-1} \exp[-(E_i - E_v)/kT] \tag{11}$$

parameter not necessarily the same as ϕ, a is a mean distance between the barriers in the protein valence bands (taking account of random orientation in the applied field) and ν' is the frequency at which a hole approaches the interprotein barrier. This frequency is not necessarily the same as ν (see above) since the latter involves the activation process from localized states at E_d as well as transfer in the bands. Thus the steady conductivity σ_o will be given by equation (12) with an activation energy clearly related to the activation energies for

$$\sigma_o = e^2 \nu' a^2 \phi' N_v (kT)^{-1} \exp[-(E_i + E_f - 2E_v)/kT] \tag{12}$$

$\sigma(\omega)$ (see equations 3, 4 and 8). From the experimental determinations of σ_o (Pethig 1978; Bone *et al.* 1978) and assuming $\phi' = 1$, $e\nu' a^2 N_v/kT$ has a value of 1.6×10^{21} for collagen complexes and similar values for the other proteins. For ν' and a we have no immediate values and the only alternative is to use f_o (equation 7) for ν' and the values of r for a. We find that $e\nu' a^2/kT$ has values between 1.8 and $7.2 \times 10^{-6}\,m^2\,V^{-1}\,s^{-1}$ at normal temperature and hence that N_v lies between 10^{26} and $10^{27}\,m^{-3}$. Assuming now that the contribution to the total hole population from the E_d levels is relatively small and equating the free hole concentration below E_v with the electron concentration in E_a states, we estimated $(E_f - E_v)$ to be about 0.15 eV. Thus E_a lies about 0.32 eV above E_v. Ladik *et al.* (1978) have argued from theoretical considerations that the interval might be 0.6 eV but the estimates derived from theoretical calculation and from experimental determinations must each contain a considerable margin of error.

The experimentally determined activation energy for σ_o is 0.72 eV and thus we find from equation (12) that the interfacial barrier $(E_i - E_v)$ is about 0.57 eV, which is significantly close to the value of W (0.52 eV) at the high frequency limit of the $\epsilon''(\omega)$ dispersion. This suggests that the lower limit of $(E_d - E_v)$ might be very small and indicates a distribution of localized E_d states extending from 0.23 eV above E_v, so that the tail of the states lies above E_f, down to the valence band edge with a maximum density at 0.1 eV (Fig. 7).

The model can also provide an explanation of the high-field effects reported by Bone & Pethig at this symposium. Close agreement with the Poole–Frenkel law for σ_o will come about by a field-lowering of the interprotein

$(E_i - E_v)$ barrier (assuming no change in E_f), especially if the barrier is Coulombic in form at its peak.

Of more significance, perhaps, is the fact that the model will also explain why the area under the dispersion curve (see p. 71) changes with field. An increase in field will lower the barrier $(E_d - E_v)$ for activation of a localized hole into the valence band; the degree of lowering depends on the nature of the hole-binding forces. All E_d levels will be shifted down and this could happen without significantly shifting E_f, which is controlled mainly by the acceptor and valence band states. When this occurs, N, the density of pair elements (equation 9), could change dramatically, since $F(E_d)$ is extremely sensitive to $(E_d - E_f)$. In general we may expect that (as in Fig. 7) most of the distribution of E_d will be located below E_f, so that when the E_d levels move down $(E_f - E_d)$ will increase and the contribution to N will decrease. Some of the states may be above E_f, however, and the contribution to N from these will, in the first instance, increase as E_d moves down towards E_f and then decrease as E_d moves on below E_f. Pethig has evidence for N first increasing with field at low values and then decreasing with field at high values, in accordance with these predictions.

CONCLUSIONS

The model proposed appears to explain the major features of conduction phenomena in dry compressed proteins but further refinement awaits the production of more experimental data, especially if they can be obtained from other types of measurements (e.g., of Hall coefficient, drift time and photoconductivity). The conclusion is that electrons, or rather their hole equivalents, can move over relatively long pathways in protein solids, but it must be emphasized that the process is of a very different order of magnitude from that associated with metallic conductors. The proteins may be weak conductors even when complexed but that is not to say that this is therefore an insignificant biological process. Indeed, with the conductivities measured, complexed proteins in fields of 10^7 V m^{-1} (which are attainable in cell membranes) will allow a flow of about 100 electrons/s along a pathway 1 μm square. This flow is sensitive to the barrier E_i which, as Fig. 7 indicates, is extremely large for dry proteins. It may be much smaller in the biological situation.

The electron pathways will be determined by protein configurations and by contacts with other protein molecules and cell components. The degree of electron movement along any one of these can be controlled in a precise and elegant way by the acceptor complex agent which does not itself take part in the electron flow. Long-range redox reactions, for example, could thus be under remote but sensitive control. Desaturation of the valence band also allows response to any microscopic electric or magnetic fields present and encourages polarization which generates forces of a structure-ordering nature. Laki & Ladik (1976) have drawn attention to these mechanical possibilities which could be important in protein structures such as those of actin and fibrin where molecular alignment is a requirement. The increasing evidence for a cell cytoskeleton consisting of actin microfilaments in an interconnected network could be of great importance in the light of an electronic theory of biology. Not only could the mechanical forces behind the construction and dismantling of the cytoskeleton be electronically controlled through polarization, but also the cytoskeleton might be an electronic network monitoring and controlling cell function.

ACKNOWLEDGEMENTS

I am greatly indebted to Professor A. Szent-Györgyi, Dr R. Pethig and Mr P. Gascoyne for many stimulating discussions, especially while in the congenial environment of the Marine Biological Laboratory, Woods Hole, Massachusetts.

References

Adamec, V. & Calderwood, J.H. (1975) Electrical conduction in dielectrics at high fields. *J. Phys. D Appl. Phys. 8,* 551-560

Austin, I.G. & Mott, N.F. (1969) Polarons in crystalline and non-crystalline materials. *Adv. Phys. 41,* 41-102

Bone, S., Lewis, T.J., Pethig, R. & Szent-Györgyi, A. (1978) Electronic properties of some protein–methylgloxal complexes. *Proc. Natl. Acad. Sci. U.S.A. 75,* 315-318

Brillouin, L. (1962) Giant molecules and semiconductors, in *Horizons in Chemistry* (Kasha, M. & Pullman, B., eds.), pp. 170-183, Academic Press, New York

Davis, E.A. & Mott, N.F. (1970) Conduction in non-crystalline systems. V. Conductivity, optical absorption and photoconductivity in amorphous semiconductors. *Philos. Mag. 22,* 903-922

Eley, D.D. & Spivey, D.I. (1960) The semiconductivity of organic substances. *Trans. Faraday Soc. 56,* 1432-1442

Fleming, R.J. (1970) Charge transfer in paraffinic hydrocarbons under the influence of external electric fields. *Trans. Faraday Soc. 66,* 3090-3102

Frenkel, J. (1938) On pre-breakdown phenomena in insulators and electronic semiconductors. *Phys. Rev. 54,* 647-648

Kertész, M., Koller, J. & Azman, A. (1977) Calculated forbidden band gap in periodic protein models indicating them to be insulators. *Nature (Lond.) 266,* 278

Kroemer, H. (1957) Quasi-electric and quasi-magnetic fields in non-uniform semiconductors. *R.C.A. Rev. 18,* 332-342

LADIK, J. (1976) Energy-band structure and charge transfer in biopolymers. *Int. J. Quantum Chem. Quantum Biol. Symp. 3,* 237-246

LADIK, J., SUHAI, S. & SEEL, M. (1978) The electronic structure of biopolymers and possible mechanisms of chemical carcinogenesis. *Int. J. Quantum Chem. Quantum Biol. Symp. 5,* 35-49

LAKI, K. & LADIK, J. (1976) A note on the electronic theory of cancer. *Int. J. Quantum Chem. Quantum Biol. Symp. 3,* 51-57

LEWIS, T.J. (1977) The dielectric behaviour of non-crystalline solids, in *Dielectric and Related Molecular Processes,* vol. 3 (Davies, M., ed.), pp. 186-218, The Chemical Society, London

McCUBBIN, W.L. (1963) Electronic processes in paraffinic hydrocarbons. *Trans. Faraday Soc. 59,* 769-774

MOTT, N.F. (1967) Electrons in disordered structures. *Adv. Phys. 16,* 49-144

MOTT, N.F. (1969) Conduction in non-crystalline materials. III. Localized states in a pseudo-gap and near extremities of conduction and valence bands. *Philos. Mag. 19,* 835-852

MOTT, N.F. & DAVIS, E.A. (1971) *Electronic Processes in Non-Crystalline Materials,* Oxford University Press, London

OTTO, P., SUHAI, S. & LADIK, J. (1977) *Ab initio* supermolecule study of charge transfer in the glyoxal–formamide and in the H_2S–formamide systems. *Int. J. Quantum Chem. Quantum Biol. Symp. 4,* 451-457

PETHIG, R. (1978) Electronic properties of protein–methylglyoxal complexes: strong evidence for energy-band conduction. *Int. J. Quantum Chem. Quantum Biol. Symp. 5,* 159-171

PETHIG, R. & SZENT-GYÖRGYI, A. (1977) Electronic properties of the casein–methylglyoxal complex. *Proc. Natl. Acad. Sci. U.S.A. 74,* 226-228

POLLAK, M. & GEBALLE, T.H. (1961) Low-frequency conductivity due to hopping processes in silicon. *Phys. Res. 122,* 1742-1753

SAYER, M., MANSINGH, A., REYES, J.M. & ROSENBLATT, G. (1971) Polaronic hopping conduction in vanadium phosphate glasses. *J. Appl. Phys. 42,* 2857-2864

SZENT-GYÖRGYI, A. (1941*a*) The study of energy-levels in biochemistry. *Nature (Lond.) 148,* 157-159

SZENT-GYÖRGYI, A. (1941*b*) Towards a new biochemistry. *Science (Wash. D.C.) 93,* 609-611

SZENT-GYÖRGYI, A. (1977) Protein radicals, regulation and cancer. *Int. J. Quantum Chem. Quantum Biol. Symp. 4,* 179-184

Discussion

Ladik: In the case of a strictly periodic protein model the valence band is about 2 eV broad; we should not call that narrow.

If I may return to the matter of disorder, earlier Professor Laki gave in his paper (pp. 33–42), the example of an ABCABC....structure. That is an exaggerated disorder; it is the worst possible situation. A second point is that the calculation he mentioned is a semi-empirical MINDO calculation which underestimates the band widths and therefore overestimates the gaps.

Eley: Professor Lewis's model stressed hole mobility, and our recent thermoelectric power work supports this conclusion for the dry biopolymers, haemoglobin and NaDNA. We now have thermoelectric power equipment which appears to give reliable results on specimens with resistances as high as 10^{13} Ω (Eley *et al.* 1977*a*). In the dry state both these biopolymers give p-type behaviour, the haemoglobin result agreeing, at least in sign, with earlier work (Cardew & Eley 1959). Adsorption of water causes a change to n-type behaviour, supporting our view of adsorbed water as an electron donor. There are problems from the quantitative point of view. If the effect of water is simply to lower the energy gap we should expect the thermoelectric power charge to parallel that for conductivity (cf. Bosman & Crevecoeur 1966). This is so for NaDNA, but for haemoglobin, adsorbed water changes the thermoelectric power much more than it does the conductivity. This we attribute to the weaker adsorption of water on haemoglobin, as a result of which it also (along with the charge carriers) diffuses down the temperature gradient. To work out the full implications of the two interacting diffusion processes we need to apply non-equilibrium thermodynamics, which we plan to do later.

A further remark on a.c. loss in haemoglobin. At very low frequencies one sees an Ω peak, attributable to electrode polarization, and at somewhat higher frequencies an α relaxation. Often such α relaxations are attributable to dipolar rotation. In this case, since the Ω and α relaxations have a similar activation energy to the steady state d.c. conduction ($\Delta\epsilon = 2.4$ eV), we are inclined to assign both to electronic conduction; i.e. therefore the α relaxation is probably a Maxwell–Wagner relaxation (Eley *et al.* 1979). In similar experiments in the intact mitochondrion where the activation energy for the α relaxation is less than that for the lower frequency Ω relaxation we have suggested that dipolar rotation in the phospholipids may be the determining factor (Eley *et al.* 1977*b*). A lot more work is necessary to establish these interpretations properly.

Lewis: One would expect that, over a considerable range of temperatures, the steady-state (d.c.) conductivity would have an activation energy slightly different from that found from a.c. measurement at the low frequency limit. In the scheme of energies shown in Fig. 6 (p. 73) one can see why that might be. Holes which are going to take part in the steady-state conductivity will involve the Fermi energy E_f, while those contributing to local hopping conductivity will involve energies such as E_d (see also equations 7 and 9).

The idea of relaxation is fashionable in polymer research and, in particular, several attempts have been made to relate the so-called α-relaxation to some gross motion of polymer chains. The model that I have proposed describes α-relaxation well but the implication then is that, instead of some molecular rearrangement occurring, only an electron (or a hole) is required to transfer.

Damadian: To settle this question of semiconductivity in proteins someone should make an amplifier from, say, a hydrated crystal protein. Has that been tried?

Eley: The d.c. dark conductivity of compressed discs of solid dry protein at

298 K is as low as $10^{-17}\ \Omega^{-1}\ cm^{-1}$, and for a compressed disc of heavy beef heart mitochondria, $7 \times 10^{-13}\ \Omega^{-1}\ cm^{-1}$. Even with a complete monolayer of water on haemoglobin (equivalent to some 3 B.E.T. monolayers) the conductivity is still only around $10^{-7}\ \Omega^{-1}\ cm^{-1}$. A major problem has always been how biological electron transfer, for example in an oxidation–reduction reaction, can occur through such highly resistive systems. If conduction goes via an energy band, there is always the possibility that the necessary energy is available to inject an electron into the conduction band, when of course the low conductivity values quoted above become less relevant.

Phillips: Professor Lewis, how do you distinguish what you called the 'interprotein barriers' from intraprotein effects? Do you conceive of these molecules as being linear chains with conduction along them and interprotein barriers between the chains? In any realistic model of protein structures it seems to me very difficult to distinguish between the forces that hold the different parts of a folded protein together and those that hold separate protein molecules together.

Lewis: I did not attempt to specify the interprotein barrier because the present electrical measurements do not provide that sort of detail about the conduction process. A compressed sample of protein must contain a variety of interprotein barriers between protein molecules or clusters of molecules. However, within each protein molecule along the main polypeptide chain, hole conduction would be possible and relatively easy, according to the energy-band calculations of Professor Ladik. This facile hole pathway might be expected to follow the main chain through any folding and turning of the molecule without losing to a significant degree its conductive properties.

The interprotein barriers, E_i, appear to be large for the dry compressed samples but, as I mentioned, they could be very much smaller or even nonexistent in a biological situation. Thus one could conceive of pathways for hole transfer over relatively long distances. In this connection I should stress the distinction I made between 'localized' transitions between E_d states situated either in adjacent protein molecules or perhaps, in some cases, within the same protein molecule, and the delocalized transitions which are emphasized above. The former give rise to a.c. and the latter to steady-state conduction.

Pethig: For the sake of accuracy (and this does not alter the essential feature of Professor Lewis's model) I should point out that only for lysozyme has the effective charge-hopping distance (mean free path) appeared to be much greater than the dimension of the protein molecule. For casein, bovine serum albumin and collagen it appears that intramolecular barriers occur.

Ladik: In the energy scheme of localized states above the valence band, do you mean that these localized states are only acceptor states (e.g. methylglyoxal), or do you mean also localized states due to different side-chains in the proteins?

Lewis: There could be two types of localized state in the band gap: (1) states derived essentially from the valence band and probably characterized in energy by the specific side-chains on the polypeptide backbone; and (2) acceptor states due to the presence of methylglyoxal or some other acceptor complexed with the protein.

Ladik: The localized states due to the different side-chains are by no means necessarily outside the band; they may be in the band or even at the bottom of the band, thereby not affecting the conductivity at all.

Eley: The idea of intramolecular energy bands separated by intermolecular energy barriers was suggested for proteins by Eley & Spivey (1960). At that time we supposed, following Evans & Gergely (1949), that the conduction bands within the molecule arose from the hydrogen-bonded $\gt C = O - H - N \lt$ system, either between chains (the β sheet of Pauling) or within the same chain, i.e. the α helix. The barriers were postulated to arise wherever the hydrogen-bond system was interrupted—that is, whether inside a globular protein molecule, between molecules, or between the crystals themselves.

Williams: Let me put in a few words here about electron transfer in biology so that the theoreticians may bear in mind what is known. Electron transfer nearly always involves two systems. The first is between metal ions. Such ions are kept in isolation from one another in protein centres. The other system includes organic molecules such as flavins and organic free radicals. The exact concentration of these lone spins is known precisely. We also know how the electrons pass from one metal ion or organic group to another, in proteins, and how far they can move (at least 1.0–1.5 nm, although some people believe they can go further). There can be long chains of such units. Many biological systems that conduct electrons for their own purposes are built up of centres of metal atoms or redox organic couples 1.0 or 1.5 nm apart, in sequences. These are 'impurity centres' in the sense of this discussion. There is no evidence, however, to suggest that these impurities are in any way associated with the band structure of the proteins. Rather, electron transfer occurs by direct tunnelling between the two centres. All the reported measurements indicate that it does not involve the energy states of the protein *per se.* In other words, I am convinced that in biological systems electrons do not use the energy states of the polypeptide (or DNA) chain; they rely on tunnelling between impurity centres. Nature then is seen to build the electronic devices without any reference to the backbone of proteins or DNA. In this way it can build conductors which are typical hop conductors as found in semiconductive devices. They are not the same as the devices based on wide bands that we have been discussing so far.

Eley: Does that apply to the whole of the electron path through the mitochondria?

Williams: Yes. The electron transfer can be followed almost step by step. The centres are closely interconnected as far as we know at present. In one protein, not in mitochondria, xanthine oxidase, the distance of hop may be as long as 3.0 nm but, unfortunately, we don't know the nature of this protein in detail. The rates of electron transfer are known: they are about $10^3\ s^{-1}$. Also, we know the activation energies for some of these processes—they are very small.

Eley: At a meeting of the British Biophysical Society in 1962 Professor J.C. Skau if I remember correctly suggested to me that electron transfer might occur across the nerve membrane, in the recovery period after the passage of an impulse. This system might merit examination from the present point of view.

Williams: If the electron transfer were to become indiscriminate in proteins, as it might, using a band of a peptide chain, biology would face a dreadful problem. Electrons which have to go to particular places would go indiscriminately. I am not saying that the introduction of substances like methylglyoxal does not introduce new functional properties of the backbone of proteins, but that there is no need to use models which are based on energy states of the backbone in the usual biological steps.

References

BOSMAN, A.J. & CREVECOEUR, C. (1966) Mechanism of the electrical conduction in Li-doped NiO. *Phys. Rev. 144,* 763-770

CARDEW, M.H. & ELEY, D.D. (1959) The semiconductivity of organic substances. Part 3. Haemoglobin and some amino acids. *Disc. Faraday Soc. 27,* 115-128

ELEY, D.D. & SPIVEY, D.I. (1960) Semiconductivity of organic substances. Part 6. A range of proteins. *Trans. Faraday Soc. 56,* 1432-1442

ELEY, D.D., LOCKHART, N.C. & RICHARDSON, C.N. (1977*a*) Apparatus for the measurement of the Seebeck coefficient in solid biopolymers. *J. Phys. E Sci. Instrum. 10,* 1220-1222

ELEY, D.D., LOCKHART, N.C. & RICHARDSON, C.N. (1977*b*) Electrical properties and structural transitions in the mitochondrion. *J. Bioenerg. Biomembr. 9,* 289-301

ELEY, D.D., LOCKHART, N.C. & RICHARDSON, C.N. (1979) Thermoelectric effects and dielectric polarisation in biopolymers. *J. Chem. Soc. Faraday Trans. I 75,* 323

EVANS, M.G. & GERGELY, J. (1949) A discussion of the possibility of bands of energy levels in protein. *Biochim. Biophys. Acta 3,* 188-197

Electronic and dielectric properties of protein–methylglyoxal complexes

STEPHEN BONE and RONALD PETHIG

Laboratory of the National Foundation for Cancer Research, School of Electronic Engineering Science, University College of North Wales, Bangor, Gwynedd

Abstract Steady-state conduction and dielectric measurements over the frequency range 10^{-5} to 10^5 Hz are reported for several proteins that have been complexed with methylglyoxal. Compared with the normal (white) proteins the brown protein–methylglyoxal complexes exhibit a marked increase in electronic conductivity and a pronounced low-frequency dielectric dispersion. The intensity of the brown colour and the electronic activity is directly related to the number of free lysine groups available to react with the methylglyoxal. It is proposed that the methylglyoxal molecules form Schiff bases with the ϵ-amino groups of lysine residues and that these Schiff bases then form a charge-transfer complex with a neighbouring peptide unit. For collagen, in particular, it is found that the electron 'holes' so formed in the polypeptide backbone are capable of long range motion in what can be interpreted as being the valence band of extended electronic states of the protein structure. The protein–methylglyoxal complexes have electronic and dielectric properties similar to those exhibited by the perylene–chloranil charge-transfer complex.

Proteins can be considered as one of the classes of condensed matter, and it is reasonable to expect that they will exhibit some of the solid-state electronic properties known to occur in conventional amorphous materials and organic polymers. Of particular relevance to this symposium is the concept that as a result of charge-transfer reactions with other molecules the ground-state energy levels of a protein may become partially desaturated of electronic charge. If a well-defined valence band of extended energy states and an associated 'tail' of localized states exists for the protein molecule, then the electron 'holes' formed by such charge-transfer reactions will be capable of extensive mobility within the *ground-state* energy levels of the protein structure. Such a process will lend to proteins a submolecular subtlety of behaviour not possible for their electron-saturated counterparts. This in essence forms the cornerstone of Szent-Györgyi's electronic concept of the 'living state', about which more should be known if we are to reach a deeper understanding of the cellular deficiencies inherent in cancer and other diseases (Szent-Györgyi 1976).

The purpose of this paper is to describe some of the electrical and dielectric measurements that we have made to investigate the possibility that, when methyglyoxal is incorporated into the structure of protein molecules, a charge-transfer reaction can result which produces electron holes capable of extensive delocalization within the protein structures. Details of the chemical preparation procedures and electronic measurement techniques used for these studies can be derived from reports of the earlier work (Pethig & Szent-Györgyi 1977; Bone *et al.* 1978; Pethig 1978).

The electrical measurements described here were made on dry, or only slightly moistened, compressed powder protein discs. Can such studies be of any biological relevance? Energy-band calculations (Suhai 1974; Ladik 1978) for β-polyglycine indicate that electronic delocalization is likely to be most favourable *along* the polypeptide backbones of protein structures rather than through the hydrogen-bonded structures that help stabilize the protein's tertiary structure. This conclusion is in agreement with that formulated by Petrov *et al.* (1977). If that conclusion is applicable to proteins generally, it means that the dehydration of proteins should not seriously alter the basic electronic properties associated with their primary structures. In their natural state many proteins are bound into hydrophobic lipid matrices, and furthermore it can be calculated (Pethig 1977) that the internal structures of proteins have an effective high frequency molecular permittivity of the order 2.6, and this is a value which will not be greatly increased on hydration. As a first step towards investigating the basic electronic properties of protein structures, measurements on dry compressed protein discs should, therefore, provide useful data. A problem that can arise from electrical measurements on compressed powders is that the desired intramolecular electronic properties may be masked by intermolecular, inter-crystallite, and electrode–interface effects. We have given such problems careful consideration.

To lend support to the opinion that a charge-transfer process is involved in the protein–methylglyoxal interaction, we compare the electrical properties of the protein samples with those of the well-characterized perylene–chloranil charge-transfer complex.

THE POSSIBLE CHARGE-TRANSFER INTERACTION

The proteins that have been studied in some detail are bovine serum albumin (BSA), casein (Hammersten preparation), collagen (bovine Achilles tendon) and lysozyme (chicken egg white). On reacting with methylglyoxal these proteins assume a stable brown colour. From the work of Takahashi (1977) on ribonuclease we can expect that, apart from reacting with terminal α-amino groups, methylglyoxal will react with the side-chains of arginine and lysine and, to a lesser extent, with cysteine. The lysine side-chains of BSA and casein have been reductively dimethylated using the method described by Lin *et al.* (1969) and it has been found that the colour and electron spin resonance (e.s.r.) signal intensities exhibited by the complexes of these modified proteins with methylglyoxal are directly related to the number of 'unblocked' lysine residues (Joy Behi, Jane A. McLaughlin, P.R.C. Gascoyne, R. Pethig & A. Szent-Györgyi, unpublished work). The concentration of free lysine residues was determined using the spectroscopic method described by Fields (1972). We have also modified the arginine side-chains of BSA using the procedure described by Smith (1977), and the modified protein still reacts with methylglyoxal to produce a stable brown complex exhibiting a larger e.s.r. signal than untreated BSA. These results suggest that the relevant reaction for our studies involves the lysine side-chains and we believe that the first step is the formation of Schiff base ($-C=N-$) linkages to the ϵ-amino groups. Otto *et al.* (1978) have shown that such a Schiff base is as good an electron acceptor as methylglyoxal itself and they also show that no change of the main polypeptide chain configurations is required for the lysine side-chain to bend back to enable a charge-transfer interaction to occur between the electron-donating 'next-door' peptide unit and the electron-accepting Schiff base. It is hoped that a crystallographer can be encouraged to investigate the validity of this interesting concept.

D.C. MEASUREMENTS

When extracted as a dry powder, the brown methylglyoxal-treated proteins exhibit much larger d.c. conductivities than those observed for the dry untreated control proteins (see Fig. 1 and Table 1). Electronic transference number measurements (Pethig & Szent-Györgyi 1977) indicate that this increased conductivity is associated with mobile electronic, rather than ionic, charges. For the temperature range investigated (200–330 K), the conductivity σ of the brown protein complexes obeys the standard equation (1)with the

$$\sigma = \sigma_0 \exp(-\Delta E/kT) \qquad (1)$$

value for the activation energy ΔE having a value typically of the order 0.65 eV for temperatures above 250 K and a value of around 0.3 eV below 250 K.

Of the proteins studied, the collagen–methylglyoxal samples have shown the largest steady-state conductivity with values up to 2×10^{-9} mho m^{-1} at 297 K. However, this conductivity value is altered after temperature cycling and good semiconductivity activation plots have been obtained only after the conductivity, measured at room temperature, has fallen by some two to three orders of magnitude below its initial high value. (Typically, at 297 K, $\sigma \approx 2 \times 10^{-12}$ mho m^{-1} and $\Delta E \approx 1.1 \pm 0.1$ eV). This would suggest that the conductivity of the dry collagen complexes is sensitive to molecular conformational changes.

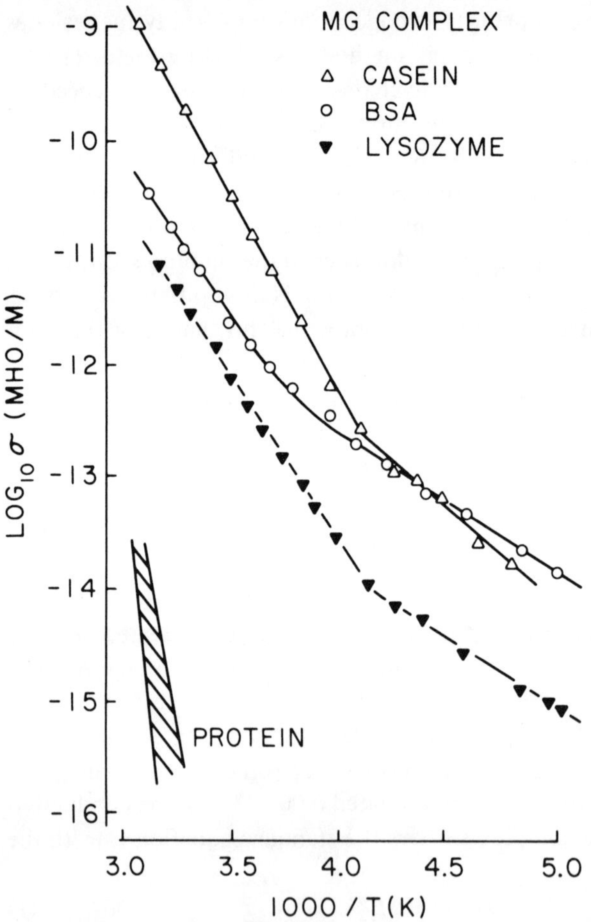

FIG. 1. The temperature variation of the dry-state conductivity of various proteins that have been complexed with methylglyoxal (MG). Also shown is the range of conductivities of the dry untreated proteins.

TABLE 1

Conductivity and dielectric data at 297 K for proteins, protein–methylglyoxal complexes, and the perylene–chloranil complex

Parameter	*Dry protein*	*Dry protein complex*	*Perylene-chloranil complex*
σ/(mho/m)	$10^{-15} - 10^{-18}$	$3 \times 10^{-10} - 2 \times 10^{-13}$	$2 \times 10^{-9} - 3 \times 10^{-12}$
ΔE eV (equation 1)			
$T \gtrsim 260$ K	1.4 ± 0.1	0.65 ± 0.06	0.73 ± 0.02
$T \lesssim 260$ K	?	0.3 ± 0.05	0.4 ± 0.02
τ s (equation 4)		$0.4 \sim 25$	$5 \sim 80$
W/eV (equation 4)		0.55 ± 0.1	0.64 ± 0.04
$\epsilon_r/(10^{-4}$ Hz)	3.3 ± 0.2	$15 - 80$	$20 - 60$
$\epsilon_r/(10^4$ Hz)	3.2 ± 0.2	4.5 ± 0.5	3.5 ± 0.1

An exception to the values given here occurs for the collagen complexes, where $\Delta E = 1.1 \pm 0.1$ eV for $T \gtrsim 260$ K.

When the conduction data for constant temperature are plotted as log(current) against $F^{1/2}$, or log(conductivity) against $F^{1/2}$, where F is the applied electric field, straight lines are obtained only for the log(conductivity) data. This result is shown in Fig. 2 for a lysozyme-methylglyoxal sample and is consistent with the sample conductivity being limited by the Poole–Frenkel effect of field-assisted thermal release of charge carriers over Coulombic potential energy barriers in the bulk material rather than by Schottky electron emission from the electrodes (Frenkel 1938; Hill 1971). For the Poole–Frenkel effect the conductivity is proportional to $\exp(\beta F^{1/2}/2kT)$ when the Fermi energy lies well above/below the electron/hole 'trapping' sites, and is proportional to $\exp(\beta F^{1/2}/kT)$ when the traps are situated close to the Fermi energy (Mark & Hartman 1968). The field-lowering coefficient β is given by equation (2)

$$\beta = (q^3/\pi\epsilon_0\epsilon_r)^{1/2} \tag{2}$$

where q is the electronic charge, ϵ_0 is the permittivity of free space and ϵ_r is the relative permittivity of the sample. The straight-line plots of log(σ) against $F^{1/2}$ plots will then have a gradient of value within the limits $\beta/2kT <$ gradient $< \beta/kT$, depending on the location of the trapping sites and the Fermi energy. For the lysozyme–complex of Fig. 2 the corresponding value for ϵ_r lies within the limits $11.1 < \epsilon_r < 44.4$. The value of 11.1 is more than twice the permittivity value obtained at the higher frequencies of measurement (see Table 1), whereas the value of 44.4 is close to that of the limiting low frequency value observed for the lysozyme–methylglyoxal sample of Fig. 2. Good

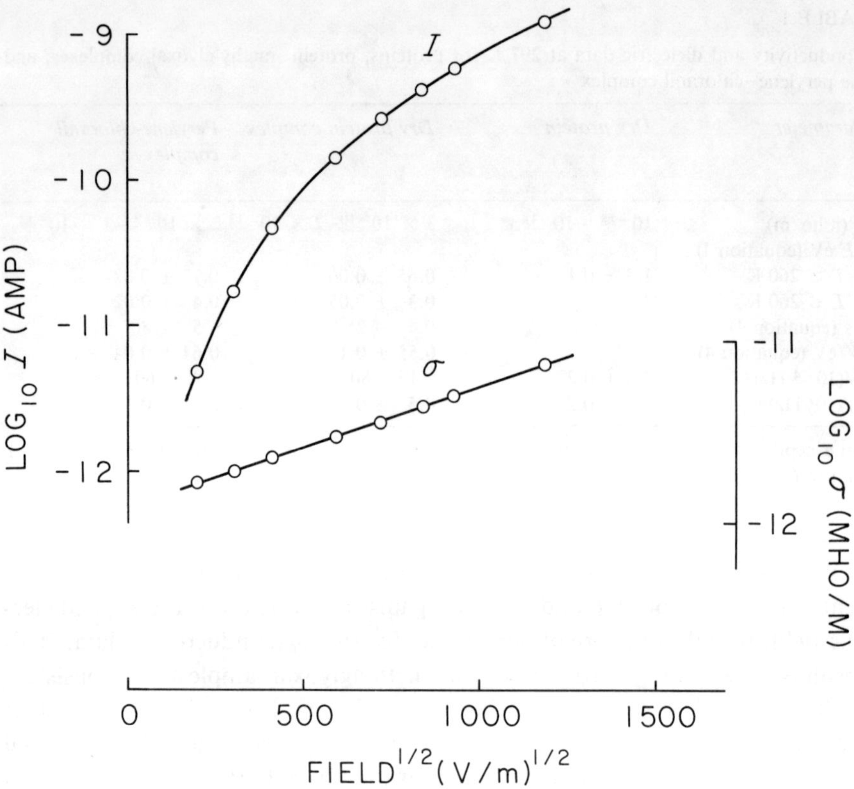

FIG. 2. The Schottky plot log(I) against $F^{1/2}$, and the Poole–Frenkel plot of log(σ) against $F^{1/2}$; plots for a dry lysozyme–methylglyoxal sample at 279 K.

linear Poole–Frenkel plots have been obtained for various BSA, collagen and lysozyme complexes, and a similar trend has been obtained regarding the most reasonable interpretation of the permittivity values derived from the gradients of these plots. This in turn implies that the trapping sites are located close to the Fermi energy, and that the charges reside in them long enough to polarize the surrounding molecular structure. The perylene–chloranil complexes have similar d.c. electrical properties to those of the protein–methylglyoxal complexes and they also show a bulk Poole–Frenkel effect (Pethig & Soni 1975). As described in the earlier reports, the protein–methylglyoxal complexes exhibit large e.s.r. signals. The spin concentrations, n_s about ($5 \pm 4 \times 10^{23}$ m^{-3}), remain relatively constant as a function of temperature and are at least 100 times greater than those observed for the untreated proteins (P.R.C. Gascoyne, personal communication). The collagen complexes tend to exhibit lower values for n_s of the order 4×10^{21} to 6×10^{22} m^{-3}. If the value for n_s is taken to represent the number of separated electronic charges produced by the charge-transfer interaction between the Schiff bases and the peptide units, then a value for the microscopic mobility μ can be estimated from the formula (3). The factor 1/2 arises because the charge accepted by the Schiff base is

$$\sigma = \tfrac{1}{2} n_s q \mu C \exp(-\Delta E/kT) \qquad (3)$$

assumed to be immobile, and the parameter C depends on the energy distribution of the localized trapping sites. Since n_s has been found to be practically independent of temperature, ΔE in equation (3) has been assumed to represent the mobility activation energy. Using the data of Table 1, together with the fact that there is a tendency for the more conducting samples to have larger n_s values and smaller ΔE values, then on average the microscopic mobility values for the BSA, casein and lysozyme complexes fall within the range $0.5 \times 10^{-5} < \mu C < 3.5 \times 10^{-5}$ m^2 V^{-1} s^{-1} and, for the collagen complexes, $0.5 \times 10^6 < \mu C < 2.5 \times 10^6$ m^2 V^{-1} s^{-1}. If the trapping sites are distributed such that their density is proportional to their energy location, the factor C has the value $\Delta E/kT$ (Davis & Mott 1970). For a discrete set of trapping levels of density N_t located at energy ΔE from the relevant band edge, then $C \approx N_b/N_t$, where N_b is the effective density of extended states at the band edge. These two examples indicate that the factor C should be expected to be larger than unity. The borderline between a conduction mechanism involving hopping charge transport in localized energy states and one involving coherent wave-like transport in well-defined bands of extended states is generally considered to occur at a mobility value of the order 1×10^{-4} m^2 V^{-1} s^{-1}. This suggests that, of the protein complexes studied, the collagen complexes may be the only ones for which charge transport in bands of extended states can occur. This may arise from the fact that the collagen molecule, with a molecular weight of the order 3×10^5 and being some 20 times larger than lysozyme, for example, has a sufficiently large number of interacting constituent atoms for the formation of well-defined energy bands to be possible.

The conductivity of the samples is very sensitive to hydration. When the water content is increased to around 20%, the conductivity of the dry samples increases by a factor of 10^6–10^7 (Joyce Eden & R. Pethig, unpublished data). We believe that this effect is primarily related to an increased permittivity which results in a reduction of the Coulombic energy binding together electron-hole pairs, and also a reduction of the inter- and intramolecular potential energy barriers limiting long-range charge mobility. We believe that the intrinsic energy-band structures of the proteins are not seriously modified by such an increase in hydration.

A.C. MEASUREMENTS

Measurements of the a.c. conductivity $\sigma(\omega)$ and complex relative permittivity ($\epsilon_r = \epsilon' - i\epsilon''$) have been made over the frequency range 10^{-5} Hz to 33 GHz. As the measurement frequency increases, the conductivity increases above the steady-state value, and the protein–methylglyoxal samples typically show evidence of a dielectric dispersion in the frequency range 10^{-3} to 10^{-1} Hz. An example of this behaviour is shown in Fig. 3 for BSA. Above 10^4 Hz the samples exhibit a frequency-dependent conductivity of the form $\sigma(\omega) = A\omega^n$ with the frequency exponent n having a value close to unity. The dielectric dispersion exhibited by the protein–methylglyoxal samples can be shown more clearly as the frequency variation of the loss factor ϵ'' (Fig. 4), or as a Cole–Cole plot of ϵ'' against ϵ' (Fig. 5). The form of these dispersions indicates that they are characterized by a wide distribution of relaxation times.

With increasing temperature the dielectric loss peaks are displaced toward higher frequencies, as shown in Fig. 4. For each loss peak the characteristic relaxation time τ is given by $1/(2\pi f_m)$, where f_m is the frequency at which the maximum dielectric loss occurs. From the temperature variations of τ (e.g. Fig. 6) the dielectric dispersions are found to obey an activated law of the

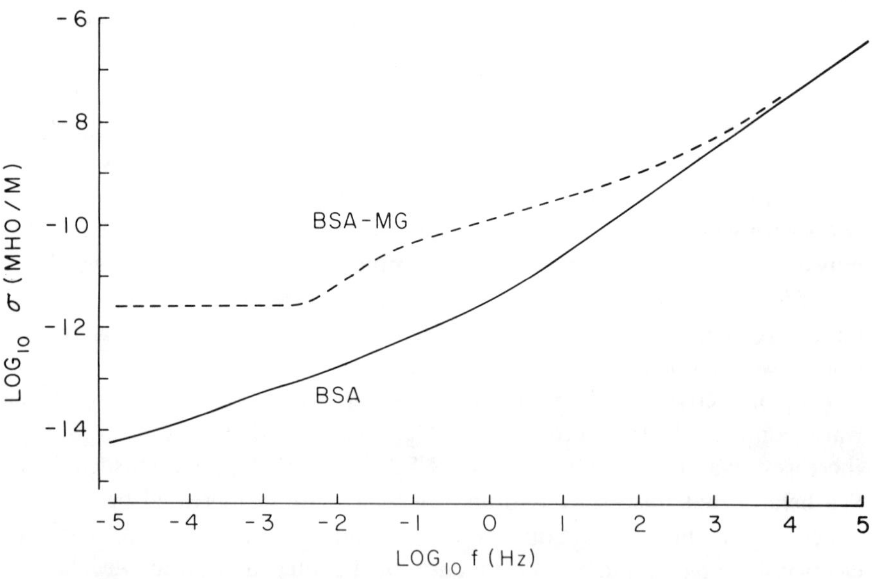

FIG. 3. The frequency variation of the conductivity of a dry BSA–methylglyoxal (MG) sample and of a dry untreated BSA sample at 297 K.

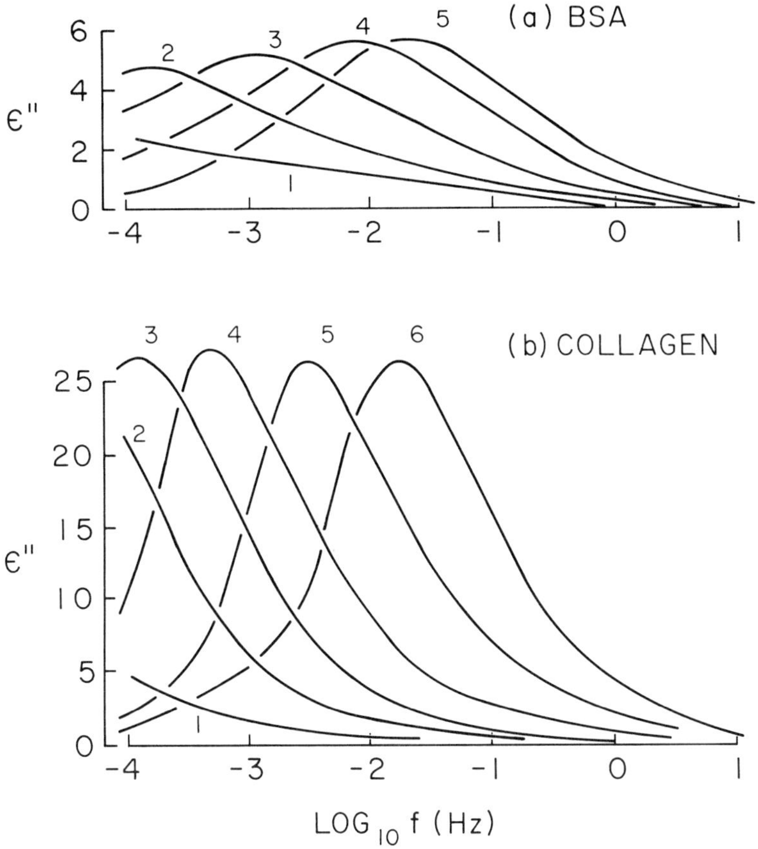

FIG. 4. The frequency and temperature variation of the dielectric loss factor ϵ'' for (*a*) a BSA–methylglyoxal sample at temperature values (1) 226 K; (2) 252 K; (3) 277 K; (4) 304.5 K; (5) 319.5 K and (*b*) a collagen–methylglyoxal sample at temperature values (1) 258 K; (2) 275.5 K; (3) 285 K; (4) 294 K; (5) 304 K and (6) 316 K.

form (4). The ranges of values obtained for τ and W for the various protein

$$\tau = \tau_0 \exp(W/kT) \tag{4}$$

complexes studied are given in Table 1(p. 87). The molar entropy of activation, ΔS, for the activated dispersion process can be estimated by assuming that τ_0 is given according to chemical reaction rate theory by equation (5), where h

$$\tau_0 = (\mathrm{h}/kT)\exp(-\Delta S/k) \tag{5}$$

is Planck's constant. At room temperature the derived value for ΔS is of the order of $-$ 10 cal/(deg. mol), which implies that the relaxation process is asso-

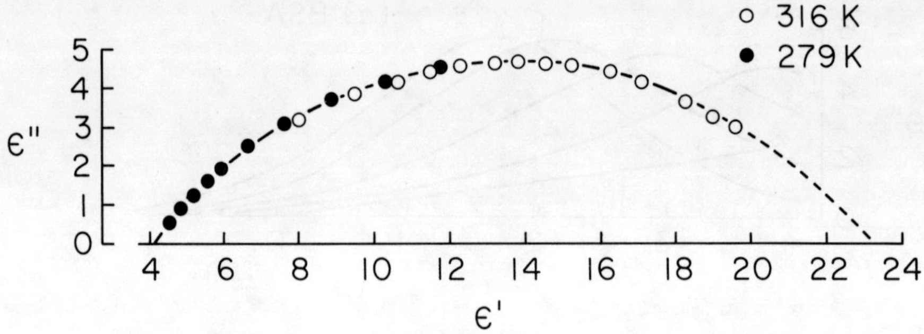

FIG. 5. The plot of ϵ'' against ϵ' for a lysozyme–methylglyoxal sample at 279 K and 316 K.

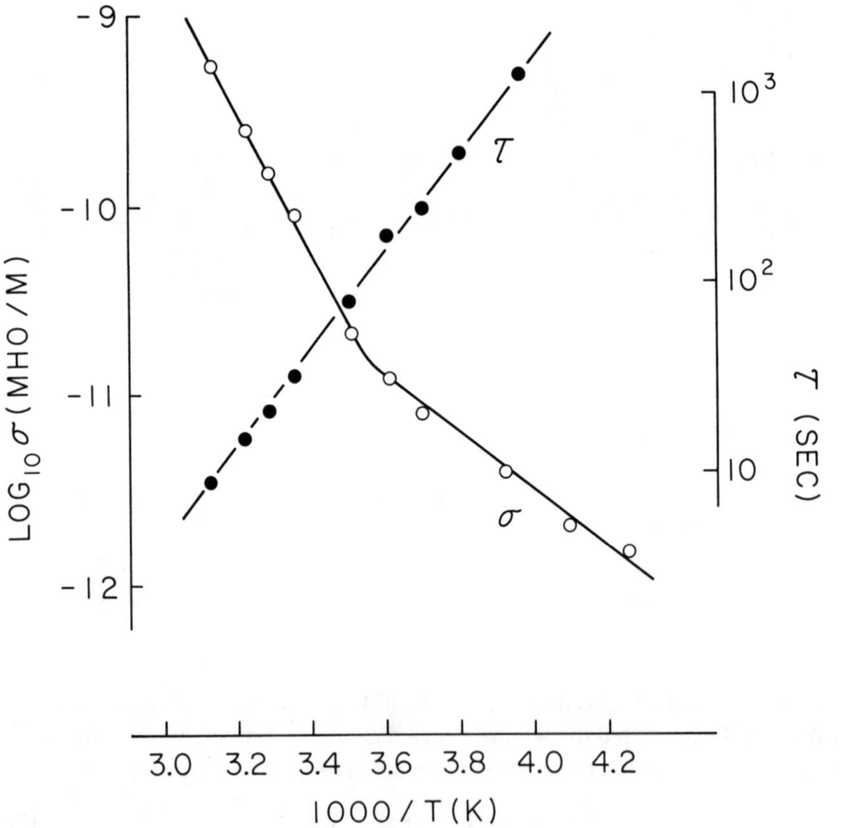

FIG. 6. The variation of the conductivity σ and dielectric relaxation time τ as a function of reciprocal temperature for the BSA–methylglyoxal sample of Fig. 1.

ciated with some degree of ordering of the molecular lattice. The values of the limiting low and high frequency relative permittivity are also given in Table 1 and it is the difference between these two quantities that gives a direct measure of the 'strength' of the dielectric dispersions. From Table 1 it can be seen that the dielectric behaviour of the protein–methylglyoxal complexes is similar to that observed for the perylene-chloranil complex (Carnochan & Pethig 1976; Bone & Pethig 1978). As was found for the latter complex, no apparent deviation from the activation law of equation (4) has been found at the temperature at which the semiconductivity activation energy ΔE of equation (1) changes. This is demonstrated in Fig. 6 for a BSA-methylglyoxal sample. The magnitude of the dielectric loss peaks has been found to vary significantly with the strength of the electric field applied to the samples. This is shown for a collagen–methylglyoxal sample in Fig. 7 in which the area under the plot of

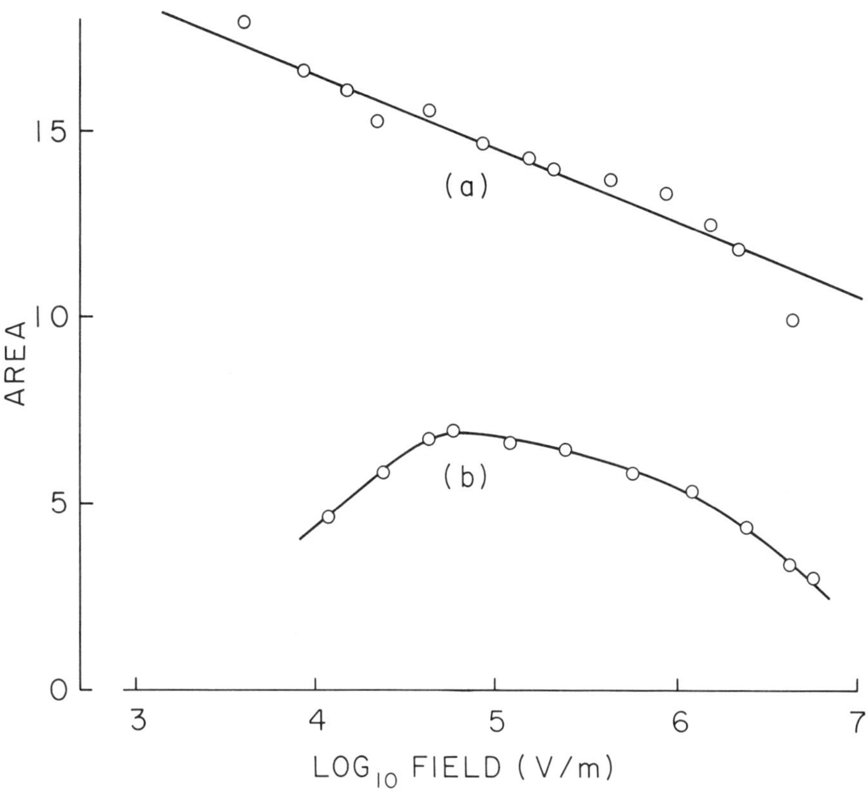

FIG. 7. The variation of the area A under the plot of ϵ'' against $\log(f)$ dielectric loss peaks as a function of the applied field for (*a*) a collagen–methylglyoxal sample and (*b*) a perylene–chloranil sample.

ϵ'' against log(f) decreases exponentially with increasing field strength. Also shown is the typical way in which the perylene–chloranil samples behave as a function of applied field strength. These results cannot be interpreted in terms of the conventional field saturation of dipolar polarizability. With increasing applied field strength the frequency f_m first increases slightly and then falls to a value slightly less than the original low field value of f_m. According to the form of equations (4) and (5) this implies that the larger field strengths induce some form of molecular ordering in the samples. With increasing hydration the value for f_m increases (Joyce Eden & R. Pethig, unpublished data) and this can be interpreted in terms of an increasing permittivity value lowering the effective value for W in equation (4).

The dielectric dispersions for the perylene–chloranil and protein–methylglyoxal samples are in general anomalously high when an interpretation is attempted in terms of conventional molecular dipole relaxation theory. With the view-point that a solid containing a system of non-interacting charges hopping over potential energy barriers will, at a constant temperature, be dielectrically indistinguishable from a solid containing a sparsely distributed set of molecular dipoles, the dielectric properties of the perylene–chloranil complexes have been interpreted in terms of a model involving the hopping of electrons between trapping centres (Carnochan & Pethig 1976). The maximum contribution to the dielectric dispersion occurs where the factor $N_t(E)f(1-f)$ is a maximum, in which $N_t(E)$ is the trap concentration and f is the probability of their electron occupancy at energy E. The dispersion strength will alter according to changes of the Fermi energy and, although sharp changes can occur for the semiconduction activation energy ΔE as a result of even a small shift of the Fermi energy, no such sharp changes should occur for the relaxation activation energy W. This can be used as an explanation for the temperature dependencies of σ and τ shown in Fig. 6.

The dielectric dispersions exhibited by the protein–methylglyoxal complexes have also been interpreted in terms of a hopping electron model (Bone *et al.* 1978; Lewis 1978). According to this model, the area A under the plot of ϵ'' against log(f) is directly proportional to the density of hopping charges and to the mean distance r travelled by the charges between hopping over potential barriers. If the density of mobile charges is taken as $n_s/2$, then the distance is given by equation (6), where ϵ is the relative permittivity of the test samples at

$$r^2 = (27.6AkT\epsilon_0/n_s\pi q^2)[2/(3+\epsilon)] \qquad (6)$$

frequency f_m and is included to take account of the well-known Lorentz correction for the local field. Values derived for r for various protein complexes are given in Table 2, and make use of e.s.r. results kindly supplied by P.R.C.

TABLE 2

Values derived for the mean hop distance r of equation (6) for various protein–methylglyoxal complexes

Methylglyoxal complex	$n_s(m^{-3})$	ϵ	A	r/nm
BSA	4×10^{23}	43	53	8.5
Casein	8×10^{23}	22	24	5.5
Collagen	4×10^{21}	18.5	19	74.5
Lysozyme	6×10^{23}	25	28	6.5

n_s, electron spin concentration; ϵ, relative permittivity at the frequency of maximum dielectric loss; A, area under the dielectric loss peak.

Gascoyne. The values derived for r imply that the mobile charges are able to migrate along lengthy pathways between the dominant potential barriers or trapping sites. The much longer conduction path determined for the collagen complex is in line with the indication from the d.c. measurements that energy band conduction may be possible for the collagen structure.

We are led to consider that the protein–methylglyoxal compressed powders represent an example of one of the well-known (Maxwell–Wagner) models of an inhomogeneous dielectric. For our case the model can be imagined to consist of conducting protein regions separated by poorly conducting barriers. Koops (1951) used a similar model to describe the dielectric properties of ferrites, and it is interesting that when his formulae are used the dispersions typically exhibited by the protein–methylglyoxal samples can be described by assigning to the conducting regions a small spread of conductivity values centred around a value of 10^{-10} mho m^{-1}, with the poorly conducting barriers having an effective conductivity of 10^{-12} mho m^{-1} and an effective thickness one-tenth of that of the conducting regions. (The limiting high and low frequency permittivity values are predicted to be 4.9 and 45, respectively, with τ = 4 s and a d.c. conductivity of 10^{-11} mho m^{-1}.) An explanation of the effect shown in Fig. 7 of the dispersion strengths decreasing with increasing electrical field can also be approached in terms of this model, since the limiting low frequency permittivity is predicted to decrease if the resistivity and width of the poorly conducting barriers are reduced by the applied field.

CONCLUSIONS

The interaction of methylglyoxal with several proteins produces brown complexes that exhibit an increased free-electron spin density and electrical conductivity by comparison with normal untreated proteins. The presence of lysine side-chains appears to be essential for the brown colouration and the increased e.s.r. and electrical activity in the proteins. Although the electronic conduction properties at the molecular level will have been largely masked by effects associated with intermolecular and inter-crystallite boundaries, the results outlined here can be taken as evidence that when methylglyoxal interacts with proteins it increases the effective electronic conductivity of the protein molecules. A possible explanation for this is that the creation of mobile charge carriers is facilitated by the action of methylglyoxal molecules forming charge-transfer complexes within the protein structure. It is envisaged that electronic charge is donated from valence electronic states of the protein structure to electron-accepting states associated with the methylglyoxal molecules. An analogy can be made with the doping of giant molecules (crystals) of silicon with electron-accepting impurity atoms, and the first experimental demonstration of the possibility that such doping or charge-transfer effects occur in protein structures appears to have been that by Davis *et al.* (1960) and by Eley & Snart (1965). Although it may only be taken as indirect evidence of a charge-transfer state, the protein–methylglyoxal samples exhibit similar electronic and dielectric properties to those of the perylene–chloranil complex whose charge-transfer character has been well established. For the collagen samples there appears to be evidence for the concept of long-range motion of holes in extended valence band states, and support for this has come from preliminary measurements (T.E. Cross & R. Pethig, unpublished data) of high-field effect and Hall effect measurements at 10 and 33 GHz where intermolecular boundary effects are largely eliminated. The results for collagen have encouraged us to extend our studies to other structural proteins such as fibrin and tubulin.

The creation of charge-transfer-induced delocalized charges in biomacromolecules can be envisaged to produce a whole range of sub-molecular electronic subtleties whose relevance has largely been overlooked by the biological sciences. These electronic effects could involve not only simple electronic conductivity, but also such phenomena as solitons, charge-density waves, spin-density waves and the other types of collective electronic modes currently under extensive study in organic conductors and polymeric charge-transfer systems (e.g. see Pal *et al.* 1977). The challenge is now to identify and explore those areas where such electronic subtleties could be of direct relevance to cancer research. It seems certain that in this way new knowledge will be gained of value to the medical sciences in general.

ACKNOWLEDGEMENTS

We thank Dr Albert Szent-Györgyi for his stimulating ideas and encouragement, Joy Behi and Jane McLaughlin for their valuable preparative work, and P.R.C. Gascoyne for kindly supplying the e.s.r. data. Thanks are also due to W.M. Arnold, P. Carnochan, T.E. Cross, Joyce Eden, J.C. Fothergill, R.E.J. Jones and Professors J. Ladik and T.J. Lewis for valuable discussions.

References

BONE, S. & PETHIG, R. (1978) Temperature dependence of the low frequency dielectric dispersion in the perylene–chloranil complex. *J. Chem. Soc. Faraday Trans. I 74,* 720-726

BONE, S., LEWIS, T.J., PETHIG, R. & SZENT-GYÖRGYI, A. (1978) Electronic properties of some protein–methylglyoxal complexes. *Proc. Natl. Acad. Sci. U.S.A. 75,* 315-318

CARNOCHAN, P. & PETHIG, R. (1976) Low frequency dielectric dispersions in the perylene–chloranil charge transfer complex. *J. Chem. Soc. Faraday Trans. I 72,* 2355-2363

DAVIS, E.A. & MOTT, N.F. (1970) Conduction in non-crystalline systems V. Conductivity, optical absorption and photoconductivity in amorphous semiconductors. *Philos. Mag. 22,* 903-922

DAVIS, K.M.C., ELEY, D.D. & SNART, R.S. (1960) Semiconductivity in proteins and haemoglobin. *Nature (Lond.) 188,* 724

ELEY, D.D. & SNART, R.S. (1965) Conduction in chloroplast components. *Biochim. Biophys. Acta 102,* 379-385

FIELDS, R. (1972) The rapid determination of amino groups with TNBS. *Methods Enzymol. 25B,* 464-468

FRENKEL, J. (1938) On pre-breakdown phenomena in insulators and electronic semiconductors. *Phys. Rev. 54,* 647-648

HILL, R.M. (1971) Poole-Frenkel conduction in amorphous solids. *Philos. Mag. 23,* 59-86

KOOPS, C.G. (1951) On the dispersion of resistivity and dielectric constant of some semiconducdors at audiofrequencies. *Phys. Rev. 83,* 121-124

LADIK, J.J. (1978) Electronic structure of biopolymers, in *Quantum Theory of Polymers* (André, J.-M., Delhalle, J.K. & Ladik, J.J., eds.), pp. 257-278, Reidel, Dordrecht & Boston

LEWIS, T.J. (1978) Conduction in protein and methylglyoxal-protein complexes. *Int. J. Quantum Chem. Quantum Biol. Symp. 5,* 149-158

LIN, Y., MEANS, G.E. & FEENEY, R.E. (1969) The action of proteolytic enzymes on *N,N*-dimethyl proteins. *J. Biol. Chem. 244,* 789-793

MARK, P. & HARTMAN, T.E. (1968) On distinguishing between the Schottky and Poole-Frenkel effects in insulators. *J. Appl. Phys. 39,* 2163-2164

OTTO, P., LADIK, J., LAKI, K. & SZENT-GYÖRGYI, A. (1978) Internal charge transfer in proteins to the Schiff bases of their lysine side chains. *Proc. Natl. Acad. Sci. U.S.A. 75,* 3548-3550

PAL, L., GRÜNER, G., JANOSSY, A. & SOLYOM, J. (eds.) (1977) *Organic Conductors and Semiconductors,* Springer-Verlag, Berlin

PETHIG, R. (1977) Some dielectric and electronic properties of biomacromolecules, in *Dielectric and Related Molecular Processes,* Vol. 3, pp. 219-252, Specialist Periodical Report of the Chemical Society, London

PETHIG, R. (1978) Electronic properties of protein–methylglyoxal complexes: strong evidence for energy-band conduction. *Int. J. Quantum Chem. Quantum Biol. Symp. 5,* 159-171

PETHIG, R. & SONI, V. (1975) High field electrical conduction in some organic charge-transfer complexes. *J. Chem. Soc. Faraday Trans. I 71,* 1534-1544

PETHIG, R. & SZENT-GYÖRGYI, A. (1977) Electronic properties of the casein–methylglyoxal complex. *Proc. Natl. Acad. Sci. U.S.A. 74,* 226-228

PETROV, E.G., UKRAINSKII, I.I. & KHARKYANEN, V.N. (1977) Donor–acceptor model of electron transfer through proteins. *Preprint of the Academy of Sciences of the Ukrainian S.S.R., Kiev, ITP-77-38E,* 1-47

SMITH, E.L. (1977) Reversible blocking at arginine by cyclohexanedione. *Methods Enzymol. 47,* 156-161

SUHAI, S. (1974) Theoretical investigation of semiconductive properties in proteins. I. Electrical conductivity, charge carrier mobilities, and free paths in β-polyglycine. *Biopolymers 13,* 1731-1737

SZENT-GYÖRGYI, A. (1976) *Electronic Biology and Cancer,* Dekker, New York

TAKAHASHI, K. (1977) Further studies on the reactions of phenylglyoxal and related reagents with proteins. *J. Biochem. (Tokyo) 81,* 403-414

Discussion

Brandt: Your arguments in favour of the reaction of lysine rather than arginine with methylglyoxal are convincing, yet many proteins have thiol groups hidden in their interior. Treatment with methylglyoxal may itself expose these internal SH groups. Treatment of the proteins with sodium dodecyl sulphate to reduce the SH groups before comparison of the methylglyoxal–protein complex with the non-complexed protein might provide better proof of the involvement of lysine.

Pethig: Yes, we plan to do various experiments with proteins whose SH groups have been modified.

Ladik: In the light of Dr Pethig's results let me take up Professor Williams's earlier point (p. 81). Conductivity, τ, may be expressed (equation

$$\tau = e(n\mu_e + p\mu_h) \tag{1}$$

1) in terms of the number of electrons, n, the electron mobility, μ_e, the number of holes, p, and the mobility of the holes, μ_h. A large gap will affect the conductivity but if the bands are broad, as they are claimed to be in a strictly periodic polypeptide, and therefore the mobilities are high, if holes are created in the valence band as electrons are taken out from it by acceptors, there will be conductivity. This is a fact which is due to basic physical principles. I would not say that just because in mitochondria one finds tunnelling between impurity centres one should generally dismiss the idea of conductivity in the bands of the backbone. Not enough calculations have been made, on the other hand, on the effects of disorder on the band widths. If it turns out that because of the disorder the bands are too narrow, conduction along the backbone becomes questionable. The possibility of conductivity along the backbone should be kept in mind, however.

Williams: I accept that it is possible, but does it happen in that way?

Swartz: If one is going to propose theoretical interpretations of experimental results, it is critical to understand and define the experimental system. How pure were the proteins that you used? What were their exact structures in the preparations used for your experiments?

Williams: Such questions must be answered. For example, how many methylglyoxal molecules were bound to each lysozyme molecule? Discussion of possible mechanisms depends on such knowledge.

Pethig: With regard to the possible involvement of a polymeric form of methylglyoxal, we were somewhat dismayed by the results of Bonsignore *et al.* (1977); they described a yellow polymer of methylglyoxal and we had been preparing yellowish-brown protein-methylglyoxal complexes. We polymerized methylglyoxal according to their method and found the conductivity of the yellow polymer to be considerably lower than that of the coloured protein. This was encouraging as it indicated to us that the high conductivity of the coloured proteins was not likely to have been influenced by the presence, if any, of a polymeric form of methylglyoxal. Also, our results with ^{14}C-labelled material and methylglyoxal weight-uptake measurements indicated an insufficient number of reacted methylglyoxal molecules to form a polymeric system that had much consequence. The methylglyoxal content was typically of the order of 10% of the total protein weight. Our results indicate interactions of methylglyoxal with arginine and lysine residues, and perhaps with cysteine, but we have yet to find proper evidence for an extensive involvement of the methylglyoxal polymer.

Williams: Are you saying that 10% by weight of the lysozyme complex was methylglyoxal?

Pethig: The figure has varied from protein to protein, and also according to the pH of the incubating protein-methylglyoxal suspension. For bovine serum albumin, for example, at pH 4.5 the methylglyoxal content is around 8% (dry protein weight basis) and at pH 7 the content is around 12%. Casein appears to bind less methylglyoxal. The results for lysozyme appear to be similar to those for the albumin.

Williams: That is, at least 10 groups per protein molecule?

Pethig: Lysozyme contains six lysine groups and together with the terminal NH_2 group we have seven possible (Schiff base) reaction sites.

Williams: As the structure of lysozyme is known with precision, there is no reason why we cannot tell whether other groups than lysine have been attacked.

Pethig: Other groups are attacked (e.g. arginine and possibly cysteine) but we are interested only in the site that is relevant to the appearance of the brown colour, the e.s.r. signal and the conductivity. This site appears to be lysine.

Swartz: For your conductivity measurements do you dry and compress the protein in air?

Pethig: After separation on a filter and rigorous washing, the proteins are dried in a vacuum desiccator. The test samples are compacted in vacuum using a standard die. Unfortunately, at some point during the experimental procedures (e.g. in mounting the samples in the conductivity cell) the proteins are exposed to air. The conductivity measurements are made in vacuum or an inert atmosphere, or in a carefully controlled water vapour atmosphere. The effect of hydration on the conductivity greatly exceeds that due to exposure to oxygen, for example.

Swartz: What do the e.s.r. spectra of that material prepared for conductivity measurements look like?

Pethig: The spectra show a main signal centred at $g = 2.004$, and the intensity of the signal increases with increasing methylglyoxal content. Peter Gascoyne at Woods Hole has used a computer technique to analyse the hyperfine structure and has identified two components of the signal: one remains constant as a function of hydration and the other component can be understood in terms of proton splitting. As more water is added the latter component increases in intensity. The e.s.r. signal remains relatively constant as a function of temperature. This work has yet to be published.

Swartz: How would an unpaired electron affect the conductivity measurements?

Pethig: We want to believe that the e.s.r. signal is directly related to the transfer (unpairing) of charge from the polypeptide backbone to the methylglyoxal acceptor attached to the lysine group. In turn the measured conductivity would then arise from the long-range mobility of these unpaired charges.

Swartz: Is the change in intensity of the signal on the binding of methylglyoxal to the protein the right order of magnitude to account for that?

Pethig: Yes. Typically we find about 10^{14} spins/g or less for the unreacted control proteins and a value up to about 10^{18} spins/g for the brown protein complexes. The e.s.r. and brown colour appear to be directly related, and in general those proteins with the darkest brown colour have the highest conductivity.

Crumpton: May I pursue Professor Williams's point further? I think we need to have much more information about the reaction of proteins with methylglyoxal. For instance, the conditions used such as pH, temperature and concentration of reagents, the variety of proteins examined and whether all proteins give a brown colour and, if so, whether the brown colour is the same (i.e. are the spectra identical?). Such information is essential for a comprehensive and convincing evaluation of the hypothesis.

I would also like to make a couple of suggestions. First, it is now possible using a mixture of proteases and peptidases to hydrolyse proteins completely and by subsequent analysis of the digest to obtain an amino acid composition of the protein. Using this approach it should be possible to show unequivocally which amino acid residues of the protein react with methylglyoxal. Another, more sophisticated approach would be to diffuse methylglyoxal into a crystalline protein, such as lysozyme, of known X-ray structure and then determine the alteration in structure by difference Fourier analysis. These results should be much more authoritative than the data presented so far.

Pethig: Precise spectra of the coloured proteins are difficult to obtain because the proteins tend to be insoluble. The structural proteins in cells are the ones that Professor Szent-Györgyi has singled out and we too believe that these will be the most interesting ones to study. In that regard our results for collagen which indicate the possibility for coherent wave-like charge transport may be significant. The conductivity results for the soluble proteins albumin, casein and lysozyme appear to be understandable only in terms of the hopping of charge carriers between localized sites. The results for collagen may be related to the fact that it is the largest macromolecule we have studied. We intend to investigate other structural proteins such as tubulin and actin.

Professor Szent-Györgyi once remarked to me that the process of electron transport in the respiratory chain of mitochondria can be compared to the action of electrical power stations. We are really looking for mechanisms that underlie the subtleties of life—the things that go on at the power sockets in the kitchen, as it were. We are not necessarily looking for the movement of whole electrons between metal atoms but for partial electronic desaturation and charge mobility within protein structures. We need not consider only naive charge conduction effects but should also consider solitons, charge-density waves and spin-density waves, for example. We have hardly touched this subject yet. As our main expertise is in electrical materials science we are hesitant to start such work as enzyme digestion. I hope that suitably qualified researchers will take up some of these exciting ideas.

Williams: Amino acid analyses can be done with no difficulty.

Phillips: The idea of the lysine–methylglyoxal Schiff base folding and interacting with the main polypeptide chain in a charge-transfer complex (p. 85) looks fine on the blackboard, but have you investigated its feasibility with molecular models? One hardly needs to work out the crystal structure.

Pethig: We have looked at molecular models and Professor Ladik's group have made a more thorough study. The reaction is sterically possible.

Phillips: It will depend on the conformation of the main polypeptide chain.

Ladik: We considered a single chain, not a two-dimensional structure.

Phillips: The single chain has two variable conformational angles at the α-carbon atom which have to be defined. Is the methylglyoxal then planar, parallel to the peptide group?

Ladik: Nearly parallel. With regard to the solitons that Dr Pethig mentioned, A.S. Davidov, one of the leading Soviet theoretical solid-state physicists, pointed out that in proteins non-linear vibration of quasi-particles, the vibrational solitons, is probably extremely important in the transfer of energy.

Eley: From the experimental viewpoint it is very satisfactory, even if perhaps unexpected, that compressed discs of dry proteins do give such reproducible results. In fact, we normally use a compressed disc of dry bovine plasma albumin (Armour Fraction V) as a test substance with new students making electrical measurements on highly resistive polymers (cf. Eley & Thomas 1968). It is perhaps surprising that the different dry proteins so far examined did not show a greater variability of behaviour; thus in Spivey's work the 400K resistivity values all lay between a value of $2.6 \times 10^{11}\ \Omega$ cm for thrombin and $2.0 \times 10^{14}\ \Omega$ cm for elastin, with the electron transfer protein, cytochrome *c*, $3.8 \times 10^{11}\ \Omega$ cm. Again, different electron acceptors seem to have a similar effect in raising the conductivity of dry proteins; thus Dr Pethig's results with methylglyoxal–casein (Pethig 1978) are very similar to our chloranil–BPA results (Davis *et al.* 1960), and our unpublished work with ubiquinone–BPA and other systems (K.M.C. Davis, D.D. Eley & R.S. Snart, unpublished 1963; Eley & Snart 1965). However, more recent studies on thin films have given ferricytochrome *c* $6.5 \times 10^{10}\ \Omega$ cm and ferrocytochrome *c* $6.5 \times 10^{8}\ \Omega$ cm, at 328K (Nakahara *et al.* 1977), and this does suggest that conductivity measurements may show biological specificity. I would refer here to the postulate that biological electron transfer involves electron injection from the substrate into specific protein systems—cf. a mitochondrial model based on this idea (Eley *et al.* 1972).

On the question of single crystal work, this has not so far been achieved for proteins. In principle, measurements on single crystals of dry proteins cannot be made on globular proteins, where the crystals inevitably contain water and salts. They will be limited to fibrous proteins such as silk if these can be prepared in single-crystal form. We have, however, reported some work on single crystals of amino acids (Eley & Williams 1968), where the actual observed dark currents almost certainly arise from electrode injection, unlike proteins where thermal ionization (intrinsic conductivity) seems more likely to be responsible for the observed dark conduction.

In making measurements on electron-transfer systems such as the mitochondrial system, it is impossible, so far, to apply electrodes to the two ends of the chain of enzymes. Thus conductivity measurements on the intact mitochondrion, such as we have done (Eley *et al.* 1977), must involve a complex path of charge carriers, including not only mitochondrial membranes, but also other features not susceptible to detailed specification at present. The value of the e.s.r. technique is that it can identify the different resting places (prosthetic groups) of the electron on its passage through the mitochondrial chain. From the viewpoint of semiconduction theory, these resting places are shallow traps. Our object in making semiconduction measurements should be to relate the results to e.s.r. work. The value of a.c. measurements, especially microwave measurements, is that it should limit the length of the electron path to the biologically active material but, as I have said, other forms of a.c. loss increase the difficulties of interpretation. And in the case of microwave Hall measurements, as Trukhan first showed, the mobilities are limited by Faraday rotation in the walls of the cavity, as we have also found. Attempts to overcome this limitation have not yet been successful but we are still very interested in this problem (cf. Fletcher 1976).

Williams: What other simple polymeric materials have been tested?

Eley: At an early stage we studied dry compressed discs of polyglycine (Cardew & Eley 1959) and poly-L-tyrosine in helical and random coil forms (Eley & Spivey 1960). The resistivity (10^{12}–10^{13} Ω cm at 400K) and the energy gaps, ~ 3.0 eV, for these synthetic poly(amino acids) are very similar, actually somewhat higher than the corresponding values for typical proteins, such as native haemoglobin. The conductivity mechanisms are probably the same, and determined in my view by intrinsic thermal ionization of the polymer (cf. my suggestions earlier about charge transfer). Polyvinyl carbazole has been much investigated, as it is used in photocopying processes. At 'low' applied voltages it has a resistivity of 10^{18} Ω cm at 363K, and an 'energy gap' of 2.2–2.6 eV (activation energy 1.1–1.3 eV) (Okamoto *et al.* 1973). Thus its conductivity is somewhat less than that exhibited by proteins, although the 'energy gaps' are similar. The authors explain the behaviour at low voltages in terms of a hole injection from the gold electrode. We have discussed the possibility of electron, and hole, injection for amino acids (Eley & Williams 1968) and benzoic acid and other small molecules (Eley *et al.* 1968), but we regarded it as less likely for proteins. At high fields polyvinyl carbazole showed a diminution of activation energy and a change to non-ohmic conduction, which was attributed to Poole-Frenkel conduction from positively charged acceptor levels. We have not studied high-field effects in our work on solid proteins, so far, but this clearly requires further investigation work.

Williams: I am asking about any simple polymer backbone made from a saturated carbon system with a few side-chains and amino groups. Do they react with methylglyoxal?

Laki: We used polylysine. It is a well-defined material and its reaction with methylglyoxal can be analysed. In our case, we know that for every lysine residue, one methylglyoxal is attached to it.

Williams: But when one varies the protein instead of the side-chain one seems to get a similar answer. That suggests to me that the results reflect a property of the methylglyoxal. Therefore, one should take away the backbone to eliminate it as the origin of the effect.

Pethig: Collagen behaves differently.

Williams: That is a protein. As long as there is an amino group for methylglyoxal to react with, do you get the same answer?

Ladik: Iodine-doped polyacetylene (in some cases) showed greater conductivity than copper (unpublished observations).

Williams: But then you have eliminated the side-chains and the methylglyoxal! Surely any polymer able to form a Schiff base with methylglyoxal will give a yellowish substance.

Pethig: We are asking for a specific biological reaction that has taken perhaps 10^9 years or more to fully evolve. The folding back of the Schiff base to form a charge-transfer complex is surely specific for a lysine side-chain and a particular polypeptide configuration. For example, on being stretched the collagen complex loses its brown colour and becomes pale. Choosing another backbone might not be relevant to our main study.

Crumpton: May I suggest some suitable materials for experiment? Lysine-Sepharose 4B and AH-Sepharose 4B, which are commercially available (Pharmacia Fine Chemicals), are good mimics of biological polymers with exposed free amino groups. Polyacrylamide is another. These experiments are vital but are easy to do.

Phillips: You have repeated that stretching collagen bleaches it, Dr Pethig. However, the biological characteristic of collagen is its resistance to stretching. In what form is the collagen? Is it still in the native form or is it gelatin?

Pethig: The collagen was dissolved in 0.1 M-dichloroacetic acid before being incubated with methylglyoxal. After this treatment a gelatinous material is formed. Stretching this material lightens the brown colour; on release the brown colour returns.

Phillips: What is the birefringence of the specimen?

Pethig: We have not looked at that. Inspection of the gelatinous material under an optical microscope reveals the existence of many regions having a stranded structure. Perhaps this implies that some of the original helical structure of the collagen has been preserved. The material is certainly not completely amorphous.

Lewis: Whatever the reaction between the protein and methylglyoxal, the fact that the steady-state conductivity increases by some four orders of magnitude implies an important reaction, and some critical process must be responsible for such a large change. Many people not accustomed to considering electrical phenomena might tend to disregard such evidence, offering more conventional biochemical arguments against the hypothesis and proposing experiments more traditionally acceptable. However, it is an electrical rather than a chemical phenomenon that we are trying to understand and we should not lose sight of that fact.

Williams: The biochemists and chemists are asking whether a series of methylglyoxal–amine compounds spaced about 1.0–1.5 nm apart (as a consequence of the number of groups per protein) will give this result independent of the matrix. If you do not know that, we can deduce nothing about protein chemistry from the observations on these glyoxal-doped polymers.

References

BONSIGNORE, A., LEONCINI, G., ANDISIO, G., ZETTA, L. & FERRUTI, P. (1977) Characterization of the polymer from methylglyoxal in the presence of L(+)-lysine. *Ital. J. Biochem. 26*, 162-168

CARDEW, M.H. & ELEY, D.D. (1959) The semiconductivity of organic substances, Part 3. Haemoglobin and some amino acids. *Disc. Faraday Soc. 27,* 115-128

DAVIS, K.M.C., ELEY, D.D. & SNART, R.S. (1960) Semiconductivity in proteins and haemoglobin. Enhanced semiconductivity in protein complexes. *Nature (Lond.) 188,* 724-725

ELEY, D.D. & SNART, R.S. (1965) Conduction in chloroplast components. *Biochim. Biophys. Acta 102,* 379-385

ELEY, D.D. & SPIVEY, D.I. (1960) Semiconductivity of organic substances. Part 6. A range of proteins. *Trans. Faraday Soc. 56,* 1432-1442

ELEY, D.D. & THOMAS, P.W. (1968) Semiconductivity of organic substances. Part 14. Electrode effects in proteins. *Trans. Faraday Soc. 64,* 2459-2462

ELEY, D.D. & WILLIAMS, W.P. (1968) Semiconductivity of organic substances. Part 13. Single crystal and polycrystalline amino acids. *Trans. Faraday Soc. 64,* 1528-1536

ELEY, D.D., FAWCETT, A.S. & WILLIS, M.R. (1968) Semiconductivity of organic substances. Part 12. Electrode injection of charge carriers into crystals of small aromatic molecules. *Trans. Faraday Soc. 64,* 1513

ELEY, D.D., MAYER, R.J. & PETHIG, R. (1972) Microwave Hall mobility measurements on heavy beef heart mitochondria. *Bioenergetics 4,* 389-402

ELEY, D.D., LOCKHART, N.C. & RICHARDSON, C.N. (1977) Electrical properties and structural transitions in the mitochondrion. *J. Bioenerg. Biomembr. 9,* 289-301

FLETCHER, J.R. (1976) An improved bimodal cavity for microwave Hall effect measurements. *J. Phys. E. Sci. Instrum. 9,* 481-483

NAKAHARA, Y., KIMURA, K. & INOKUCHI, H. (1977) Electrical conductivity of cytochrome *c* anhydrous film. *Chem. Phys. Lett. 47,* 251-254

OKAMOTO, K., KUSABAYASHI, S. & MIKAWA, H. (1973) The photoconductivity of poly(N-vinylcarbazole). II. Dark conductivity in a sandwich-type cell. *Bull. Chem. Soc. Jpn. 46,* 1953-1959

PETHIG, R. (1978) Electronic properties of protein–methylglyoxal complexes—strong evidence for energy band conduction. *Int. J. Quantum Chem. Quantum Biol. Symp. 5,* 159-171

Free radicals in cancer

H.M. SWARTZ

Department of Radiology, Medical College of Wisconsin, Milwaukee, Wisconsin

Abstract It has been hypothesized that free radicals play a significant role in cancer. A historical review indicates a series of rises and falls in the acceptance of this hypothesis and it remains controversial. The strongest evidence for a critical role of free radicals in cancer is based on electron spin resonance (e.s.r.) data from lyophilized (freeze-dried) tumours. Recent results indicate that such data are artifactual in the sense that the observed signals are not directly related to free radicals existing before lyophilization. These data also indicate, however, that some of the observed changes are reproducible and may be indirectly linked to biophysical or biochemical changes that occur in tumour cells. A possible key to such a link is via antioxidants, especially ascorbic acid. It is now feasible to do experiments to definitively determine: (1) the generality of the effect of lyophilization on e.s.r. spectra of tumours; (2) the molecular nature of the free radicals observed in lyophilized tumours and normal tissues; (3) the effect of redox reactions and substances on the observed e.s.r. spectra; and (4) the occurrence of free radical changes during carcinogenesis.

The subject of the role of free radicals in cancer is complex and controversial. To facilitate understanding of this field I shall consider it first from a historical perspective in fairly general terms and then consider the most recent developments in more detail. The discussion of recent developments will emphasize the findings from our laboratory. The implications of the experimental findings will then be discussed and experimental directions for the future will be suggested.

EXPERIMENTAL FINDINGS IN ELECTRON SPIN RESONANCE STUDIES OF CANCER

Historical perspectives

A schematic representation of the historical developments in this field is shown in Fig. 1. Developments are represented as a series of peaks and valleys with the peaks representing indications of the importance of free radicals in cancer and the valleys the opposite. Five areas are differentiated: two high-points, two low points and the current situation, which is intermediate with the eventual direction not yet known.

Peak Number One. The first peak had three foundations. The first was that the unusual reactivity of free radicals might be responsible for the extraordinary changes in the properties of cells which we call cancer (Brues & Barron 1951; Fitzhugh 1953; Swartz 1972). This general theoretical consideration was supported by more specific theoretical considerations on the nature of some chemical carcinogens (Brues & Barron 1951; Pullman & Pullman 1955; Szent-Györgyi *et al.* 1960) and by experimental evidence of the production of free radicals by physical carcinogens, especially ionizing radiation. The role of free radical intermediates in chemical carcinogenesis is too large a subject to be considered here and the reader is referred to a recent review (Ts'o *et al.* 1977). It has been amply demonstrated that many carcinogens can form free radicals but it is much more difficult to determine whether or not the free radical forms of the carcinogens are the active intermediates that actually react with the target molecules. The role of free radicals in ionizing radiations seems less equivocal, there being general agreement that free radical reactions account for most of the effects of ionizing radiation.

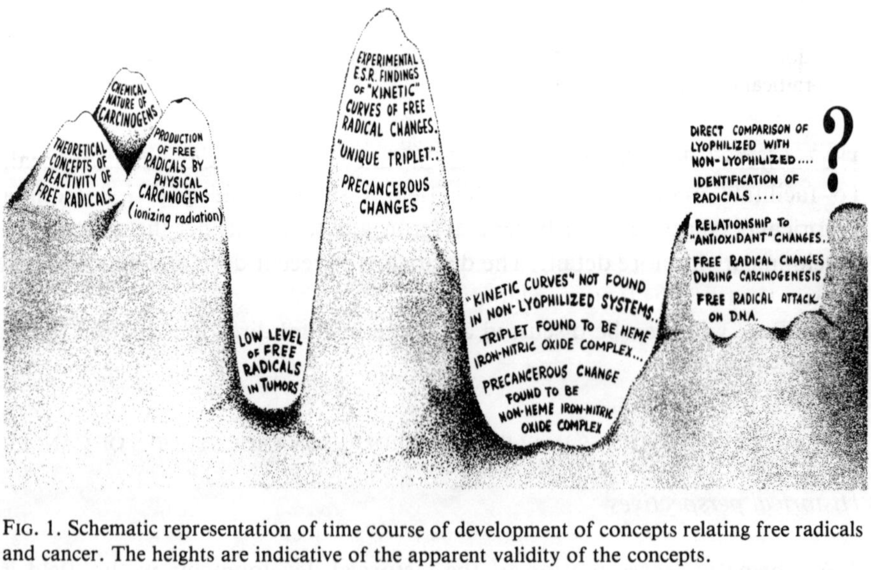

FIG. 1. Schematic representation of time course of development of concepts relating free radicals and cancer. The heights are indicative of the apparent validity of the concepts.

Valley Number One. The development of electron spin resonance spectroscopy (see Bolton 1972 and Swartz 1972 for a review of this technique) made it possible to examine tumours directly in order to study the amount and nature of free radicals in them. Although there was some variation in the early studies, the general trend was clearly that there were fewer free radicals in tumours than in comparable normal tissues (Truby & Goldzeiher 1958; Commoner *et al.* 1954; Swartz 1972). The findings of a low level of free radicals in tumours appeared to contradict the hypothesis of a critical role of free radicals in cancer.

Peak Number Two. This low state of the hypothesis was soon apparently reversed, however, by the findings of Emanuel and co-workers (reviewed by Emanuel 1976). In a series of studies on different tumours they reported a remarkably consistent pattern of free radical changes in developing tumours. A composite curve of these experiments is shown in Fig. 2. This pattern of

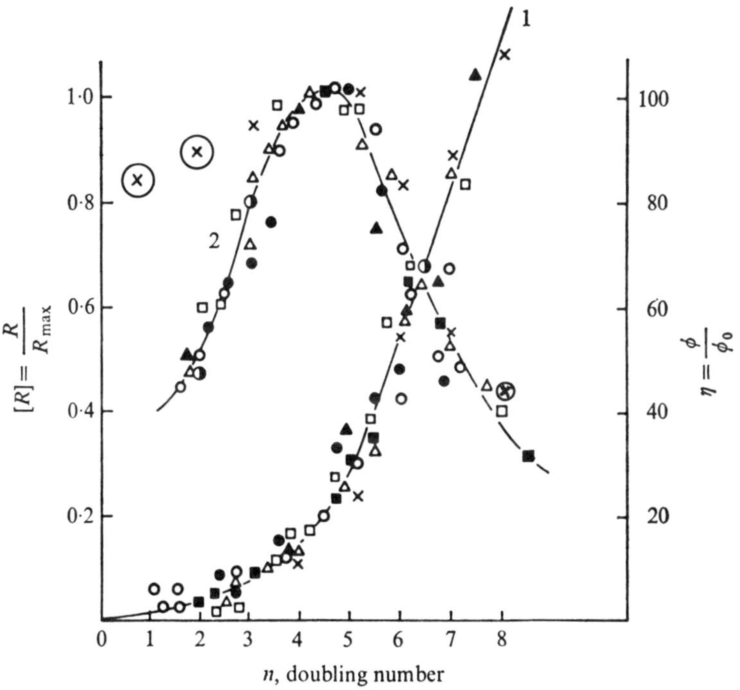

FIG. 2. Generalized kinetic curves for relative growth(1) and free radical content(2) of tumour tissue, revised from Emanuel (1976). [R], relative intensity of free radicals as determined by peak-to-peak measurements of e.s.r. spectra; η, relative weight of tumour; ϕ, weight of tumour . The symbols correspond to seven mouse tumours and one rat tumour (Walker carcinosarcoma, □). The original points for the mouse leukaemia are ×; points added (taken from Fig. 2 of the same reference; see p. 119) are designated by ⊗.

change seemed to reconcile quite adequately the experimental findings of low levels of free radicals in mature tumours with the hypothesis associating malignant changes with increased levels of free radicals. The concept that the number of free radicals increases in the early stages of cancer was very exciting, because this might be used for diagnostic purposes and free radical inhibitors might interfere with the development and growth of cancer cells. Work from Emanuel's and other laboratories in the Soviet Union supported both these possibilities. They reported the retardation or reversal of tumour growth by free radical inhibitors in animals and proposed the development of a new class of chemotherapeutic agents based on this action (Emanuel 1976). In addition to their animal experiments, they also reported elevated levels of free radicals in the leucocytes of patients with leukaemia (Pavlova & Livenson 1965; Kassirskii *et al.* 1967; Klochko *et al.* 1970). These changes paralleled the course of the clinical disease, being highest in untreated patients and returning towards normal after treatment. They also reported that in treated patients in remission, the level of the free radicals increased before any other evidence of recurrence of their disease.

At about the same time, other e.s.r.-detectable changes in tumours were reported which also appeared to be potentially useful in understanding or detecting cancer. Brennan *et al.* reported the finding of a unique e.s.r. pattern consisting of a triplet of lines in a tumour cell preparation (Brennan *et al.* 1966). A third potentially characteristic change was reported from Commoner's laboratory; they found the development of an e.s.r. signal with a g-factor of 2.035 which preceded the occurrence of liver tumours in animals fed butter yellow or other carcinogens (Vithayathil *et al.* 1965). Thus for a time there was evidence that not only were free radical changes intimately involved in cancer, as had been hypothesized, but it appeared that other changes detectable by e.s.r. also occurred in cancer. Some of these were not necessarily free radicals, because some of the properties of the triplet and the $g = 2.035$ signal suggested that they were associated with metals. At about the same time, a number of reports appeared in which free radicals were studied indirectly by measuring effects on polymerization reactions and 'antioxidant' content (Burlakova & Malochkina 1968; Goloshchapov *et al.* 1973). The results of these experiments were somewhat mixed but in general supported the hypothesis of increased levels of free radicals during the development of cancer.

Valley Number Two. It soon became apparent, however, that there were some very serious discrepancies in the literature, which led to scepticism about the role of free radicals in cancer. Premature attempts claiming that a firm link had been established and clinical use was imminent accelerated this scepticism; consequently, studies in this area are now viewed in a very questioning manner by most serious cancer investigators.

Although the kinetic curves were considered to be universal by Emanuel and co-workers, experiments from other laboratories were clearly in disagreement. For some time no exactly parallel experiments were being done in other laboratories but in a number of studies observations were made over periods of time from the administration of the carcinogens (or the injection of tumour cells) until the death of the animals from the tumour (Mallard & Kent 1964, 1966; Vithayathil *et al.* 1965; Duchesne & Van de Vorst 1970; Mailer *et al.* 1974). The general pattern was that free radical levels tended to decrease, without any early rise. Although none of these studies were done on the same tumour systems as were used by Emanuel and his co-workers, some were sufficiently similar to suggest that this was not the source of the discrepancy. The most likely explanation appeared to be in the manner of preparing the samples. Those studies that showed a dramatic early rise were done in lyophilized (freeze-dried) tissue samples while the negative studies were done in quickly frozen samples, or in unfrozen samples. When the nature of the e.s.r. signal seen in lyophilized samples was then examined, it was found that lyophilization itself induced e.s.r. signals, especially if the samples were exposed to oxygen and certain concentrations of water vapour (Heckly 1972). The suggestion that the apparent discrepancy in the results was due to the method of sample preparation was supported by experiments done in our laboratory with a leukaemia similar to one used in the Soviet studies, in which we used non-lyophilized rather than lyophilized samples. We found only a gradual decrease in the concentration of free radicals (Fig. 3) (Swartz *et al.* 1973).

There were equally disappointing findings in regard to the other e.s.r. changes that were originally felt to be characteristic of tumours. The triplet signal was soon shown to be due to a nitric oxide–haemoglobin complex which formed spontaneously in any hypoxic or anoxic tissue, especially if the tissue had necrotic areas (Maruyama *et al.* 1971). Such conditions are especially common in tumours, which readily explained why this signal is often observed in tumour tissues. The extent of the physiological changes that are required to generate the triplet signal makes it unlikely that e.s.r. would be a sensitive tool for detecting the presence of tumour cells, because the changes would occur only with relatively advanced tumours; such tumours should be readily characterized by classical means of studying tissues. Similarly, Commoner's group found that the signals at $g = 2.035$ that they had reported were due to a complex of nitric oxide with non-haem iron proteins (Woolum & Commoner 1970). There appeared to be no direct relationship between the

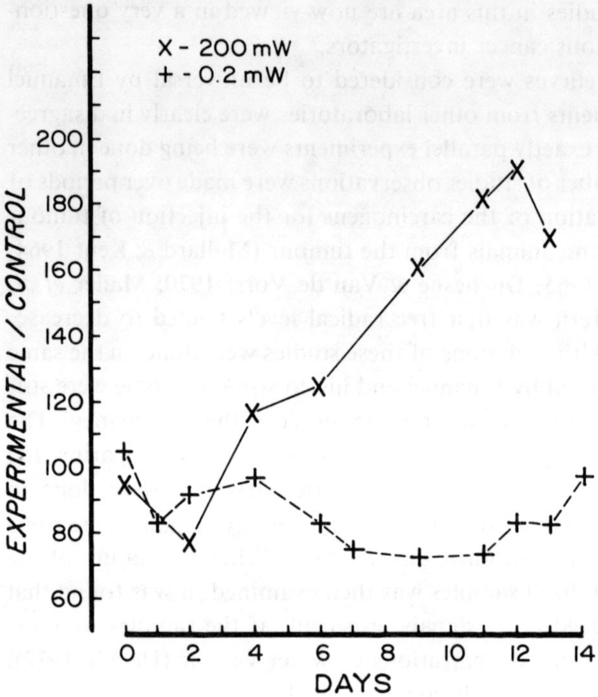

FIG. 3. Relative e.s.r. intensity (experimental/control) in $g = 2.00$ region of samples of spleen from mice injected with AKR leukaemia cells. The line-shapes and power saturation data indicate that at 200 mW the spectra were primarily due to unpaired electrons associated with a paramagnetic metal and at 0.2 mW were primarily due to organic free radicals. (See Swartz & Gutierrez 1977 for further details of the experiment.)

development of this signal and the subsequent development of tumours and indeed some data indicated that there was an inverse relationship (i.e., the larger the e.s.r. signals, the fewer the number of tumours that appeared) (Commoner *et al.* 1970).

The plateau region. This was the state of affairs when we began our current series of experiments. There have been some promising results but we cannot yet determine whether overall these will support or negate the theory, so we term this period a 'plateau'. Our initial goal was to see if we could reproduce some of the most interesting results obtained by Emanuel and his co-workers and, if so, to try to relate these to the methodological approaches used by other laboratories. If we were able to reproduce the results, we had a further goal; we would then try to understand the molecular basis for the changes observed. We have made considerable progress towards these goals and as a consequence have added another goal: to relate the e.s.r. changes to the changes that have been reported in levels of 'antioxidant activities'.

Recent experimental results

We first studied e.s.r. changes in mice injected with AKR leukaemia cells, using non-lyophilized samples (Swartz *et al.* 1973). Another leukaemia was studied by Emanuel's group and in their hands showed the 'characteristic' changes in free radical concentrations (Fig. 2, p. 109. As indicated in Fig. 3, we did not find similar changes in free radical concentrations in the spleen (which was the primary tissue containing tumour cells). Some e.s.r. changes were observed both in the spleen and in the blood but these were attributable to changes in paramagnetic trace elements, rather than free radicals. The precise origin of the change in the spleen has not been determined; the change in the blood is now known to be due to the copper-containing protein caeruloplasmin (Mailer *et al.* 1974). The results of this experiment directly confirm that even in the same type of tumour system studied by Emanuel, the characteristic change in free radicals was not reproduced when non-lyophilized samples were studied.

We next developed an experimental procedure in which the same samples could be studied both before and after lyophilization. We used a tumour system that had been used by Emanuel's group, the Walker 256 carcinosarcoma (Saprin *et al.* 1967; Swartz & Gutierrez 1977; Gutierrez & Swartz 1979). The minimum number of cells necessary to give reproducible tumours were injected in the thighs of a number of rats. At predetermined times the area containing the tumour was excised as well as the comparable area of the other leg (which had not been injected with tumour cells) and from animals in which no tumour cells had been injected. Samples were formed into cylinders of 4 mm in diameter whose length exceeded the sensitive length of the e.s.r. cavity. They were frozen in liquid nitrogen, stored and then studied at 77K. After e.s.r. studies had been done on the rapidly frozen samples, the same samples were lyophilized to constant weight without significant exposure to oxygen and then re-examined by e.s.r. spectroscopy. Typical results are shown in Fig. 4. They indicated that in this system the discrepancies between the two types of experiments could be explained by the sample preparation technique. Before lyophilization the samples of the tumour had an e.s.r. signal of lower intensity than the normal muscle with which they were being compared. After lyophilization the peak-to-peak height of the e.s.r. signal became larger and appeared to follow the first part of the curve reported by Emanuel and his group. A closer examination of the data indicated further that the lines seen

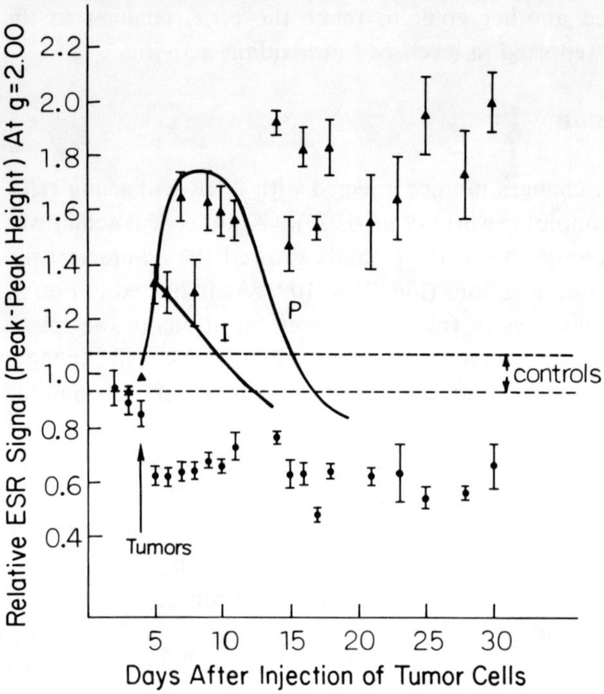

FIG. 4. Peak-to-peak heights of e.s.r. spectra from lyophilized (▲) and non-lyophilized (•) Walker 256 tumour and normal muscle samples of Sprague-Dawley rats (averages and S.E.). The solid lines are data from Saprin *et al.* (1967) on the same system, with samples derived from the periphery (P) and internal parts (I) of the tumour. We show our results for the internal part of the tumour; the periphery gave similar results if necrotic areas were avoided. Tumours were palpable by Day 4. The sample for Day 4 contains 10–15% muscle. From Day 5 the tumours were well defined and samples were 100% tumour cells. All samples exceeded the cavity's sensitive volume. The peak-to-peak heights were obtained from 9 GHz spectra at 0.01 mW output power, 8 G modulation amplitude at 77K. A DPPH standard was used for daily calibration and *g*-value calculations.

in the tumour samples were narrower than those seen in the muscle and when this was taken into account the total number of free radicals was the same in both types of samples (Table 1).

To help determine the generality of the relationship that we saw in the experiments with the Walker tumour we studied breast tumours induced by 7, 12-dimethylbenzanthracene (DMBA) in the rat (Swartz *et al.* 1978; Gutierrez *et al.* 1979). This potent carcinogen was administered by stomach tube to 52-day-old female rats. Some animals were killed at predetermined intervals after being given the carcinogen and others were biopsied at predetermined in-

TABLE 1

Comparison of e.s.r. peak-to-peak heights and integrated signal intensities for Walker 256 tumours

LYOPHILIZED SAMPLES

Days after injection of 10^4 cells	*No. of samples*	*Linewidth ΔH (gauss)*	*Relative peak-to-peak height*	*Relative integrated intensity*
Muscle (control)	12	13.1 ± 0.2[a]	14.4 ± 1.0[a]	48 ± 1.6[a]
6	2	10.5 ± 0.4	20.2 ± 0.8	52 ± 10
7	3	9.6 ± 0.3	24.4 ± 0.2	42 ± 2
9	3	10.3 ± 0.9	28.4 ± 3.8	48 ± 9
10	3	10.0 ± 0.3	24.9 ± 1.9	42 ± 0.6
15	4	10.3 ± 0.2	24.9 ± 1.1	41 ± 3
17	3	9.8 ± 0.2	24.7 ± 1.0	41 ± 4
21	3	9.7 ± 0.3	21.8 ± 1.5	40 ± 1
28	2	9.5 ± 0.4	16.9 ± 0.8	58 ± 3
30	2	9.0 ± 0.0	33.7 ± 1.3	48 ± 0.5

NON-LYOPHILIZED SAMPLES

Muscle (control)	9	13.6 ± 0.2	13.5 ± 0.2	34 ± 4
2	3	13.0 ± 0.6	13.9 ± 1.1	42 ± 4
6	3	13.3 ± 0.3	8.6 ± 0.6	25 ± 7
15	3	13.3 ± 0.3	8.4 ± 0.6	18 ± 6
17	3	13.3 ± 0.3	6.6 ± 0.5	12 ± 1
28	3	14.0 ± 0.6	8.3 ± 0.4	35 ± 4

Integrated intensities of e.s.r. spectra observed in developing Walker 256 carcinoma. (Data from Swartz & Gutierrez 1977.)

[a] Mean ± S.E.

tervals after the appearance of tumours. With each biopsy, control samples were also taken from grossly normal breasts from the same animal and from animals not given DMBA. In order to be able to study the small sample that one obtains with a rat breast, we used a higher-frequency e.s.r. spectrometer (35 GHz). At this frequency, in addition to free radicals, one can also readily observe paramagnetic manganous ions. The examination procedure was similar to that used for the Walker tumour; the samples were first studied after rapid freezing in liquid nitrogen and were then restudied after lyophilization. Because of the small size of the samples, observations were made in 1-mm (inside diameter) quartz tubes.

The e.s.r. spectra of the frozen tumours had twice the intensity of the normal breast tissue (Fig. 5 and Table 2). This increase was different from that reported for any other tumour system studied by any method of preparing the tissue samples. The increase did not correlate with a simple change in the

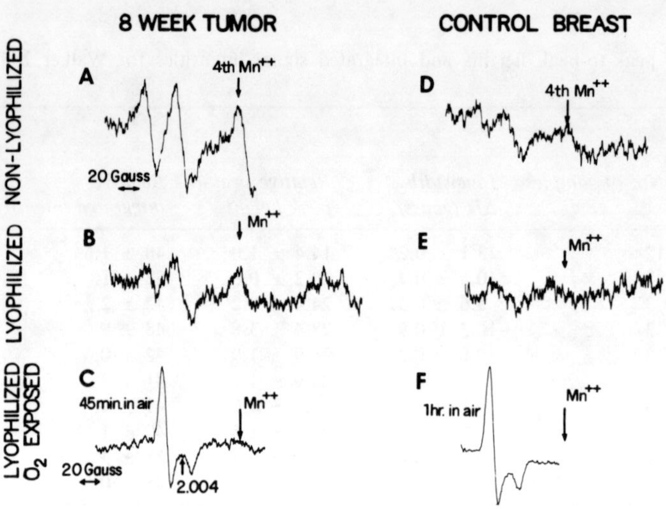

FIG. 5. E.s.r. spectra of a 7,12-dimethylbenzanthracene-induced breast tumour in rats, eight weeks after its detection by palpation. A control sample is shown for comparison. The conditions are: microwave frequency 35 GHz, modulation amplitude 8 G at 0.06 mW microwave output power, temperature −150°C. *g*-values were calculated from a published value of $g = 2.0012$ for Mn^{2+}. (From Gutierrez *et al.* 1979.)

number or type of cells; this was shown by studying signal intensity in relation to quantitative microscopic determinations of cell types.

In contrast to the findings with the Walker tumour system, the DMBA-induced breast tumours showed similar trends in both lyophilized and non-lyophilized samples. For both types of preparation the magnitude of the

TABLE 2

Relative peak-to-peak heights in mammary tumours induced in rats by 7,12-dimethylbenzanthracene

Days after tumour detection	*Free radicals[b] in lyophilized tissues*	*Free radicals[b] in frozen tissues*	*Mn^{2+}[c] in lyophilized tissues*	Mn^{2+}[c] *in frozen tissues*
Normal tissue (control)	12 ± 1 (12)[a]	15 ± 0.7 (47)[a]	17 ± 3 (12)[a]	58 ± 4 (36)[a]
0–4	24 ± 2 (6)*	30 ± 3 (10)*	45 ± 7 (6)*	67 ± 5 (10)
9–14	22 ± 1 (6)*	34 ± 2 (13)*	44 ± 15 (6)*	95 ± 12 (13)*
21	30 ± 4 (7)*	34 ± 5 (7)*	26 ± 7 (7)	96 ± 11 (7)*
56	28 ± 3 (6)*	32 ± 2 (10)*	38 ± 8 (6)*	108 ± 15 (10)*
91–119	25 ± 3 (6)	26 ± 2 (8)*	14 ± 1 (6)	83 ± 8 (8)

[a] Mean ± S.E. Number in parenthesis is number of samples.
[b] Data obtained at 0.06 mW at 35 GHz.
[c] Data obtained at 6.0 mW at 35 GHz.
*Significantly different from controls at 95% confidence level, Mann-Whitney test.

signal in the tumour was approximately twice that seen in the normal breast and it remained at this level throughout the experiment. There was only one point (samples obtained 9–14 days after tumours were detectable) in which the trend of the lyophilized samples varied significantly from that seen in the non-lyophilized samples (Table 2).

In contrast to the abrupt change in free radical levels observed as the DMBA-induced tumours increased in size, the amount of paramagnetic manganous ion (Mn^{2+}) increased gradually over the period of the experiment. The relative changes in manganous ion concentration were larger but much more irregular in the lyophilized samples. The basis of this difference is not clear, but it indicates an additional experimental uncertainty that is introduced when samples are lyophilized.

The effect of oxygen on the e.s.r. signals seen in lyophilized samples was studied in detail in the experiment on DMBA-induced breast tumours. Oxygen increases the e.s.r. signal considerably and changes its shape to that of a π-electron type radical with axial symmetry (Fig. 5*F*). Some of the tumour samples had a somewhat similarly shaped signal before the introduction of oxygen. As discussed later (p. 122), there is some evidence that, in view of the possible origin of the oxygen-related e.s.r. signal, the signal seen in these tumours is due in part to ascorbic acid; if so, then the signal seen in the tumour samples may provide significant insights into the nature of the biochemical changes that we are observing.

From our experiments as described to this point we can state that (1) in suitably controlled conditions one can get reproducible data from lyophilized samples when exposure to oxygen is avoided; (2) the signals in the lyophilized samples have no regular relationship to those seen before lyophilization; (3) increases in the peak-to-peak height of the lyophilization-induced signal occur similar to those reported by Emanuel and co-workers; (4) in at least one system the increase in free radicals in tumour tissue relative to control tissues was apparent and not real and was due to a narrowing of the line (Swartz & Gutierrez 1977; Gutierrez & Swartz 1979); and (5) the introduction of oxygen to lyophilized samples resulted in characteristic line-shape changes and increased intensity, indicating that it might be possible to identify the molecular basis of the signal.

IMPLICATIONS OF THE RESULTS AND SUGGESTIONS FOR FUTURE EXPERIMENTS

Has the basic hypothesis really been tested?

If one reconsiders the basic hypothesis, which is that free radical reactions are important for the transformation of cells from the normal to the malignant state, and when one also considers the experiments done so far, it is reasonable to conclude that the hypothesis has not really been tested. This seems quite clear for those experiments in which malignant cells have been transplanted into animals. In such experiments one really does not study malignant transformation at all but only the multiplication of cells that are already malignant. The transplanted cells presumably do not change their basic characteristics during the experiments except for those changes related to poor nutrition resulting from development of large masses of tumour cells without adequate blood supplies. There may be other important aspects of cancer and the organism's response to cancer that can be studied in such an experiment; we shall return to these shortly. This does not affect our conclusion that these experiments do not give us information on free radicals in the development of cancer.

At first consideration the other type of experiment, in which a carcinogenic agent has been administered, appears to be an appropriate test of the hypothesis, but this also is probably not so. The malignant changes presumably take place in only a few cells and probably occur soon after the carcinogen is given. By the time a detectable tumour is present the malignant transformation process has been long completed, so that even the smallest tumours reflect stages well after the development of the cancer. It is probably not possible to get the information we are seeking by looking in animals at stages prior to the development of detectable tumours. Only one or at most a few cells become malignant and the tumours develop from these. Most of the cells are not transformed and so our samples must inevitably be composed of the 99% or more of cells that do not undergo malignant change. (It is possible that all the cells undergo the same premalignant chemical changes and that those that become malignant do so because they undergo a second alteration; if so, e.s.r. studies might be a valid way to investigate the biophysical nature of the first step.)

Interpretation of results obtained in e.s.r. studies of cancer

This discussion will focus on the results of the studies in which tissues or cells are lyophilized. These are the experiments that have been interpreted as indicating important and useful correlations between free radicals and cancer (Emanuel 1976). We have already discussed the evidence that such studies do not directly reflect free radical levels *in vivo* and our discussion here will therefore centre on other interpretations and other uses of these data.

Reproducibility of the 'universal kinetic curve' of Emanuel. Our results indicate that an early apparent increase in the intensity of the free radical signal in tumours can be reproducibly observed in at least some tumour systems. Close examination of all the data, however, indicates that the universality may be less than originally thought. In at least some of the systems the 'increase' is actually due to a change in shape of the e.s.r. signal and not a true increase in the free radical content (Swartz & Gutierrez 1977; Gutierrez & Swartz 1979). The summary of data presented by Emanuel presents an amazingly consistent pattern of changes in eight tumour systems (Fig. 2, p. 109; Emanuel 1976). In the same paper, however, he gives detailed data on one of these tumour systems (La leukaemia) and these data do not correspond with the general curve: these points have been added to the figure taken from Emanuel's review (1976).

The drop in free radical content in mature tumours has not been observed in the two systems we have studied; a possible reason for this discrepancy is described in the following section.

Meaning of the kinetic curve. Even with the above comments, the data from Emanuel's group are impressive in their extent and consistency and it seems desirable to try to relate them to other experimental findings and considerations. Although the following interpretation is somewhat speculative, I feel it is likely to be valid because it is based on an examination of their published papers, on extensive discussions held with their group during a recent seven-week visit to Emanuel's laboratory, and on data from my own laboratory.

I ascribe the early 'rise' in free radical content to two factors: a different type of lyophilization-induced radical occurring in the tumour, and an increasing percentage of tumour cells in the samples during the first few days of the experiments. The latter increase seems inevitable for data obtained in the first 1–4 days because of the small number of tumour cells present and the impossibility of isolating them from the associated normal cells with the simple biopsy procedures that were used. If this interpretation is correct it indicates that although the apparent quantitative changes are not significant, there are qualitative changes and these may be significant because they must be due to a biochemical difference between the tumour cells and the normal cells. The decrease in free radicals observed in larger tumours may simply reflect a decreased number of viable tumour cells resulting from the haemorrhage and necrosis that are common in such tumours. Emanuel's group has reported a difference between internal and peripheral parts of the tumour that is consistent with this interpretation (Saprin *et al.* 1967). Our failure to find a decrease in large tumours is probably the result of our experimental procedures, which sought to include only tissue that appeared viable in our samples.

Meaning of the observed changes. On the basis of the lack of close correspondence with the spectra of non-lyophilized samples, I assume that the spectra of the lyophilized samples do not directly reflect free radicals existing *in vivo*. They may, however, indirectly reflect these if: (1) the intensity of the signal seen in lyophilized tissues is related to the concentration of free radicals *in vivo* because the free radicals act as promotors or inhibitors of reactions that produce free radicals in lyophilized materials; or (2) there are common factors that simultaneously affect the levels of free radicals *in vivo* and the lyophilization-induced free radicals (for example, antioxidants).

Another possibility is that the level of lyophilization-induced free radicals is affected by intracellular conditions that affect the action of carcinogens *in vivo*. Again, such a link might be provided by the overall level of antioxidants or of a specific antioxidant.

A third possibility is that the lyophilization-induced free radical seen in tumours is a reflection of some aspect of the host's response to the presence of the tumour.

All these possibilities theoretically could be tested; they are specifically separated out here to encourage such experimental tests of their validity. The more we recognize that the observed free radicals do not directly reflect free radicals in *in vivo,* the more the probability of making appropriate experimental tests increases.

Usefulness of results of e.s.r. studies of cancer

Again I primarily consider the spectra observed after lyophilization. Other changes do not seem to be specific for tumours. Even for the lyophilization-induced signal the changes observed do not seem to be universal or readily quantified. There is, however, a suggestion that these changes may be reproducible. If so, they may be useful either empirically or for theoretical considerations of the nature of cancer.

The empirical use of the e.s.r. findings is probably limited to readily obtainable samples such as blood, urine or exfoliated cells. Useable changes have not been reported in the latter two types of specimens. Changes in free radicals in blood have been reported, again principally by Emanuel and coworkers (Pavlova & Livenson 1965; Saprin *et al.* 1966; Kassirskii *et al.* 1967; Klochko *et al.* 1970). (I omit from discussion here changes in blood proteins that contain paramagnetic trace elements.) Emanuel's group present data suggesting excellent correlations of free radical levels in leucocytes with the stage and treatment of leukaemia. They have never followed up these exciting results, however, and I am unaware of other groups that have reproduced them.

Perhaps the most productive area of use is in the theoretical implications of the e.s.r. changes. As discussed above, the observed changes may reflect some aspect of tumour metabolism, tumour growth, intracellular conditions or host response. It seems quite possible that if there is a general link between the observed e.s.r. changes and some fundamental properties of malignant cells, the link is via factors related to redox conditions in the cell. This could tie the observed e.s.r. changes into some very interesting but highly speculative considerations about the nature of cancer. These include the theories of Dr

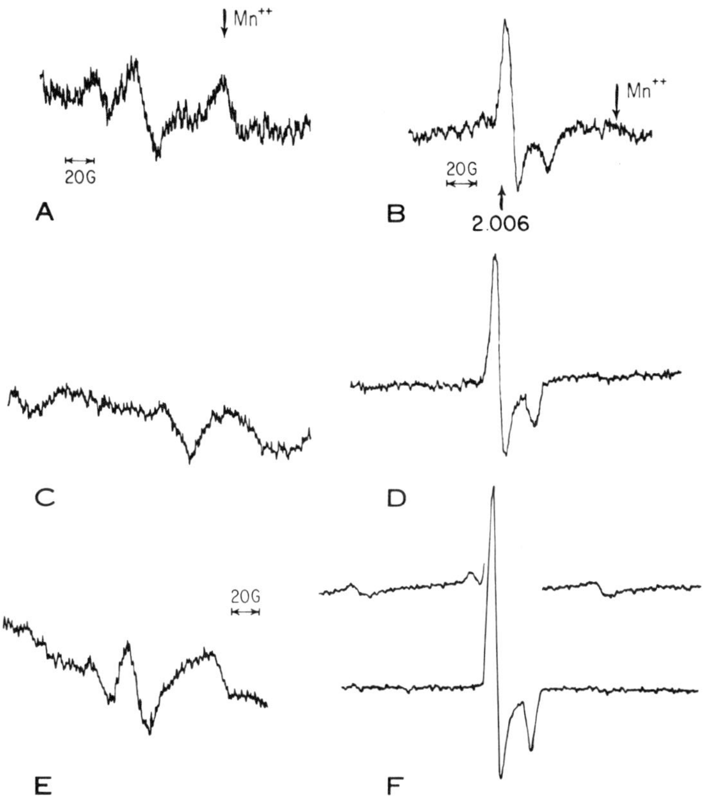

FIG. 6. E.s.r. spectra of lyophilized samples at 35 GHz after exposure to air for 45 min. A. 7, 12-Dimethylbenzanthracene induced rat breast tumour (before exposure to air). B. The same tumour after exposure to air. C. Bovine serum albumin. D. Bovine serum albumin plus 10^{-4} M-ascorbic acid. E. Rat liver microsomes (1 ml contained microsomes from 0.5 g liver). F. Rat liver microsomes plus 10^{-4} M-ascorbic acid. E.s.r. conditions similar for all spectra except for amplification; temperature – 150°C, incident microwave power 0.06 mW, modulation amplitude 8 G. In F a spectrum obtained at 60 mW is also shown to indicate the position of the second, third and fourth peaks of Mn^{2+}. The position of the fourth Mn^{2+} peak is also indicated in A and B.

Albert Szent-Györgyi and theories and findings relating to changes in lipids, membranes and antioxidant activity.

In this regard, an especially attractive hypothesis is that the e.s.r. signal seen in lyophilized tissues is due to an ascorbyl radical immobilized on a solid matrix such as a protein (Naktinis & Cerniauskiene 1974). We have some evidence that supports this concept and are now doing experiments to test rigorously for this possibility. Fig. 6 shows e.s.r. spectra seen in lyophilized tumour tissue before and after exposure to oxygen, and lyophilized and oxygen-exposed serum albumin and microsomes with and without ascorbic acid. The data are not yet sufficient to permit firm conclusions to be drawn.

Experiments that need to be done

There seem to be several experiments that need to be done to resolve some of the questions, suggestions, and speculations that have been raised here. These experiments include: (*a*) Direct comparison of samples of different tumours before and after lyophilization to determine the reproducibility and generality of reported e.s.r. spectra of lyophilized tumours and their relationship to spectra seen before lyophilization; (*b*) Determination of the molecular nature of the free radicals observed in lyophilized tumour and normal tissues, with emphasis on the possible role of ascorbic acid; (*c*) Determination of the relationship of redox and 'antioxidant activity' changes *in vivo* and free radical changes in lyophilized normal and malignant cells: and (*d*) A real test of the hypothesis that free radicals play a significant role in carcinogenesis, to be obtained by studies of a synchronized tissue culture system that undergoes malignant change with a high percentage of transformation.

ACKNOWLEDGEMENT

This research was supported in part by grants CA13341 and RRO1008 from the National Institutes of Health.

References

BOLTON, J.R. (1972) Chapter 1 in *Biological Applications of Electron Spin Resonance* (Swartz, H.M., Borg, D.C. & Bolton, J.R., eds.) Wiley, New York

BRENNAN, M.J., COLE, T. & SINGLEY, J.A. (1966) A unique hyperfine ESR spectrum in mouse neoplasms analyzed by computer simulation. *Proc. Soc. Exp. Biol. Med. 123,* 715-718

BRUES, A.M. & BARRON, E.S.G. (1951) Biochemistry of cancer. *Annu. Rev. Biochem. 20,* 343-366

BURLAKOVA, Ye.B. & MALOCHKINA, Ye.M. (1968) Changes in the antioxidative activity of mouse liver lipids on induction of hepatoma with *o*-aminoazotoluene. *Biofizika 13,* 443-448

COMMONER, B., TOWNSEND, J. & PAKE, G. (1954) Free radicals in biological materials. *Nature (Lond.) 174,* 689-691

COMMONER, B., WOOLUM, J.C., TERNBERG, J.L. & SENTURIA, B.H., Jr (1970) The effects of α-acetylaminofluorene and nitrite on free radicals and carcinogenesis in rat liver. *Cancer Res. 30,* 2091-2097

DUCHESNE, J. & VAN DE VORST, A. (1970) Free radicals in normal and pathological surviving tissues. *Bull. Cl. Sci., Acad. R. Belg. 56,* 433-448

EMANUEL, N.M. (1976) Free radicals and the action of inhibitors of free radical processes under pathological states and aging in living organisms and in man. *Q. Rev. Biophys. 9,* 283-307

FITZHUGH, A. (1953) Malignant tumors and high polymers. *Science (Wash. D.C.) 118,* 783

GOLOSHCHAPOV, A.N., ALESENKO, A.V., BOGDANOV, G.N. & BURLAKOVA, Ye.B. (1973) Physio-chemical characteristics of lipids and proteins during the development of transplantable tumours. *Biofizika 18,* 1047-1051

GUTIERREZ, P.L. & SWARTZ, H.M. (1979) Paramagnetic changes in cancer: the growth of Walker 256 carcinoma studied in frozen and lyophilized tissues. *Br. J. Cancer, 39,* 24-34

GUTIERREZ, P., SWARTZ, H.M. & WILKINSON, E.J. (1979) Paramagnetic changes in cancer: 7-12-dimethylbenzanthracene (DMBA) induced tumours studied in non-lyophilized and lyophilized tissues. *Br. J. Cancer 39,* 330-336

HECKLY, R. (1972) Chapter 5 in *Biological Applications of Electron Spin Resonance* (Swartz, H.M., Borg, D.C. & Bolton, J.R., eds.), Wiley, New York

KASSIRSKII, I.A., EMANUEL, N.M., KLOCHKO, E.V., PINZUR, V.I. & KRUGLYAKOVA, K.E. (1967) *Problemy Gematologii i Perelivaniya Krovi 8,* 11

KLOCHKO, E.V., KOVAL'CHUK, L.V., KRUGLYAKOVA, K.E., SEITS, I.F., LUGANOVA, I.S., BLINOV, M.N. & EMANUEL, N.M. (1970) Free radicals and energy exchange in leukocytes during leukosis. *Dokl. Acad. Nauk SSSR 190,* 476-479

MAILER, C., SWARTZ, H.M., KONIECZNY, M., AMBEGAONKAR, S. & MOORE, V. (1974) Identity of the paramagnetic element found in increased concentrations in plasma of cancer patients and its relationship to other pathological processes. *Cancer Res. 34,* 637-642

MALLARD, J.R. & KENT, M. (1964) Differences observed between electron spin resonance signals from surviving tumour tissues and from their corresponding normal tissues. *Nature (Lond.) 204,* 1192-1194

MALLARD, J.R. & KENT, M. (1966) Electron spin resonance of surviving rat tissues. *Nature (Lond.) 210,* 588-591

MARUYAMA, T., KATAOKA, N., NEGASE, S., NAKADA, N., SATO, H. & SASAKI, J. (1971) Identification of 3-line ESR signal and its relationship to ascites tumors. *Cancer Res. 31,* 179-184

NAKTINIS, I. & CERNIAUSKIENE, L. (1974) Nature of the free radical EPR signal recorded from dried animal tissues. *Biofizika 19,* 1039-1044

PAVLOVA, N.L. & LIVENSON, A.R. (1965) ESR spectra of human normal and leukemia blood. *Biofizika 10,* 169-171

PULLMAN, A. & PULLMAN, B. (1955) *Cancérisation par les Substances Chimiques et Structure Moléculaire,* Masson, Paris

SAPRIN, A.N., KLOCHKO, E.V., KRUGLYAKOVA, K.E., CHIBRIKIN, V.M. & EMANUEL, N.M. (1966) Kinetics of the change in the content of free radicals in mouse organs in experimental leukosis. *Proc. Acad. Sci. USSR 167,* 222-224

SAPRIN, A.N., MANENKOV, A., NAGLER, Ye.A., KAZNACHEYEV, Yu.S. & EMANUEL, N.M. (1967) Course of change in the content of free radicals in developing Walker carcinoma and the action of Thio-TEPA. *Biofizika 12,* 1099-1102

SWARTZ, H.M. (1972) Electron spin resonance studies of carcinogenesis. *Adv. Cancer Res. 15,* 227-252

SWARTZ, H.M. & GUTIERREZ, P.L. (1977) Free radical increases in cancer: evidence that there is not a real increase. *Science (Wash. D.C.) 198,* 936-938

SWARTZ, H.M., MAILER, C., AMBEGAONKAR, S., ANTHOLINE, W., MCNELLIS, D. & SCHNELLER, S. (1973) Paramagnetic changes during development of a transplanted AKR/J leukemia in mice as measured by electron spin resonance. *Cancer Res. 33,* 2588-2595

SWARTZ, H.M., ANTHOLINE, W. & REICHLING, B. (1978) Paramagnetic changes during development of DMBA induced mammary tumors in Sprague-Dawley rats. *Phys. Med. Biol. 23,* 235-244

SZENT-GYÖRGYI, A., ISENBERG, I. & BAIRD, S.L. (1960) On the electron donating properties of carcinogens. *Proc. Natl. Acad. Sci. U.S.A. 46,* 1444-1449

TRUBY, L.K. & GOLDZEIHER, J.W. (1958) ESR investigations of rat liver and rat hepatoma. *Nature (Lond.) 182,* 1371-1372

TS'O, P.O., CASPARY, W.J. & LORENTZEN, R.J. (1977) The involvement of free radicals in carcinogenesis, in *Free Radicals in Biology* (Pryor, W.A., ed.), vol.3, pp. 251-303, Academic Press, New York

VITHAYATHIL, A., TERNBERG, J.L. & COMMONER, B. (1965) Changes in ESR signals of rat liver during chemical carcinogenesis. *Nature (Lond.) 207,* 1246-1247

WOOLUM, J.C. & COMMONER, B. (1970) Isolation and identification of a paramagnetic complex from the livers of carcinogen treated rats. *Biochim. Biophys. Acta 201,* 131-140

Discussion

Williams: Before we discuss this paper I should appreciate a little clarification. As I understand cancer (cf. my Note on Cancer, pp. 19-31), there are two levels to the problem. (1) Normal cells are converted into some state, which we don't understand–a transformed cell of some sort. (2) This cell then multiplies to give *n* cells, which represent a tumour. We can introduce more steps into this scheme but, first and foremost, we have to separate carcinogenesis, the transforming step (1), from the maintained state of growth (2). In the second state a cancer cell behaves unlike any other body cell—it is autonomous, and it grows in a formless manner. All the cells which are native to the system grow with physical form. This is because the surface of a tumour cell is, in some way, distinctive and, therefore, its characteristics as a cell are distinctive from those of all normal cells. This is step (2). Professor Slater will discuss carcinogenesis (pp. 301-321) and the way in which transient radicals affect the first step (1), but Dr Szent-Györgyi's theory of cancer is concerned not with that process, but with the maintained state of cancer cells in tumours. He proposes that cancer cells contain special states of proteins which alter their general nature and he is therefore referring not to carcinogenesis but to the maintained state, or the autonomy, of cancer cells. Consequently, let us limit this discussion to the nature of the maintained cell. This is in line with Dr Swartz's point that he is looking specifically at the nature of the total cell population in the tumour and that we should not confuse this with carcinogenesis. In this maintained state is there a free radical population and, therefore, some special protein which is altering the metabolism of that cell in a special way?

Dianzani: Carcinogenesis is, as you say, a multistep process. Many cancerous cells may occupy living organisms without giving any clinical sign of cancer. Between carcinogenesis and the production of clinical cancer all the defences of the organism, especially the immunological cells, may act. These cells may produce several types of free radicals when contacting their targets. Macrophages, for example, are known to be good producers of prostaglandins. During prostaglandin production and release, free radicals are generated. During phagocytosis by macrophages or leucocytes, superoxide anions are released from these cells into the medium. Macrophages and leucocytes probably play a role in the organism's reaction against cancer cells, so the possibility exists that free radicals that may be detected in cancer tissues are not due to cancer cells themselves, but are only the result of contamination by inflammatory cells. This contamination is expected to be rather high shortly after transplantation of the tumour. So, macrophages or other types of inflammatory cells may be responsible for an eventual increase in e.s.r. signal. The number of inflammatory cells actually present in the tumour tissue may change with time, as new processes, especially necrosis of tumour cells, increase the extent of the inflammatory process. Thus it seems important to study the problem more deeply by using a more critical model. The synchronized cell culture system suggested by Dr Swartz would probably be a good model.

With regard to the concentration of antioxidants in tumour cells, several years ago Lash (1966) and, in my laboratory, Ugazio *et al.*(1968, 1969) showed that antioxidants are plentiful in several hepatomas. Lipid peroxidation by these tumour homogenates, as measured by the production of malonaldehyde, appears to be minimal in comparison with that of normal liver homogenate. In addition, when a homogenate from an ascites tumour is added to a homogenate of normal liver, the usual production of malonaldehyde by the latter homogenate is severely depressed, suggesting that the tumour tissue may be rich in antioxidants.

Ladik: One may translate Professor Williams's 'carcinogenesis' and 'maintenance' into the words 'cause' and 'consequence'. I am not sure what Professor Szent-Györgyi says is the cause or the consequence. One can argue that carcinogens affect DNA by binding to it and releasing some blocking proteins which frees the carcinogenic information that transforms the cells; that would be the cause. The consequence of that may be, for instance, that methylglyoxalases cease functioning and, therefore, proteins become insulators. But one can argue the converse. Some methylglyoxalases may be knocked out and thus proteins become insulators. In this case the interactions between DNA and proteins decrease (the dispersion polarization forces are much smaller between insulators than between semiconductors). In this case the fact of a protein becoming an insulator, and therefore carcinogenesis starts, is the cause, leading to the consequence that some undesired part of the genetic information in DNA becomes free. What is cause and consequence is rather like the hen and the egg problem.

Williams: I disagree. The first process is the changing of the genetic material (the chromatin, i.e. not just the DNA).

Ladik: Let us rather say a change in the *accessible part* of the DNA.

Williams: Yes. I am trying to say that carcinogenesis is action at one level; a change in the total genetic material. If one wants to say that the initiation of cancer is due to changes in ill-defined proteins and not changes in genetic material, especially DNA, one has to explain how such change is inherited.

Ladik: Let us assume that DNA contains the 'carcinogenic' information but that it is blocked by proteins, so it is not accessible in differentiated cells. If the conduction properties of proteins are changed somehow (assuming for the moment that this is the cause), the interaction between protein and DNA becomes weaker at that place and the undesirable part of the DNA is unblocked (that is the 'consequence'). That is one possible argument; I don't think that one can claim that the protein 'story' is a consequence only.

Williams: But it isn't heritable; even if the protein is damaged, the new DNA would produce new, undamaged, protein.

Ladik: The new DNA may code for new enzymes which affect the biochemical reactions in the cell in such a way that proteins become insulators again. One can put forward many such mechanisms.

Bondy: To turn this argument upside down, may I discuss why cells *don't* become cancerous? In the course of differentiation each cell line must develop a certain pattern for the use of its DNA. Each time the cell divides, this pattern must be reproduced until such time as there is further differentiation and a new activation of DNA occurs. So far as I am aware, this differentiation pattern depends on changes not in the DNA but in the proteins which surround, expose, protect it. If so, each cell division requires the daughter cells to reestablish the pattern of differentiation. The critical matter, then, consists of the particular proteins that are involved in this 'steering' of the DNA and not the DNA itself. If some activity or event produces a clone of cancer cells, there must be a continuous aberration of this differential organization or exposure of the DNA for that clone to be maintained. What happens to the protein of the cancer cell to make it abnormal is not what happens to the DNA. What happens to the DNA is secondary to the organizing principles which are, so far as I am aware, most likely to be proteins.

Kornberg: This argument is becoming almost theological, and appears to be philosophically incapable of resolution! One difference between a cancer cell and an untransformed cell is that the cancer cell is not subject to, and does not respond to, the same regulatory signals as does the transformed cell. For example, in a cancer cell differentiation does not occur to the same extent as in normal cells, growth is uncontrolled, and this invasive phenomenon, which we recognize as the growth of the tumour in place of the normal cells and causes the damage, is associated with altered cell-to-cell interaction. It is true that the changes that go on must be changes in the behaviour of the enzyme proteins, the integrated action of which causes cell regulation to be maintained normally. But it is also true, if we can extrapolate from the experience with microorganisms, that many of these regulatory phenomena are enshrined not only in the genes that specify the structure of enzymes but also in the genes that specify the expression of the genetic potential—regulatory genes. Therefore, both have to be involved. Clearly there is a protein or a series of proteins that must be altered to cause a change in the over-all pattern but clearly also this change must be maintained which, therefore, must underlie the change of the genetic material. Or one could argue that by altering an enzyme which, for example, repairs DNA, one might cause a situation whereby a genetic mistake could no longer be corrected. There is no logical barrier to putting the first cause anywhere.

Williams: It is difficult to make firm statements about the condition that is cancer but we can say that certain steps initiate and, normally, they differ somehow from some of the events subsequently seen in maintained cancer cell populations. It is no good confusing what could initiate tumours with the maintained states of cancer cells, which are different in their metabolism. As I understood Dr Szent-Györgyi's hypothesis, a cancer cell is a cancer cell not just because something has happened but because there is a maintained state in a cancer cell which differs from the maintained state in all other cells and it is that maintained state, not carcinogenesis, which he wants to inspect.

Szent-Györgyi: That is correct.

Pethig: The concept of conducting proteins could be relevant for both the states (initiation and maintenance) being discussed. If polarization forces are important in keeping the DNA and histone together, then any mechanism that reduces the conductivity, and hence polarizability, of the histone molecule would tend to weaken the binding strength, lead to detachment of the histone from the DNA, and thence to an initiation of carcinogenesis. In the maintained state, such features as membrane fluidity and cellular contact inhibition may be influenced by the conductivity (polarizability) of the transmembrane proteins (e.g. glycophorin).

Ladik: Let us assume that initially a nucleohistone is a conductor and that therefore there is a strong interaction between DNA and the histone. When something attacks the histone the interaction is weakened, enabling the histone to leave and expose a new part of the DNA. That exposed part of DNA, which would not normally be transcribed, will now be. That is possible and I would not call it the maintained state.

Phillips: But in the daughter cells the histones again have to be disengaged from the DNA.

Ladik: As I said earlier, one can imagine that the newly accessible information in DNA finally gives rise to such biochemical reactions which 'damage' the histones also in the daughter cells. In this way the change can become hereditary.

Phillips: Unless the histone is again defective in the daughter cell, that piece of DNA will not be copied.

Ladik: Not quite so, because that part of the DNA could have been copied as tumour virus which then contains the undesired genetic information. It is not certainly so, but it is possible.

Slater: Professor Williams, you talk about the maintained state as if cancer cells, once formed, are a stable well-defined entity, but cancer cells often change their morphological appearance and biological character. I support what Professor Dianzani said about the complexity of what we know as clinical cancer. We sometimes mislead ourselves by devising models that are too simple—and it would be difficult to get a model less simple than the one you have outlined! Once initiated, tumours often develop in different ways. The results of analysing tumours that have progressed considerably from their original form are very difficult to interpret in relation to the initial change (or changes) that produce the biological disturbance we know as cancer. To analyse what has changed in the initiation process we must look at the very early stages in carcinogenesis. Turning to Dr Swartz's comments, I agree that an essential role for free radicals in chemical carcinogenesis has not been established unequivocally, although many carcinogens are known to be activated metabolically to free-radical intermediates. To illustrate an extra complexity that we see in this field, I should say that most of the carcinogens that form free radicals, and which we study by e.s.r. techniques, are generally highly cytotoxic. So, superimposed on their carcinogenicity is an intense cytotoxicity that can produce a variety of metabolic disturbances quite distinct from disturbances involved in carcinogenesis. Therefore, in trying to interpret the e.s.r. signals in terms of carcinogenic actions, we have to bear in mind the tremendous complexity of the other toxic reactions of free radicals on cells. For example, aflatoxin B_1 at a moderate dose produces a widespread centrilobular necrosis in the liver (for refs. see Patterson 1978): only a relatively few cells will be transformed but many cells will be killed.

Swartz: I agree. One of my basic points was that to study something as complex as a tumour by simply obtaining e.s.r. spectra was reasonable 25 years ago when the theory was first put forward. In view of our knowledge today it is not very reasonable because we know we are not likely to derive any real knowledge. Such data won't help us to discover whether a carcinogen proceeds by a one-electron step. To find that out one has to use a less complex system, and observe it at the proper time. Furthermore, even if free radicals are observed, they can have other actions besides carcinogenesis.

Brandt: I too feel uncomfortable with your model, Professor Williams. Once a tumour has been induced, one often finds not only different cell types but also the difference between benign and malignant growth. Some cancers are benign; others remain benign for many years before becoming malignant and invasively dangerous. So, many cells have within them the capability in their genetic material to go from a benign or even a normal state to a malignant state. We know very little about that process, even at the level of the surface proteins, the protein channels going through the membranes—both the cell membrane and the nuclear membrane. A model as simple as yours makes me a bit uncomfortable.

Williams: I apologize for that model; I know and accept the criticisms but I believe that only when one does an experiment to test a simple hypothesis can one obtain an answer. If subsequently one says that the experiment was wrong because the material was bad, that is the fault of the experimenter. As I understood things, you as experimenters are studying cancer according to the hypothesis that there is something abnormal about a tumour cell which might be recognized from the e.s.r. signals of the cells. In that case all you can do is to study the population of transformed cells in their maintained state; you cannot study carcinogenesis instead. From such experiments it seems to me that you have not got an answer. The number of steps in carcinogenesis is immaterial; the tumour never shows abnormalities of the kind that was introduced in the original theory.

Swartz: I would like to reiterate three points in closing. First, if it is carcinogenesis we are trying to study, then the e.s.r. studies of tumours that have been done are simply not pertinent because they have not made observations at the proper time under the proper circumstances. But, second, if we are seeking to test a theory that says that free-radical levels in tumour cells should be different from those of normal cells, then we can state that the available evidence is contrary to this conclusion. Finally, and perhaps most importantly, we should understand that regardless of the theory we are testing, tumours are so complex that simple measurements of levels of free radicals are unlikely to be productive.

References

LASH, E.D. (1966) The antioxidant and prooxidant activity in ascites tumors. *Arch. Biochem. Biophys. 115,* 332-336

PATTERSON, D.S.P. (1978) in *Biochemical Mechanisms of Liver Injury* (Slater, T.F., ed.), pp. 403-442, Academic Press, London

UGAZIO, G., GABRIEL, L. & BURDINO, E. (1968) Ricerche sugli inibitori della perossidazione lipidica presenti nelle cellule dell'epatoma ascite di Yoshida. *Boll. Soc. Ital. Biol. Sper. 44,* 30-34

UGAZIO, G., GABRIEL, L. & BURDINO, E. (1969) Perossidazione lipidica e processi proliferativi. *Atti Soc. Ital. Patologia 11,* 325-341

Whole-body nuclear magnetic resonance scanning: n.m.r. studies of tumour cells

RAYMOND DAMADIAN, LAWRENCE MINKOFF and MICHAEL GOLDSMITH

State University of New York, Downstate Medical Center, Brooklyn, New York

Abstract A technique, field-focusing nuclear magnetic resonance (n.m.r.) spectroscopy (FONAR), is described for doing n.m.r. scans in large samples. The method utilizes a shaped d.c. magnetic field that confines the n.m.r. signal-producing region of the sample to a small volume called the resonance aperture. The aperture contains the required values of the H_0 field to fully bracket the band of the r.f. pulse. The magnet system and r.f. pick-up coil that achieved the first human n.m.r. scan are discussed.

In 1971, one of us conceived the idea of whole-body nuclear magnetic resonance (n.m.r.) scanning and tested the idea to see if the nuclear resonance signal would detect disease non-invasively (Damadian 1971). In 1972 he developed the field-focusing n.m.r. (FONAR) technique for whole-body scanning (Damadian 1972). Others followed (Hinshaw 1976; Mansfield & Maudsley 1977; Kumar *et al.* 1975). Most recently, Bene obtained proton signals from internal human organs using the earth's magnetic field (Bene *et al.* 1977). We wish now to report the achievement of our original goal with the successful completion of the first whole-body n.m.r. scan.

The practice of medicine today is largely rooted in the anatomical descriptions of Vesalius and his intellectual successors who advanced the anatomical data base of diagnosis and treatment from gross to microscopic description. Thus, common medical diagnoses such as cirrhosis of the liver, glomerulonephritis, or Hodgkin's sarcoma connote alterations in the microscopic architecture of the diseased organ.

The intuitive driving force behind the clinical application of biochemistry, however, has been the prospect of one day converting the practice of medicine from an anatomical to a chemical footing. In the clinical setting, technological advances over the past two decades have made it possible to extract considerable chemical information from a sample of blood. However, the chemical disturbance in a diseased organ can only be inferred from clues deposited in the blood by that organ. No technique exists for going non-invasively and directly to the affected organ for its chemistry.

The non-invasive determination of the chemistry of diseased organs and tumours in humans imposed requirements that could not be met by the existing n.m.r. technology. In the conventional n.m.r. experiment the nuclear induction signal from a sample is detected by the 'pick-up' coil surrounding the sample without knowledge of how the signal-producing domains within the sample (usually small in size, e.g. 5 g) are distributed. Singling out organs inside the human body for direct inspection by n.m.r. or detecting internal malignant deposits required the development of new techniques for focusing the n.m.r. signal within the interior of the sample (which is much larger in size, e.g. human liver 1500 g) (Damadian 1972).

The forced precessions of a nuclear magnetization under a radio-frequency (r.f.) driving field (Bloch *et al.* 1946; Purcell *et al.* 1946) provided the basis for achieving this 'in-sample' focusing. But the detectability of the electromotive force induced in a coil by these precessions is subject to some restrictions. The FONAR method (Damadian 1972, 1977; Damadian *et al.* 1976) for 'in-sample' focusing originates in these restrictions. Transitions between Zeeman levels that give rise to the nuclear induction signal are constrained to occur between neighbouring levels of spin magnetic energies. Sufficient coupling of the nuclear spins to the radiation field to produce a signal detectable by radiofrequency spectroscopy occurs only when the stringent Bohr frequency condition, $h\upsilon = (\mu H_0/I)$, is satisfied; no signal being generated by the spin system when the oscillator frequency, υ, dictated by the Zeeman level separation, $\mu H_0/I$, is incorrect. Thus for any choice of frequency of the r.f. driving field there is one value of the d.c. static field, H_0, that will produce a resonance. In actual practice, shaping of the static field across the sample confines the signal-producing region of the sample to a small volume, called the resonance aperture, that contains the correct values of H_0 to bracket the band of the r.f. pulse (Damadian *et al.* 1976). The resonance aperture is constructed by calculating the series expansion of the axial magnetic induction, B_z, in spherical coordinates.

To achieve our objective of scaling up the n.m.r. technique for scanning large samples it was necessary for us to design and construct our own superconducting magnet and cryogen that would operate with the FONAR designs. The magnet, two halves of a Helmholtz pair, was wound on a machined aluminium former with a 53-inch inner diameter using a specially designed winding machine built at this laboratory. Each magnet-half contained a sweep coil and z-gradient coil in addition to the main magnet windings. The main magnet consisted of five layers of 12 mil core diameter superconducting wire and 47 layers of 22 mil core copper-clad niobium-titanium wire. The magnet, according to our computer calculations of the field mesh, is capable of 5000 gauss, although our present series of tests have generally been done with the 'supercon' in persistent mode at either 500 or 1000 gauss. The characteristics of the finished magnet are an inductance of 61.8 henrys, a stored magnetic energy (at 930 gauss) of 2.97×10^4 joules, a stability of better than seven parts in 10^7 over one hour, a weight of 120 lb (without dewar), and a maximum field of 5000 gauss. So far there has been no n.m.r.-detectable drift in the magnet's field when it is operating in the persistent mode.

The cryogen is a nitrogen-jacketed, vacuum-insulated aluminium dewar in three adjoining sections: the magnet hoop, gooseneck and storage can. The hoop contains the magnet solenoid bolted into a doughnut-shaped stainless-steel (SS type 304) liquid helium can that was welded closed with a 300 Amp Airco TIG Heliwelder. Concentric with the magnet can is a larger aluminium (6061-T6) can that doubles as a nitrogen-cooling stage and radiation shield. The outer concentric vacuum jacket is a welded cylinder of 0.5 inch 6061-T6 aluminium. Radiational losses from the dewar were minimized with superinsulation (aluminized mylar) and a single layer of aluminium tape (Emerson and Cuming) on the liquid helium can.

For the first test of the fully assembled FONAR apparatus we used a simulated human chest ('the phantom'), consisting of a 13.5 inch-diameter cylindrical container of $NiCl_2$-doped water with three air-filled methacrylate tubes 5, 3 and 1 inches in diameter, for 'lungs'. The r.f. pulses were delivered to the sample using a tape-wound 14-inch single-coil probe powered by a variable frequency Seimco model RD spectrometer operating at 2.18 MHz and delivering 10 W of power over 60 microseconds. 90° pulses were repeated with a period of 800 microseconds. The n.m.r. images shown in this paper are stored video records of the maximum peak-to-peak amplitude of a constant 5-kc off-resonance beat pattern of the phase-detected proton signal.

Fig. 1*a* is a schematic illustration of the phantom. Fig. 1*b* demonstrates the off-resonance proton signals obtained (without signal averaging) from representative locations in the phantom. The experiment of Fig. 1*b* demonstrates major advantages of the FONAR method over other methods, in that *(a)* FONAR is direct and *(b)* the FONAR signal is visible at each location of the scanning aperture. These capabilities permit us to visualize the n.m.r. behaviour of each region of the anatomy as the scan proceeds, rather than having to await a computer reconstruction of the data, as in non-focusing methods, before information can be obtained. Furthermore, at the

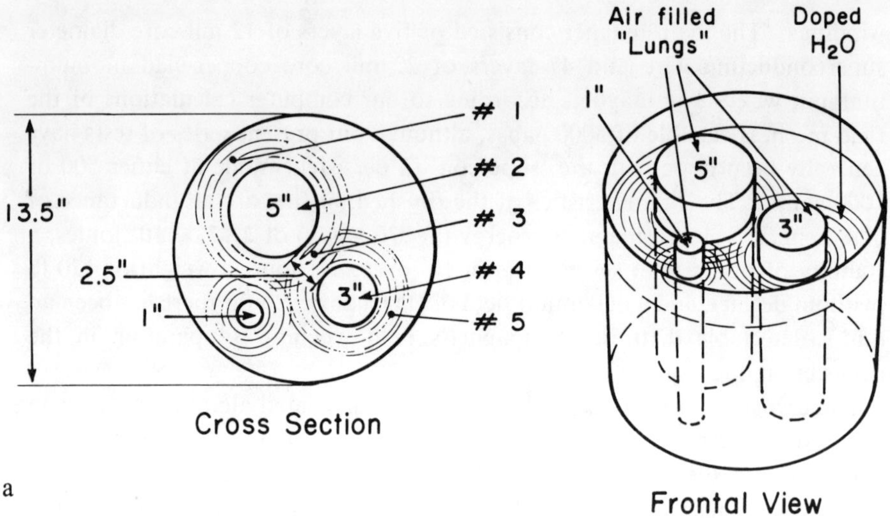

a

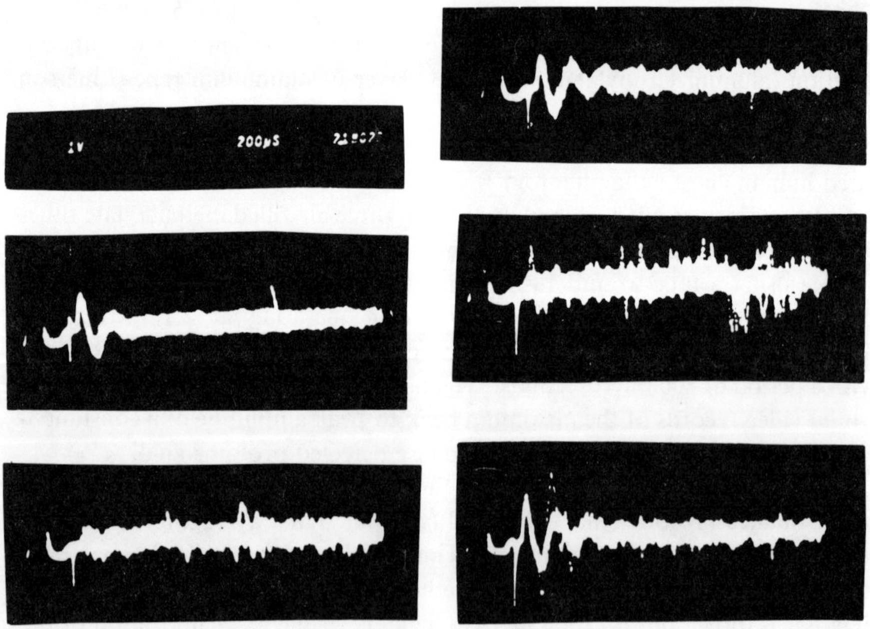

b

FIG. 1 (*a*) Schematic of the simulated (phantom) human chest used to obtain the FONAR image shown in Fig. 1*c*. The phantom consisted of a cylindrical polypropylene tank (13.5 inches in diameter) filled with an aqueous solution of $NiCl_2$ and containing three air-filled lucite cylinders with ↗

c
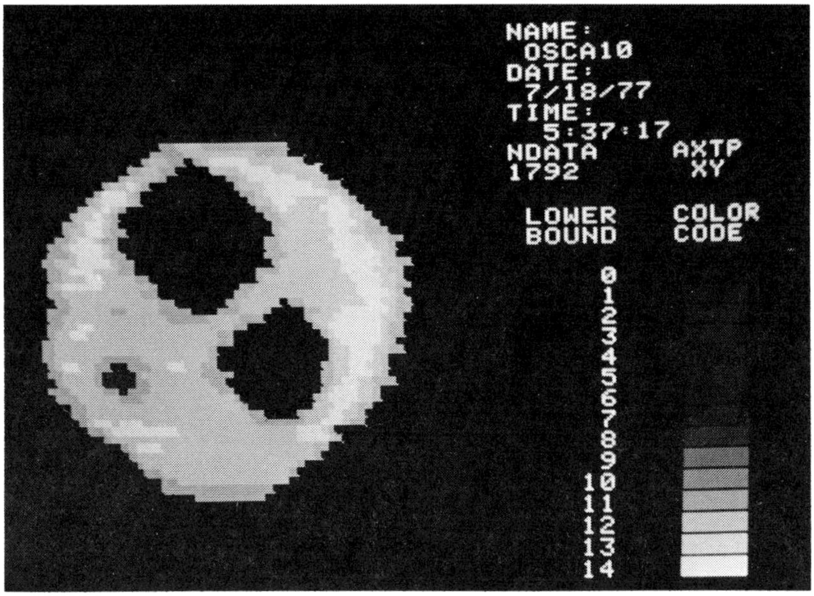

dimensions as indicated, serving as 'lungs'. The numbered regions in the drawing correspond to the position of the FONAR spot for the n.m.r. signals shown in Fig. 1*b*. Note that the FONAR process easily detected the smallest structure in the phantom (1-inch 'lung') with a 14-inch exploring coil.

(*b*) Off-resonance ^{1}H (proton) n.m.r. signals (without signal averaging) from each of the numbered regions of the phantom shown in (*a*).

(*c*) Cross-sectional image by focused n.m.r. (FONAR) of simulated chest. Zero proton signal is coded black, while three shades of grey represent the various signal intensities from doped H_2O. The image is a black-and-white photo of an original 14-colour video display.

completion of the scan the resonance aperture can be directed back to the coordinates of a suspicious locus for more detailed examination. Fig. 1*c* is the completed image of the phantom obtained in 30 minutes and displayed on a 256 × 256 pixel array.

Fig. 2*b* is a cross-sectional FONAR image of the live human chest at the level of the eighth thoracic vertebra (Fig. 2*a*). The scan, which took 4½ hours to complete, visualized the heart and mediastinum, outlined a left lung cavity smaller than the right, as it should be at this level, detected a depression in spin density in the midline across the back that could correspond to the lowered proton density of the vertebral body, and encountered a high signal-producing region immediately anterior to the vertebral body and slightly to the left side of the thorax, which corresponds to the location of the descending aorta. We estimate the resolution of this image to be approximately ¼ inch.

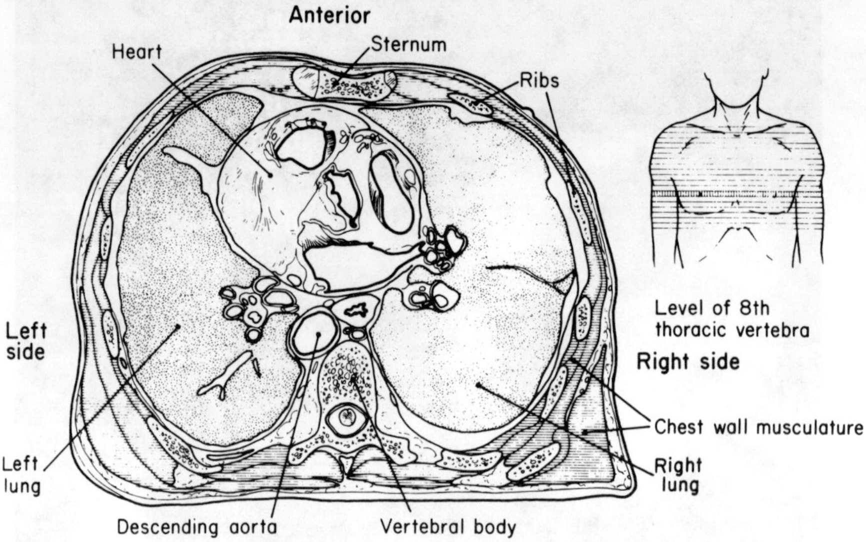
Anterior
Heart
Sternum
Ribs
Level of 8th thoracic vertebra
Left side
Right side
Chest wall musculature
Left lung
Right lung
Descending aorta
Vertebral body
Posterior

a

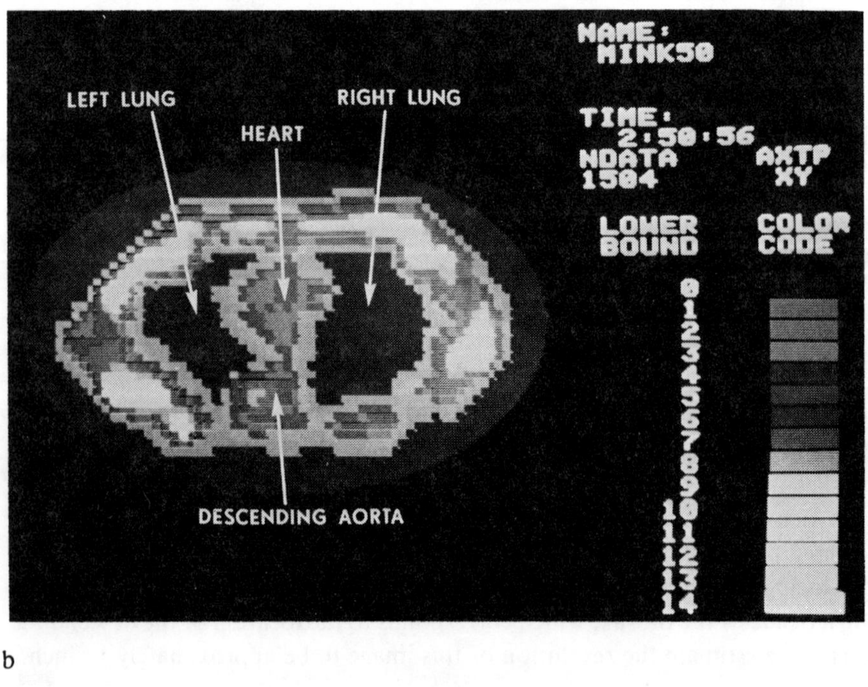
NAME: MINK50
LEFT LUNG
RIGHT LUNG
HEART
TIME: 2:50:56
NDATA 1504
AXTP XY
LOWER BOUND
COLOR CODE
0
1
2
3
4
5
6
7
8
9
10
11
12
13
14
DESCENDING AORTA

b

In the laboratory mouse a tumour surgically implanted in the anterior thorax has been successfully imaged with the FONAR method (Damadian *et al.* 1976; Damadian 1977). In humans, however, the largest structure that it has so far been possible to image by n.m.r. has been the finger (Mansfield & Maudsley 1977). We believe that the accomplishment of the first whole-body chemical image of a live human is not inconsistent with the concept that whole-body n.m.r. scanning will be useful in medicine.

References

BENE, G.J., BORCARD, B., HILTBRAND, E., MAGNIN, P. & SECHEHAYE, R. (1977) Nuclear magnetography in a weak field. Preliminary results. *C.R. Hebd. Séances Acad. Sci. Sér. B 284*(8), 141-143

BLOCH, F., HANSEN, W.W. & PACKARD, M. (1946) The nuclear induction experiment. *Phys. Rev. 70*, 474-485

DAMADIAN, R. (1971) Tumor detection by nuclear magnetic resonance. *Science (Wash. D.C.) 171*, 1151-1153

DAMADIAN, R. (1972) Apparatus and method of detecting cancer in tissue, U.S. Patent 3,789,832, filed 17 March 1972

DAMADIAN, R. (1977) Nuclear magnetic resonance: a noninvasive approach to cancer. *Hosp. Pract. 12*, 63-70

DAMADIAN, R., MINKOFF, L., GOLDSMITH, M., STANFORD, M. & KOUTCHER, J. (1976) Field focusing nuclear magnetic resonance (FONAR): visualization of a tumor in a live animal. *Science (Wash. D.C.) 194*, 1430-1432

HINSHAW, W.S. (1976) Image formation by nuclear magnetic resonance: the sensitive-point method. *J. Appl. Phys. 47*, 3709-3721

KUMAR, A., WELTI, D. & ERNST, R.R. (1975) Nuclear magnetic resonance Fourier zeugmatography. *J. Magn. Resonance 18*, 69-83

MANSFIELD, P. & MAUDSLEY, A.A. (1977) Medical imaging by nuclear magnetic resonance. *Br. J. Radiol. 50*, 188-194

PURCELL, E.M., TORREY, H.C. & POUND, R.V. (1946) Resonance absorption by nuclear magnetic moments in a solid. *Phys. Rev. 69*, 37-38

←

FIG. 2 (*a*) Schematic of the human chest at the level of the eighth thoracic vertebra.
(*b*) FONAR cross-section of the live human chest at the level of the eighth thoracic vertebra. Proton signal intensity is coded, with black assigned to zero signal amplitude, white to signals of strongest intensities, and intermediate grey scales to intermediate intensities. Top of image is anterior boundary of chest wall. Left area is left side of chest. Proceeding from anterior to posterior along the midline, the principal structure is the heart seen encroaching on the left lung field (black cavity). Left lung field is smaller than the right lung (black cavity to right of midline), as it should be (see schematic of the human chest at the thoracic level of the FONAR image in *a*). More posteriorly and slightly left of midline is a grey circular structure corresponding to the descending aorta.

In the body wall, beginning at the sternum (anterior midline) and proceeding around the ellipse, alternation of high intensity (white) with intermediate intensity (grey) could correspond to alternation of intercostal muscles (high intensity) with rib (low intensity), as shown in the schematic. The image is a black-and-white photo of the original 14-colour video display.

Discussion

Williams: This is a very remarkable method for whole-body scanning. In particular, using n.m.r. you are able to distinguish tumour cells from non-tumour cells. That, of course, is one of your original objectives, but I think it should be made clear that this is not electron spin resonance but nuclear magnetic resonance and that what is being detected has nothing to do with the idea that Dr Szent-Györgyi put forward originally—that one would be able to detect changes in conductivity, perhaps of free radical concentrations between normal and tumour cells. Here you are detecting something strange about the nuclei, and particularly the nuclei of water molecules.

Damadian: Actually, even that is not true. When we searched for an explanation of this peculiar, prolonged relaxation event there was virtually nothing in the literature to give any clue to what was happening except Szent-Györgyi's statement that the dielectric constant of water in malignant tissue ought to be high, signifying a reduced degree of the structure and order of solvent water. I was at a loss then for any explanation of our original observation and it was Szent-Györgyi's comments on water in his monograph on bioenergetics (1976) that we turned to for insight. Since then we have found this interpretation to be correct. The chief cause of the prolonged spin relaxation of water in tumours is the decreased order of solvent water molecules inside malignant tissue. This view has been contested and it has been argued that the prolonged relaxation time doesn't mean an implicit disordering in the water, but that it results from the increased water content of the malignant cells. An increased bulk water content of a transformed cell, however, implies an overall increase in the free water content of the cell and therefore in the average motional freedom of the cell water molecules. Either way, there is an increased structural disorganization or increased water content, disordering the system.

I should add that relaxation measurements have also been made in at least two laboratories establishing relaxation time prolongations in many tumours where the water content was measured and found to be unchanged.

Williams: So we are dealing here with a totally different aspect from anything said earlier about the nature of a tumour cell; this is probably the most pertinent point for our discussion.

Brandt: Dr Damadian, if you took an animal and filled one lung with water, would you get the same type of scan as if there were a tumour in that lung? The human parallel to this would be in pneumonia or pleurisy, for example, which also might interfere and give an erroneous signal of a tumour.

Damadian: This brings up an essential point, namely what the human scans we have so far produced are detecting. The first four human scans as we performed them detect only the presence of signal-producing protons in excess of what ought to be present in the normal organ. In the pictures of the thorax that you saw, pulmonary oedema, abscess and inflammation would all introduce signal-generating protons into the lung fields and show up as shadows that could be confused with tumour as they do in X-ray. This, however, gives me the opportunity to illustrate the virtue of the n.m.r. technology over other modalities. With n.m.r. we are not limited to one picture. The technique possesses the potential of gathering a wide variety of chemical information. Once we see something suspicious on a scan, for example, we can focus the FONAR spot directly on the lesion and make a detailed *chemical* enquiry of it. We can begin to do the kinds of things that a chemist would be delighted to be able to do on a liquid sample. We measure the proton relaxation time of the lesion in detail, determining its T_1, T_2, $T_{1\varrho}$. With enough signal/noise we could obtain the high resolution ^{31}P spectra of the lesion. In due course we could build up the kind of characterization which we hope would lead us to unambiguous diagnostic decisions. This illustrates my point about the technique delivering tissue chemistry rather than shadow anatomy, thereby giving us the prospect of transforming medicine from the practice of anatomy where it is rooted today to the practice of chemistry.

Brandt: I would certainly agree completely with the aim of trying to make medicine into a quantitative science rather than a store of qualitative observations.

Pethig: I should like to attempt to relate the concept of the dielectric constant to Dr Szent-Györgyi's bioelectronic ideas. There could be a direct link, because Dr Szent-Györgyi earlier showed us what happens to some of his charge-transfer reactions as a function of the dielectric constant of the solvent. At very low and high dielectric constant values the reactions did not proceed very well; the optimum relative dielectric constant value appeared to be around 60. We could interpret some of Dr Damadian's results in terms of a loosening of the water structure around proteins. In the region of the protein/aqueous medium interface we can imagine the effective dielectric constant going from a value of about 4 (near to that for ice) towards a value of 80 (normal water) as the water structure became more loosened. The stability and strength of the charge-transfer interaction involving the bending back of the lysine side-chain (as I discussed in my paper, pp. 83-98) would be influenced by such a drastic change of the local dielectric constant. Perhaps there could be a link at that level between some of Dr Damadian's results and the bioelectronic concepts.

Williams: There *could* be, but there is no good evidence yet that particular proteins are affected in any way. It is water which is either more abundant or in some way more disorganized in tumours, but not necessarily specifically around any protein.

Damadian: One can't be certain yet, of course, but simultaneous with the malignant cell's imbibition of water is the major architectural disorganization you described earlier, Dr Williams, in your second stage of tumour transformation (p. 124). It is nearly impossible in my judgement to separate the two—water overloading and architectural distortion—so that a major protein or nucleoprotein disorganization must accompany the swelling of the malignant cell.

Eley: Have you considered estimating the differential hydration of normal and tumour cells, using the effect of freezing the aqueous suspension and estimating the water bound to the proteins by the intensity of the narrow proton n.m.r. signal that remains after freezing (Kuntz *et al.* 1969)? Dr S.A. Attiga, Dr M.J. Hey and I recently (unpublished) found this technique useful in establishing that 15 molecules of water were bound to disodium cromoglycate (Intal), of which nine H_2O molecules were bound strongly and six less strongly. The nine strongly bound molecules agree with that absorbed by crystals from the vapour at the point above the crystal lattice collapse (Cox *et al.* 1971). It corresponds to one H_2O molecule per possible hydrogen-bonding centre in the molecule, if we assume that water bound to the Na^+ ions is ice-like and fails to give a narrow n.m.r. signal. So it appears that the technique which was developed for biopolymers is also useful with small molecules, and is perhaps worth examining for tumour cells.

Damadian: We have not done this, but it has been done on normal tissues by other groups. I have not been entirely satisfied with the interpretation of the results, because one ends up with a range of estimations of bound water varying from 5 to 35%, depending on initial assumptions. In fact, I have rather lost my interest at present in the endless molecular arguments on structured cell water and the numerous unlikely hypotheses generated to discredit it.

Slater: Dr Damadian, what order of resolution are you obtaining at the moment with your technique, especially in the abdomen, where the diagnosis of a tumour with your technique may be expected to be much less sensitive than in the thorax?

Damadian: We obtained a resolution of about 4–5 mm in the abdomen and thorax scans.

Slater: And how long does it now take to accumulate one of these scans?

Damadian: The first chest scan took 4½ hours; we are down to 15 minutes now. It is strictly a matter of detector signal-to-noise. Our efforts are concentrated on techniques for improving this.

Kornberg: Your most elegant pictures show that you can detect the presence of invasive cells, or cells that have invaded the tissue, by your technique, and your data also showed a large difference in the relaxation time that you observed with tumour tissue. What I am not clear about (and Dr Pethig's interpretation in particular rests on there being a fundamental difference between tumour and normal tissue) was whether you find similar pictures when the lungs, for example, are invaded by cells that are not tumorous but nevertheless present a mass which would otherwise not be present. It is important for the purpose of the discussion which motivates this symposium to establish that the difference you observe is peculiar to tumour tissues and not to any other tissue that may also contain more water than normal tissues.

Damadian: We have a lot of information on that, although not yet in the live human body. We have completed an n.m.r. study of 1000 tissues taken at surgery, biopsies of all types. We found that we could separate malignant tissues from their corresponding normals with almost total discrimination if we used a combination of n.m.r. parameters. Taking the pathologies one at a time, for inflammation there was no confusion with tumour tissue. In the case of benign tumours we sometimes confused malignant with benign tumour tissue. In at least one pathological condition we were able to make discriminations that microscopists could not. Pathologists, for example, have difficulty assessing the potential malignancy of a common breast disease, fibrocystic disease of the breast. We were able to get clear separation between normal breast tissue, fibrocystic tissue and malignant tissue. Furthermore, we were able to assign a malignancy index number to each of the fibrocystic cases, which we think provides us with a means of assessing which cases have the greatest potential for becoming malignant.

References

Cox, J.S.G., Woodard, G.D. & McCrone, (1971) Solid-state chemistry of cromolyn sodium (disodium cromoglycate). *J. Pharm. Sci. 60,* 1458-1465

Kuntz, I.D. Jr, Brassfield, T.S., Law, G.D. & Purcell, G.V. (1969) Hydration of macromolecules. *Science (Wash. D.C.) 163,* 1329-1331

Szent-Györgyi, A. (1976) *Electronic Biology and Cancer,* Dekker, New York

Quantum chemical investigations of charge-transfer interactions in relation to the electronic theory of cancer

COLIN THOMSON and JOHN R. BALL

Regional Workshop of the National Foundation for Cancer Research, Department of Chemistry, University of St. Andrews, Scotland

Abstract The results of *ab initio* 'supermolecule' calculations of the charge transfer between formamide and methylglyoxal, dimethylglyoxal and ethylglyoxal are compared for several different relative conformations of the constituent molecules. The extent and sign of the charge transfer is similar for all three molecules; the ketoaldehyde acts as an electron acceptor only for the stacked conformation.

Similar calculations on α-hydroxytetronic acid as a model for ascorbic acid show that it can act as either an acceptor from formamide or a donor to glyoxal.

In 1941 Szent-Györgyi proposed that the subtlety of many biological processes was due to electronic effects, especially electron delocalization. This idea, that it is necessary to look for explanations of biological phenomena at the submolecular level, has been rather slow to gain acceptance in the biological sciences, although Szent-Györgyi has in several publications (Szent-Györgyi 1960, 1968, 1972, 1976, 1978) further developed this view and, during the past 10 years or so, has proposed that the understanding of the molecular basis of neoplastic disease can only be expected when we understand the electronic properties and interactions of the molecules involved in the control of cell proliferation.

Recent advances in theoretical chemistry and solid-state physics, however, have led to theoretical calculations which strongly indicate that Szent-Györgyi's ideas are on sound theoretical ground, as emphasized elsewhere in this symposium (Ladik 1979), and that the detailed study of biology in electronic terms promises to yield a much greater insight into the subtlety of biological processes than attempts to interpret biological phenomena in purely molecular terms.

The basic ideas in so far as they relate to the subject matter of this paper are as follows.

Proteins (and other macromolecules), by virtue of their regular repeating structure, have energy bands which are normally completely filled in an *isolated* protein. The filled valence bands are separated from the unfilled conduction bands by a substantial energy gap ($\Delta E \approx 11$ eV; Ladik *et al.* 1978). Consequently, such proteins cannot conduct: they are insulators.

However, in biology, if we consider, for example, the structural proteins in a typical cell, the saturated valence band may become desaturated and, therefore, conducting if an electron (or electrons) can be removed from the filled bands by a molecule which is a strong electron acceptor. A molecule of this type would have to be closely associated with the particular protein and have the specific function of controlling the extent of the desaturation. (This model is analogous to that involved in semiconductors.) The result is that the acceptor molecule creates a hole in the valence band. The resulting unfilled band can lead to enhanced interactions between proteins (Laki & Ladik 1976). The extent of the charge transfer need not be large for this to occur.

The possibility of semiconduction by proteins and other biological macromolecules leads us to believe that electronic processes such as this are of vital importance in intercellular communication and in the control of cell proliferation.

Szent-Györgyi suggested (1976, 1977) that the insoluble structural proteins are particularly important in these processes and one of us (C.T.) believes that the structural proteins actin, myosin and tubulin have a key role in these phenomena (C. Thomson, unpublished work), but so far the relevant experimental measurements have not been made.

There have been, however, several experimental studies (reviewed by Szent-Györgyi 1976) of proteins 'doped' with an electron acceptor, and the most recent electrical measurements (Bone *et al.* 1978; Pethig 1978) seem to provide decisive evidence for protein semiconduction, at least under certain experimental conditions. However, it is clearly important to determine the nature of the molecules which act as electron acceptors and could be expected to desaturate the protein energy bands.

It is in this area that theoretical chemists can provide useful information. We summarize the current views on charge transfer, below.

CHARGE-TRANSFER INTERACTIONS

The desaturation of the energy bands by transfer of an electron to an acceptor molecule is a particular case of the phenomenon known as charge transfer (Kollman 1977; Slifkin 1971), in which one molecule (D) donates an electron to another molecule (A) which accepts the donated electron. This is symbolized in equation (1). The resulting charge-transfer (or donor/acceptor)

$$D + A \rightarrow (D^+ A^-) \tag{1}$$

molecule is stabilized by electrostatic and other interactions. For this process to occur in the ground state, the donor molecule must have a doubly occupied orbital which is relatively weakly bonding and the acceptor molecule must have available a low-lying unoccupied orbital which can accommodate the transferred electron. This is depicted symbolically in Fig. 1. The gain in energy by stabilization of the complex must clearly offset the energy necessary to effect the transfer. The complex may be stabilized in many ways, especially in a biological situation: we return to this point later.

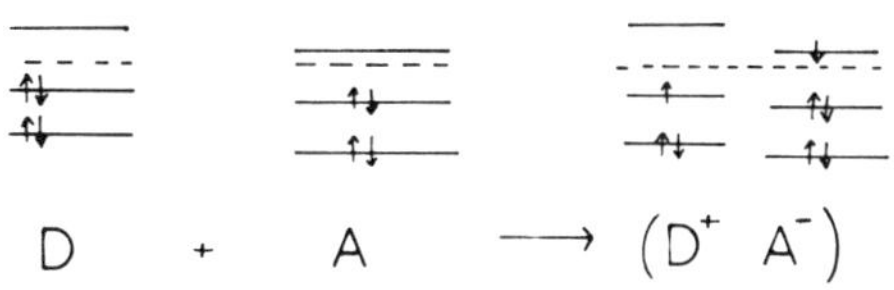

FIG. 1. Donor and acceptor orbitals in charge transfer. Energy levels above dashed line are anti-bonding: those below dashed line are bonding.

Experimental evidence for charge-transfer interactions with organic molecules has been reviewed in detail elsewhere (Slifkin 1971), but this type of complex usually involves charge transfer in an excited state which is close to the ground state. Experimentally, this phenomenon is usually studied by u.v. spectroscopy. Charge transfer can also be studied by electron spin resonance (e.s.r.) spectroscopy when free-radical species are involved (Kon & Szent-Györgyi 1973; Kimura & Szent-Györgyi 1969).

For charge transfer in the ground state the energies of the two orbitals involved must be similar. The resultant (D^+A^-) complex consists of two radicals, D^+ and A^-, in close proximity. The spins in this complex can be paired, in which case the total spin $S = 0$ and the complex is diamagnetic, or unpaired, in which case there are two possibilities: (1) if the spins are strongly coupled, $S = 1$, and we have a triplet state; (2) if the spins are not coupled, we have a biradical. It has recently been emphasized (Szent-Györgyi 1976) that the $(D^+ A^-)$ complex need not behave as a singlet or triplet but that, for suitable D and A, one can obtain two radical species D^+ and A^- in which the spins are uncoupled but the electrostatic interaction is sufficient to preserve the stability of the complex. Such complexes are paramagnetic and e.s.r. spectroscopy can be used to detect them.

Szent-Györgyi and his co-workers (Kimura & Szent-Györgyi 1969; Kon & Szent-Györgyi 1973) have described simple charge-transfer complexes of this type between amines and carbonyls and in other systems. Electrical measurements on such species have also been reported (Pethig & Szent-Györgyi 1977; Bone *et al.* 1978).

QUANTUM MECHANICAL STUDY OF CHARGE-TRANSFER INTERACTIONS

The study of these phenomena by quantum chemical methods dates back to the work of Mulliken (1950) but studies of the weaker charge transfer of the type we are concerned with necessitated the development of methods for computing in detail the electronic structure of the isolated molecules and their interaction. Although the latter may be computed by perturbation theory, it is now realized that the charge redistribution which occurs when donor and acceptor interact can be better calculated by the 'supermolecule' approach. In this method, the wave function of the isolated molecules is compared with that of the 'supermolecule' formed by placing the interacting molecules at the appropriate relative geometries.

Extensive theoretical work of this kind has been done, especially for hydrogen-bonded complexes (Kollman 1977), but rather less for the type of donor–acceptor interactions with which we are concerned.

The interaction energy between the molecules, ΔE, is defined in equation (2), where the E_i are the energies of the isolated molecules and E_{super} is the

$$\Delta E = E_{\text{super}} - \sum_i E_i \tag{2}$$

energy of the supermolecule. ΔE contains various contributions, including a charge-transfer component.

Since the interaction energies are small ($\approx$ 20 kJ mol^{-1}) it might be expected that highly accurate wave functions are needed. Although this is true in principle, in practice for the large molecular combinations with which we are concerned accurate calculations are not practicable or likely to be in the near future. We are not, however, primarily interested in accurate values for ΔE or the charge shifts, but are more interested in the order of magnitude of the charge transfer and how it varies in a series of chemically related molecules.

For obtaining information of this type, many studies have shown the value of using wave functions computed with the self-consistent-field molecular orbital method (SCF-MO). This is the simplest type of single-configuration wave function and, in the limit of an infinite basis set, gives a well defined limit, the Hartree-Fock wave function (Schaefer 1973).

Different quality SCF wave functions are available depending on the size of the basis set in which the molecular orbitals are expanded. In general, each molecular orbital ϕ_i is expanded as a linear combination of basis orbitals, as in equation (3). The simplest expansion which has been shown to give reliable

$$\phi_i = \sum_{j=1}^{N} C_{ij}\chi_j \tag{3}$$

results is the minimal basis set, in which the basis functions comprise those atomic functions occupied in the ground state of the atoms. For first row atoms this means the χ_j are 1*s*, 2*s* or 2*p* functions (1*s* only on H). The molecular integrals are usually evaluated by expanding the χ_j in a set of *n* Gaussian orbitals, and the simplest minimal basis set, extensively developed by Pople and his co-workers, is the STO−3G basis (Pople 1977), using three Gaussian orbitals per Slater orbital (STO).

The results we describe have all been obtained using STO−3G basis sets in an *ab initio* calculation of the SCF wave function. Although semiempirical calculations have been used in this field, notably by Abdulnur (1976, 1977), we have preferred the use of *ab initio* methods. The SCF wave function is then computed in terms of the fundamental physical constants: there are no other parameters, but the geometry of the molecule is, of course, one molecular parameter in the calculations.

QUANTUM MECHANICAL RESULTS AND THEIR ANALYSIS

The results of the calculation are the total energy for a particular nuclear configuration, and the molecular wave function Ψ expressed in terms of the molecular orbitals ϕ_i. For an SCF wave function, in the case of a closed shell molecule (subscript cs), we have for a $2N$ electron system the wave function given by equation (4) in the usual Slater determinant notation, whereas for an

$$\Psi_{cs} = |\phi_1 \bar{\phi}_1 \cdots\cdots \phi_N \bar{\phi}_N| \tag{4}$$

open shell (subscript os) (radical) containing ($2N + 1$) electrons the wave function is given by equation (5) where the ϕ_i and $\phi_i{}'$ refer to the molecular

$$\Psi_{os} = |\phi_1 \bar{\phi}_1{}' \cdots\cdots \phi_N \bar{\phi}_N{}' \phi_{N+1}| \tag{5}$$

orbitals for α-spin and β-spin electrons. In the restricted Hartree-Fock method the ϕ_i and $\phi_i{}'$ are identical but for radicals it is often convenient to use the unrestricted method in which the ϕ_i and $\phi_i{}'$ are different, and this is especially useful when computing the spin densities in radicals.

The orbital energies and the composition of the orbitals in terms of the basis functions are available from the calculations. It is convenient, however, to analyse the orbitals in terms of the charges on the various atoms using the Mulliken population analysis (Mulliken 1955). This method is subject to a certain degree of arbitrariness in the calculation of the overlap contribution to the charges, but in general it yields reasonable qualitative estimates of the charge distribution, particularly useful in comparing charge shifts.

The extent of charge transfer (ΔQ) is assessed by computation of the total charge on the 'supermolecule' fragments (subscript 'super') compared with that on the isolated molecules (subscript 'isol'), as in equation (6) where Q_i is

$$\Delta Q = \Sigma\{(Q_i)_{\text{super}} - (Q_i)_{\text{isol}}\} \tag{6}$$

the charge on the ith atom in each case. Examination of various regions of the molecule can also indicate where the charge transfer is most important.

The analysis of the interaction energy ΔE in terms of polarization, electrostatic interactions, dispersion and charge-transfer components has been discussed previously (Morukuma 1971; Kollman 1977) and recent applications to charge-transfer complexes have been reported (Lathan *et al.* 1975; Lucchese *et al.* 1975). The calculations reported on the TTF–TCNQ complex (Cavallone & Clementi 1975) are similar to those reported in this paper. These complexes have semiconducting properties at high temperatures. However, the extent of the charge transfer is much larger than in the cases studied here.

CHARGE TRANSFER AND KETOALDEHYDES

Returning now to the problem of desaturation of the proteins, Szent-Györgyi has emphasized the possible role of ketoaldehydes as acceptors (Szent-Györgyi 1977). The evidence for the involvement of these molecules has been reviewed elsewhere (Szent-Györgyi 1976), and indeed the recent work has shown that methylglyoxal is detectable in liver (G. Fodor, unpublished work 1978), but it is still not definitely established if one or several ketoaldehydes are involved—that is, whether there is a unique electron-acceptor molecule in these situations. In this light, we have studied the charge transfer to several ketoaldehydes using as a model donor the formamide molecule. We hoped to determine whether substitution in these molecules made any significant difference to the extent of charge transfer.

QUANTUM CHEMICAL CALCULATIONS OF CHARGE TRANSFER BETWEEN FORMAMIDE AND VARIOUS KETOALDEHYDES

We here describe *ab initio* SCF-MO calculations made using the Gaussian 70 system of programs (Hehre *et al.* 1970), modified to improve the SCF convergence. The calculations were run on the University of St. Andrews IBM 360/44 and on the University of Cambridge IBM 370/165.

We have used the experimental geometry of the donor, formamide (Costain & Dowling 1960). The geometries of the ketoaldehydes studied, methylglyoxal, ethylglyoxal and dimethylglyoxal, were obtained by geometry optimization of bond lengths and bond angles using both the semi-empirical PCILO method (Diner *et al.* 1968) and also Gaussian 70 for methylglyoxal. All distances are quoted in Å (0.1 nm), and energy values in atomic units (hartree).

The dimer conformations studied were of two types: first, the in-plane hydrogen-bonded conformations and, secondly, the stacked form in which the two molecules are in parallel planes separated by an interplane distance R'. The conformations studied are described more fully under the results for the individual molecules.

No attempt was made to optimize exhaustively the relative molecular positions, but it is believed that the conformations studied should be those most likely to result in appreciable charge transfer.

We should point out that glyoxal and methylglyoxal exist in aqueous solution in equilibrium with the hydrated forms (Schauenstein *et al.* 1977) and also with the relevant enol forms of the ketones. Calculations on the enol forms will be reported elsewhere (J.R. Ball & C. Thomson, unpublished work).

It is expected that the ketoaldehydes will react with the terminal NH_2 groups of lysine in proteins to give Schiff bases which are in favourable geometrical positions to accept electrons from the protein backbone (Otto *et al.* 1978). However, these equilibria readily result in the free ketoaldehyde being regenerated. Although calculations on the acceptor properties of the Schiff bases are being made (S. Suhai & J. Ladik, unpublished work), it is expected that calculations on the relative acceptor properties of the different ketoaldehydes should give an indication of which, if any, of these would be the best electron acceptors.

RESULTS OF KETOALDEHYDE/FORMAMIDE CALCULATIONS

Methylglyoxal/formamide

As the geometry of methylglyoxal is not known experimentally, we first optimized the geometry using the STO-3G basis set with the program Gaussian 70, keeping the C—H bond lengths fixed at 1.09 Å. The computed geometry is depicted in Fig. 2.

We should point out in passing that all calculations refer to the *trans* forms of the aldehydes. For glyoxal itself the *trans* form is predicted theoretically (Dykstra & Schaefer 1975) to be more stable, in agreement with experiment (Paldus & Ramsay 1967). The same is true for methylglyoxal, where the *trans* form is calculated to be more stable by 10 kJ mol^{-1}.

The computed values of the total energy of methylglyoxal and of formamide, together with the energies of the highest occupied (HOMO) and lowest unoccupied molecular orbitals (LUMO), are given in Table 1.

Fig. 3 shows the computed charge distributions of methylglyoxal and formamide for the above geometries and in Fig. 4 we present the atomic charges for the different 'supermolecule' configurations we have studied, the values listed referring to our minimum energy configuration.* We have varied the distance $O_4 \ldots H_{10}$ in the planar conformations and the distance R' between the molecular planes for the stacked conformation D. Table 2 gives the total energies for the distances corresponding to the energy minimum as R or R' is varied. It should be emphasized that this minimization is done for

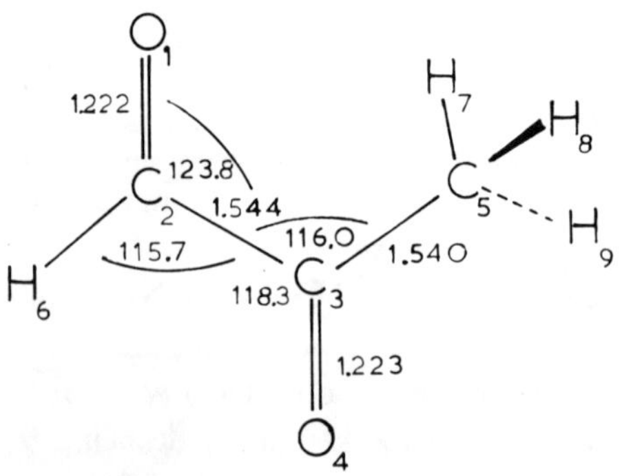

FIG. 2. Computed optimized geometry of methylglyoxal, assuming R(C—H) = 1.09 Å and a tetrahedral CH_3 group.

*The charges are quoted for configurations C, B and D. Those for configuration A were very similar to those for C.

TABLE 1

Total energy and energies of the highest occupied (HOMO) and lowest unoccupied molecular orbitals (LUMO) for the isolated molecules

Molecule	*Energy, E_T*[a]	ϵ(HOMO)	ϵ(LUMO)
Methylglyoxal	−262.1714	−0.314	+0.205
Formamide	−166.6775	−0.307	+0.343
Ethylglyoxal	−300.7188	−0.333	+0.177
Dimethylglyoxal	−300.7615	−0.302	+0.211

[a]All energy values in hartree.

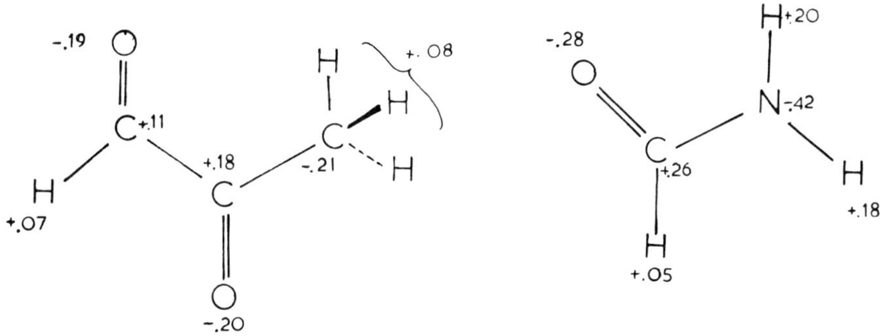

FIG. 3. STO-3G charge distribution in methylglyoxal and formamide.

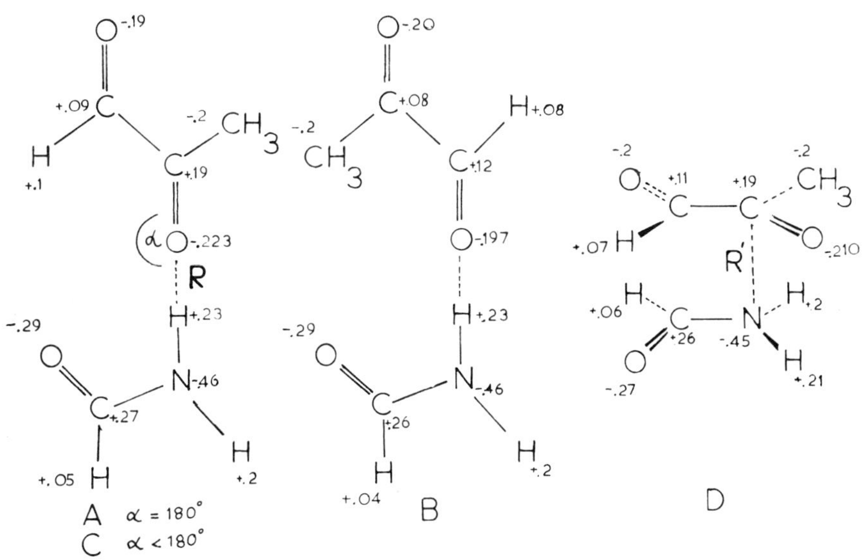

FIG. 4. 'Supermolecule' configurations and charges on the atoms for the methylglyoxal/formamide system at the configurations of Table 2.

TABLE 2

Comparison of total energy, energies of the HOMO and LUMO, the interaction energy ΔE and the net charge Q on the ketoaldehyde for the lowest energy 'supermolecule' configurations (see Fig. 4)

	Methylglyoxal			
	A	*B*	*C*	*D*[a]
Energy	−428.8607	−428.8597	−428.8651	−428.8552
HOMO	−0.295	−0.298	−0.304	−0.313
LUMO	+0.195	+0.196	+0.201	+0.205
ΔE	−0.012	−0.011	− 0.016	−0.006
Q_{KA}	+0.19	+0.017	+0.012	−0.001
	Ethylglyoxal			
	A	*B*	*C*	*D*[a]
Energy	−467.4081	−467.4082	−467.4101	−467.4206
HOMO	−0.295	−0.297	−0.297	−0.317
LUMO	+0.168	+0.169	+0.170	+0.177
ΔE	−0.012	−0.012	−0.014	−0.024
Q_{KA}	+0.019	+0.017	+0.022	−0.001
	Dimethylglyoxal			
	A		*C*	*D*[a]
Energy	−467.4510		−467.4504	−467.4438
HOMO	−0.296		−0.297	−0.301
LUMO	+0.202		+0.202	+0.211
ΔE	−0.012		−0.011	−0.005
Q_{KA}	+0.018		+0.018	−0.002

[a]Stacked form at $R = 3.2$ Å.
Energy values are in hartree.
Charges are in units of electronic charge (e).

consistency: it does not necessarily correspond to any physical configuration but enables us to compare the series of molecules in a well-defined manner. Table 2 also lists the energies of the HOMO and LUMO, the net charge on the ketoaldehyde and the interaction energy ΔE. The extent of the charge transfer is very small but, more important, the direction is very sensitive to the precise configuration. For configuration C, small variations in the distance between the molecules R and the C_3—O_4—H_{10} angle change the total energy about 0.005 hartree and also change the sign of the charge transfer. For the lowest energy configuration, the transfer is from methylglyoxal to formamide except for the stacked configuration D, where it is to the ketoaldehyde. The magnitude of the charge transfer is less than that found in calculations on glyoxal/formamide (Otto *et al.* 1977), but is not very different from the experimentally determined charge transferred from casein to methylglyoxal (Pethig 1978).

It is not clear whether the difference between our results and those of Otto *et al.* is due to the use of standard geometries in their case (Otto *et al.* 1977). The use of the optimized geometries of the ketoaldehydes throughout our work should render comparisons between the results meaningful but, since the charge transfer is so small and only to the methylglyoxal in the stacked case, it is evident that effects of solvent and other extraneous factors *in vivo* could make a large difference to the absolute value of the charge transfer.

Dimethylglyoxal/formamide

Dimethylglyoxal (biacetyl) is a diketone, not a ketoaldehyde, and we were interested to see how it compared with glyoxal and methylglyoxal in the magnitude and direction of the charge transfer.

The geometry adopted was essentially the optimized geometry of methylglyoxal, except for the replacement of H_6 by the second CH_3 group. (Geometry optimization calculations on this molecule would involve expensive computer time and were not considered worthwhile.)

For comparison with the methylglyoxal/formamide calculations, we studied the equivalent configurations A and C and the stacked form at the same distance R'. The results are given in Table 2. It is evident that the charge transfer is similar to that found for methylglyoxal, in the same direction and only to the diketone in the stacked form D. (Details of the individual atomic charges are not given here to save space, but can be obtained from the authors.)

Ethylglyoxal/formamide

One reason for studying the ethyl derivative as part of this series was the fact that various substituted derivatives have been used in cancer chemotherapy, particularly the ethoxy-substituted derivative KETOXAL, 3-ethoxy-2-oxobutanal (Schauenstein *et al.* 1977). We were interested therefore, to investigate the electron-acceptor properties of the substituted ethyl derivatives and, as a first step in this work, we report here our results for ethylglyoxal/formamide.

Of the several possible configurations, we restrict our discussion to the analogues of the configurations A to D for methylglyoxal, since explicit calculations show these to be lower energy configurations than the other alternatives. For purposes of comparison we have again varied the distances R and R' and used the PCILO-optimized geometry for ethylglyoxal. Table 2 lists the relevant quantities from the supermolecule calculations. Although the LUMO value for the isolated molecule seems to indicate that it is possibly a better electron acceptor than the other two molecules, the calculations show that this is not the case, the results showing a close parallel to those of methylglyoxal and dimethylglyoxal.

CHARGE TRANSFER AND ASCORBIC ACID

One of us (C.T.) has proposed (unpublished) the possibility that the ascorbic acid molecule or one of its metabolic derivatives is involved in the charge-transfer processes. Subsequent e.s.r. experiments (P. Gascoyne, unpublished work) showed that ascorbic acid enhances the e.s.r. signal observed when methylglyoxal interacts with amines (and proteins). The early idea stemmed from the work of Cameron & Pauling (1974) who had suggested the use of ascorbic acid in cancer therapy for different reasons. The results they obtained suggested a possible link between Szent-Györgyi's theory and ascorbate therapy, and we have undertaken a programme to study theoretically this important molecule, the ion and free radicals derived from it, and their interaction with other biologically important molecules.

We have studied the interaction between formamide and ascorbic acid, and between glyoxal and ascorbic acid, using the same basis sets as in the ketoaldehyde studies. Our intention was to see whether this molecule acted as an electron donor or an electron acceptor. We have also examined in some detail the structure of the various possible free radicals derived from ascorbic acid, and these results will be reported elsewhere. There has been one previous *ab initio* study of the ascorbic acid molecule (Carlson *et al.* 1976). The HOMO and LUMO for this molecule at the experimental geometry are given in Table 3. Comparison with Table 1 indicates that ascorbic acid might act as a donor towards the ketoaldehydes, but could be an acceptor from formamide.

We realized that we could save a considerable amount of computer time by studying the model compound α-hydroxytetronic acid (3-hydroxytetrahydrofuran-2,4-dione) (Fig. 5)—that is, ascorbic acid without the side-chain. Calculation of the wave function of this molecule gave almost identical charge distribution and energy levels to those for ascorbic acid itself (Table 3) and in

TABLE 3

Total energy, HOMO and LUMO for ascorbic acid and the ascorbic acid model, α-hydroxytetronic acid[a]

Molecule	*Energy, E_T*	*ε(HOMO)*	*ε(LUMO)*
Ascorbic acid	−672.0857	−0.250	+0.234
Ascorbic acid model[b] (α-hydroxytetronic acid)	−447.27493	−0.248	+0.240

[a]3-Hydroxytetrahydrofuran-2,4-dione.
[b]Computed at experimental geometry of ascorbic acid.
All energy values in hartree.

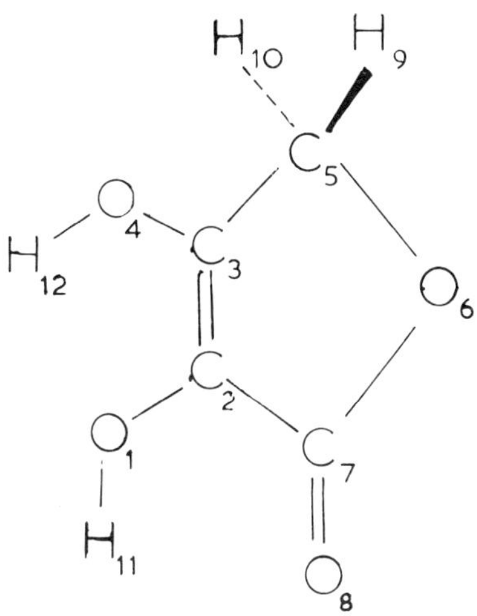

FIG. 5. α-Hydroxytetronic acid (3-hydroxytetrahydrofuran-2,4-dione, shown here in its enol form): a model for ascorbic acid.

all subsequent supermolecule calculations we used this molecule as an ascorbic acid model.

The first question to be answered is whether the model molecule acts as an electron donor or as an electron acceptor. To investigate this question, we have made *ab initio* ‘supermolecule’ calculations of the energy and charge distribution in the model/formamide system and in the model/glyoxal system, the latter since the methylglyoxal calculations were computationally too expensive to do at present. We have investigated only a limited number of possible configurations, in particular the in-plane hydrogen-bonded structures and the stacked forms. These are depicted in Figs. 6 and 7.

α-Hydroxytetronic acid/formamide

We used throughout these calculations the experimental geometry of the ascorbic acid ring (Hvoslef 1968) and of formamide. The in-plane hydrogen-bonded configuration with two hydrogen bonds was studied at several distances both with linear $H_{13}...O_8=C_7$, and also bent. The lowest energy conformation with an energy of -613.97289 hartree at an angle $\alpha = 135°$ showed a transfer of charge to the furan ring of the model of about 0.01 (Table 4). Hence in this configuration the molecule can accept electrons. The interaction energy ΔE was 0.020 hartree.

Several possible stacking configurations of the stacked form could be visualized. However, the configuration with the formamide nitrogen under C_7 and the C—N bond parallel to the C_2—C_7 bond in the model was adopted.

For comparative purposes we adopted an interplane distance of 2.8 Å. In this configuration there was a much smaller charge transfer, but it was still from the formamide to the tetronic acid, as in the in-plane case. As in the ketoaldehyde calculations, the charge transfer is very sensitive to the precise geometrical configuration.

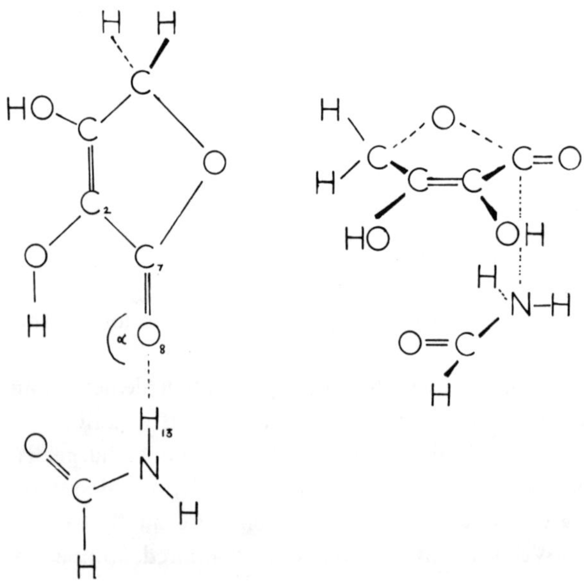

FIG. 6. 'Supermolecule' configurations studied for the ascorbic acid model/formamide system.

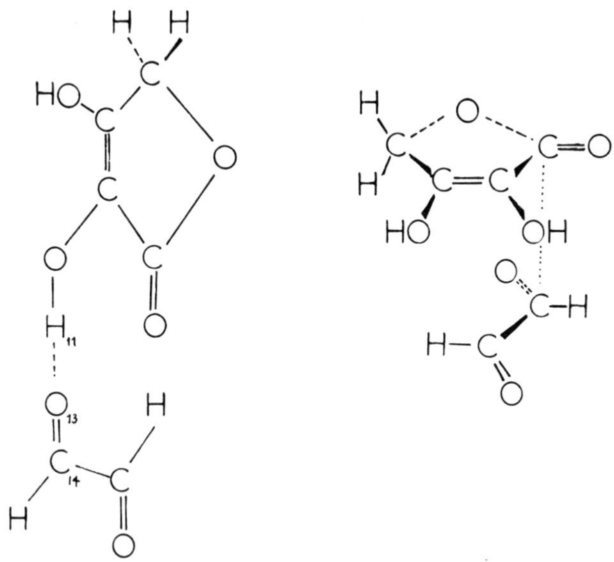

FIG. 7. 'Supermolecule' configurations studied for the ascorbic acid model/glyoxal system.

α-Hydroxytetronic acid/glyoxal

We expected that the use of glyoxal instead of methylglyoxal (for the reason given above) would not affect our conclusions, since previous work (C. Thomson, unpublished work; J. Ladik, unpublished work; Otto *et al.* 1977) has shown that glyoxal and methylglyoxal show similar charge-transfer behaviour, and this is also in accord with the calculations on the other keto-aldehydes reported here.

TABLE 4

Total energy, HOMO and LUMO, the interaction energy ΔE and the charges on the ascorbic acid model (α-hydroxytetronic acid) and formamide or glyoxal

Complex	E	ϵ_{HOMO}	ϵ_{LUMO}	Q_F	$Q_{(Glyoxal)}$	ΔE
Model/HCONH$_2$						
In plane	−613.97289	−0.240	+0.237	+0.010		−0.020
Stacked[a]	−613.95285	−0.241	+0.240	+0.003		−0.0004
Model/glyoxal						
In plane	−670.86216	−0.242	+0.197		+0.008	−0.024
Stacked[a]	−670.82799	−0.254	+0.194		−0.008	+0.011

[a]At 2.8 Å distance between planes.
Energy values in hartree.
Charges are in units of electronic charge (e).

Calculations were made for only two conformations, one in-plane, and the other stacked (Fig. 7). We varied the interplane distance and also R in the in-plane studies, but in only a limited way. The results of the calculations are given in Table 4. The lowest energy configuration at an angle $C_{14}...O_{13}...H_{11}$ of 167.5° at a distance $R(H_{11}-C_{13})$ = 2.14 Å has a substantial charge transfer to the model molecule from the glyoxal (about +0.008 e). However, in the stacked form, the ascorbic acid model acts as a donor. Thus in this case the direction of the transfer is once again very dependent on the precise geometry.

CONCLUSIONS

The purpose of our calculations was two-fold. First, using a consistent geometry for the ketoaldehydes and the *ab initio* STO-3G method, we wanted to compare the donor/acceptor properties of the different ketoaldehydes with respect to formamide.

We conclude that there is only a very small transfer to the ketoaldehyde in the stacked forms only and that it is very sensitive to the precise geometry. These calculations emphasize that it is not possible in general to draw conclusions about the extent and direction of charge transfer from calculations in which the geometrical parameters of the isolated molecule are kept fixed at standard values. At least some variation of these is necessary, although it is doubtful if full optimizations are worthwhile, for the reasons given previously (Otto *et al.* 1977). The calculations do not include possible effects of solvent or ions on the charge transfer and in future work these should be studied. They do not, however, lead us to expect that the different ketoaldehydes will differ much in their donor/acceptor properties.

The second aim was to establish the possible charge-transfer behaviour of ascorbic acid. With the model compound α-hydroxytetronic acid, the calculations show that this molecule should behave as an electron acceptor with formamide but could be either a donor or an acceptor with glyoxal.

The involvement of ascorbic acid itself, however, is questionable, since it is readily ionized and also it is easily oxidized to the ascorbic acid free radical. We have, therefore, begun a more detailed study of the properties and charge-transfer interactions involving the ascorbate free radical and the ascorbate ion, and the early results reported here should therefore be regarded as preliminary. Nevertheless, we believe that the role of ascorbate in the charge-transfer phenomena (Szent-Györgyi 1978) and in cancer treatment (Cameron & Pauling 1974) is based on firm ground, and that these calculations do show the possibility of electron-transfer processes involving this molecule.

Clearly, much work needs to be done on this problem, both theoretical and experimental. We believe, however, that quantum chemists can and must contribute to this area of research, since in the last analysis the detailed molecular and intermolecular interactions which biologically significant molecules undergo are describable in detail only by the methods of quantum mechanics, and it is now feasible to apply such methods to the study of these important problems.

ACKNOWLEDGEMENTS

We thank Mr F.C. Salisbury and the NFCR for financial support, and especially Dr Szent-Györgyi for stimulating our interest in this problem and for useful discussions and correspondence. We thank Drs Ladik and Abdulnur for preprints of their work and for useful discussions. Finally, we thank the operators and staff of the Computing Laboratories at the Universities of St. Andrews and Cambridge for the large amount of computer time allocated to this project, and for their help in running the calculations.

References

ABDULNUR, S.F. (1976) The interaction of glyoxals with proteins and DNA in relation to cancer. *Int. J. Quantum Chem. Quantum Biol. Symp. 3*, 59-64

ABDULNUR, S.F. (1977) The interaction of methylglyoxal with methylamine and the triplet state. *Int. J. Quantum Chem. Quantum Biol. Symp. 4*, 217-223

BONE, S., LEWIS, T.J., PETHIG, R. & SZENT-GYÖRGYI, A. (1978) Electronic properties of some protein–methylglyoxal complexes. *Proc. Natl. Acad. Sci. U.S.A. 75*, 315-318

CAMERON, E. & PAULING, L. (1974) The orthomolecular treatment of cancer: I. The role of ascorbic acid in host resistance. *Chem.-Biol. Interact. 9*, 273-283

CARLSON, G.L., CABLE, H. & PEDERSEN, L.G. (1976) An ab-initio study of ascorbic acid. *Chem. Phys. Lett. 38*, 75-78

CAVALLONE, F. & CLEMENTI, E. (1975) Electronic structure of TTF–TCNQ complex. *J. Chem. Phys. 63*, 4304-4307

CONSTAIN, C.C. & DOWLING, J.M. (1960) Microwave spectrum and molecular structure of formamide. *J. Chem. Phys. 32,* 158-165

DINER, S., MALRIEU, J.P., JORDAN, F. & GILBERT, M. (1968) Fully localized bond orbitals and the correlation problem. *Chem. Phys. Lett. 2*, 319-323

DYKSTRA, C.E. & SCHAEFER, H.F. III. (1975) Electronic structure of dicarbonyls. The ground state of glyoxal. *J. Am. Chem. Soc. 97*, 7210-7215

HEHRE, W.J., LATHAN, W.A., DITCHFIELD, R., NEWTON, M.D. & POPLE, J.A. (1970) Gaussian 70. *Program No. 236, Quantum Chemistry Program Exchange*, University of Indiana, Bloomington

HVOSLEF, J. (1968) The crystal structure of L-ascorbic acid, 'vitamin C'. I. The X-ray analysis. *Acta Crystallogr. B24*, 23-35

KOLLMAN, P. (1977) Hydrogen bonding and donor-acceptor interactions, in *Modern Theoretical Chemistry*, vol. 4 (Schaefer H.F. III., ed.), pp. 109-152, Plenum Press, New York

KIMURA, J.E. & SZENT-GYÖRGYI, A. (1969) Donor-acceptor interactions of nitrogen. *Proc. Natl. Acad. Sci. U.S.A. 62*, 286-288

KON, H. & SZENT-GYÖRGYI, A. (1973) Charge transfer between amine and carbonyl. *Proc. Natl. Acad. Sci. U.S.A. 70*, 3139-3140

LAKI, K. & LADIK, J. (1976) A note on the 'electronic theory' of cancer. *Int. J. Quantum Chem. Quantum Biol. Symp. 3*, 51-57

LADIK, J. (1979) Electronic structure of proteins and DNA: solid state aspects, this volume, pp. 51-58

LADIK, J., SUHAI, S. & SEEL, M. (1978) The electronic structure of biopolymers and possible mechanisms of chemical carcinogenesis. *Int. J. Quantum Chem. Quantum Biol. Symp. 5*, 35-49

LATHAN, W.A., PACK, G.R. & MORUKUMA, K. (1975) Molecular orbital studies of electron donor-acceptor complexes II. Carbonyl cyanide-benzene complex and dispersion energy contribution. *J. Am. Chem. Soc. 97*, 6624-6628

LUCCHESE, R.R., HABER, K. & SCHAEFER, H.F. III. (1975) Charge transfer complexes NH_3-O_3, NH_3-SO_2 and $N(CH_3)_3-SO_2$. *J. Am. Chem. Soc. 97*, 7617-7620

MORUKUMA, K. (1971) Molecular orbital studies of hydrogen bonds. III.C=O....H–O hydrogen bond in $H_2CO...H_2O$ and $H_2CO...2H_2O$. *J. Chem. Phys. 55*, 1236-1244

MULLIKEN, R.S. (1950) Structure of complexes formed by halogen molecules with aromatic and with oxygenated solvents. *J. Am. Chem. Soc. 72*, 600-608

MULLIKEN, R.S. (1955) Electronic population analysis on LCAO-MO wave functions. *J. Chem. Phys. 23*, 1833-2346

OTTO, P., SUHAI, S. & LADIK, J. (1977) Ab-initio supermolecule study of charge transfer in the glyoxal-formamide and in the H_2S-formamide systems. *Int. J. Quantum Chem. Quantum Biol. Symp. 4*, 451-457

OTTO, P., LADIK, J., LAKI, K. & SZENT-GYÖRGYI, A. (1978) Internal charge transfer in proteins to the Schiff bases of their lysine side chains. *Proc. Natl. Acad. Sci. U.S.A. 75*, 3548-3550

PALDUS, J. & RAMSAY, D.A. (1967) The 4550 Å band system of glyoxal. I. Rotational analyses of the (O.O) bands for $C_2H_2O_2$, C_2HDO_2 and $C_2D_2O_2$. *Can. J. Phys. 45*, 1389-1412

PETHIG, R. (1978) Electronic properties of protein–methylglyoxal complexes – strong evidence for energy band conduction. *Int. J. Quantum Chem. Quantum Biol. Symp. 5*, 159-171

PETHIG, R. & SZENT-GYÖRGYI, A. (1977) Electronic properties of the casein–methylglyoxal complex. *Proc. Natl. Acad. Sci. U.S.A. 74*, 226-228

POPLE, J.A. (1977) *A priori* geometry predictions, in *Modern Theoretical Chemistry*, vol. 4 (Schaefer, H.F. III., ed.), pp. 1-27, Plenum Press, New York

SCHAEFER, H.F. III. (1973) *Electronic Structure of Atoms and Molecules*, Addison-Wesley, New York

SCHAUENSTEIN, E., ESTERBAUER, H. & ZOLLNER, H. (1977) *Aldehydes in Biological Systems* (Gore, P.H., transl.), Pion/Academic, London

SLIFKIN, M.A. (1971) *Charge Transfer Interactions of Biomolecules*, Academic Press, New York

SZENT-GYÖRGYI, A. (1941) The study of energy levels in biochemistry. *Nature (Lond.) 148*, 157-159

SZENT-GYÖRGYI, A. (1960) *Introduction to a Submolecular Biology*, Academic Press, New York

SZENT-GYÖRGYI, A. (1968) *Bioelectronics*, Academic Press, New York

SZENT-GYÖRGYI, A. (1972) *The Living State*, Academic Press, New York

SZENT-GYÖRGYI, A. (1976) *Electronic Biology and Cancer*, Dekker, New York

SZENT-GYÖRGYI, A. (1977) The living state and cancer. *Proc. Natl. Acad. Sci. U.S.A. 74*, 2844-2847

SZENT-GYÖRGYI, A. (1978) *The Living State and Some Observations on Cancer*, Dekker, New York

Discussion

Eley: The *ab initio* method introduced some years ago by S.F. Boys represented a watershed in quantum chemistry. Many calculations, however, had been made before then by such methods as Hückel's or MINDO/2. Must we now neglect the conclusions of such studies?

Thomson: Even though *ab initio* calculations of SCF wave functions are expensive compared to semi-empirical calculations, one has a well-defined benchmark, namely the Hartree–Fock wave function. In the semi-empirical calculations—which may be extremely useful for comparing a series of molecules of a particular type—one always has to use approximations of some kind, to simplify the calculations, and/or introduce various parameters. Consequently, if one finds serious discrepancies between experimental results and calculations one can never be sure whether aproximations in the method or the particular choice of parameters provide the reason for the discrepancy. Hückel theory is a case in point: it should have been dead and buried long ago, because although Hückel designed it for calculations on hydrocarbon systems, it has been extensively used for much more complicated systems. As soon as one introduces other elements besides carbon and hydrogen, one has to cope with many more parameters which have to be derived by fitting to selected experimental data of one kind or another. It is easy with many parameters to obtain agreement with experiment but it is doubtful if one learns anything more about the physical situation this way. As calculations using less approximate methods these days take about the same time as Hückel calculations, the use of the latter in my opinion is no longer warranted.

Secondly, many of these early methods treated only π-electrons. We now know from *ab initio* calculations that the π- and σ-levels are mixed: it is not a valid approximation to neglect the σ-electrons and treat only π-electrons. Semi-empirical methods that include σ- and π-electrons have been developed, such as the CNDO, INDO, and MINDO methods, and Dr Abdulnur has done much of the earlier work on glyoxal/formamide interactions with CNDO calculations. It seems to me, however, that it is preferable nowadays to use methods which are approximate but whose limitations are well-defined and appreciated from extensive previous work. Then one has a better idea of how to tackle any problems that arise.

Eley: Are all your assumptions explicit? In the course of time can they be refined, given sufficient computer time?

Thomson: Yes, our *ab initio* calculations have as input the molecular geometry but the SCF wave function is computed in terms of the fundamental physical constants: there are no other experimental parameters. For beyond SCF calculations, the same applies, but we have one important limitation: we do not yet understand the relativistic problem well. But that should not worry us until we do *accurate* calculations on compounds containing phosphorus, sulphur and especially elements of the third row and up.

Ladik: Much work is now being done not only on relativistic Hartree–Fock calculations (which have been completed for all atoms) but also on the relativistic interaction operators. In my opinion, this will be the central problem rather than correlation for the next 10 years.

March: One of the difficulties facing us with big molecular systems is that however we solve the wave equation, we always derive the electron density in the language of delocalized electron clouds. As the molecules under consideration become larger, we cannot continue to do that. Consequently we need a model, but inevitably that model brings us back to localized building fragments of the delocalized electron distribution. Can calculations such as your *ab initio* ones be reinterpreted usefully in terms of these localized fragments in such a way that we eventually avoid the necessity of solving the wave equation?

Thomson: Not with the present way of doing SCF calculations, because one has to calculate the wave function first. If we could determine the density functional directly, we might be in a position to avoid solving the wave equation. for the larger molecules, however, the biggest problem with these molecular systems is that the time for this kind of calculation increases as the fourth power of the number of basis functions. Thus we have an immensely complicated data-handling problem for large systems. If we could avoid this N^4 problem, maybe using some of the localized features of the electron density, it would be a major achievement. At present, however, to investigate interesting large molecules such as haemoglobin, or the porphyrin ring fragments, is exceedingly time-consuming, and in my opinion we shall not make much progress until we have different methods for these calculations.

Ladik: We have also been studying charge transfer. We found that the amount of transferred charge in a supermolecule calculation depends strongly on the basis set; for example, introduction of *d* functions doubles the amount of charge transferred. It depends on correlation also. There is a still more difficult conceptual problem, namely Mulliken's population analysis, which is a poor tool. We and others are trying to plot the whole density and integrate over the subspaces of the different molecules. The problem there is how to divide the space. Sometimes I am inclined to believe that charge transfer is an ill-defined quantity and perhaps it is impossible to define it with a single number. Instead of this, the charges in the density distributions of the molecules have to be considered.

Slater: You have told us, Dr Thomson, that one of your intentions in these studies was to stimulate the experimentalists. However, experiment has shown that the side-chain of vitamin C can undergo enzymic oxidation–reduction (Tolbert *et al.* 1976). I understand the practical computational reasons for using the simpler α-hydroxytetronic acid as a model but won't you miss something important by not including the side-chain in your theoretical considerations?

Thomson: I don't think we shall, so far as the electronic properties of the molecule are concerned, because the energy levels associated with the saturated side-chain are much lower in energy than those associated with the ring system. The top three orbitals are associated with the enone part of the molecule (see Fig. 5, p. 155; $C_3 = C_2 - C_7 = O_8$). Introduction of the side-chain should not make much difference to the donor/acceptor properties of the molecule—although there may be other differences with respect to its chemical behaviour, solubility, etc. May I turn the question round? What are the biological effects of α-hydroxytetronic acid? Are they very different to those of ascorbic acid? Apart from a report in the 1930s that it is not anti-scorbutic, nothing seems to be known (Haynes & Plimmer 1960). Our calculations suggest that it would be useful as an acceptor. I would be surprised if it did not enhance the e.s.r. signal in the experiment done by Professor Szent-Györgyi and Peter Gascoyne, referred to in my paper; namely methylamine + methylglyoxal. It is significant that our theoretical calculations point to a molecule that would not be an obvious choice for biologists, and which should be looked at for its biological effects. This illustrates the value of applying these quantum chemical techniques to biological systems.

Slater: One reason for my question was that provisional data indicate that the side-chain of ascorbic acid shows some very interesting biological reactivity. Lipid peroxidation in rat liver microsomes can be stimulated by ascorbate over the concentration range 5–500 μmol/l in the presence of Fe^{2+}. We have found that the methylglyoxal–ascorbate complex behaves identically with ascorbate on a molar basis in this reaction (P. Brunswick & T.F. Slater, unpublished data). If the complex does not dissociate rapidly at neutral pH in aqueous solution, then the peroxidation data suggest an involvement of the ascorbate side-chain alcohol group in the redox reactions associated with microsomal peroxidation.

Swartz: The values for the charge transfer between the model compound and glyoxal (Table 4, p. 157) are similar, yet you claim that the model is a donor in the stacked configuration and an acceptor in the in-plane configuration.

Thomson: As with the ketoaldehydes, the charge transfer is sensitive to the precise geometry. We have not investigated other possible stacked or in-plane configurations; we have not done enough calculations to be able to generalize. However, the value of -0.008 e for the stacked complex is substantial (as with the ketoaldehyde); in the in-plane form the ketoaldehydes are not acting as acceptors. Another point is that this system is probably not the most significant one *in vivo*. As Professor Szent-Györgyi pointed out, the ascorbate *radical* is probably the active species and, according to some preliminary unpublished calculations, the transfer from the model ascorbate *radical* to glyoxal is about 10 times greater than for the neutral model. We have, however, to investigate the role of the radical in more detail, and this work is under way.

Swartz: Does that relate to the known chemistry of ascorbate and the ascorbate radical? Data on reaction rates (which are a different parameter but perhaps related) indicate that the ascorbate radical anion is a relatively unreactive molecule.

Thomson: I haven't really looked into this.

Purcell: To take up Professor Slater's point in that regard, I must stress that the side-chain is important: ascorbic acid and α-hydroxytetronic acid react differently biochemically. With regard to the quantum mechanics, your answer that the energy levels are the same is not adequate. As you said, the properties of a molecule can be described if the Schrödinger equation is solved rigorously. But since you cannot account for the differences you must have the wrong basis set, inadequate calculations, or some other problem.

Thomson: I do not accept this. If we are discussing the behaviour of the ascorbic acid molecule or ion in a conventional chemical reaction, it is not the same as describing the donor/acceptor properties in Professor Szent-Györgyi's sense. The spectroscopic behaviour, for instance, is determined by the ring system. In talking about small shifts in electron density, these are primarily determined by the ring system.

References

HAYNES, L.J. & PLIMMER, J.R. (1960) Tetronic acids. *Q. Rev. Chem. Soc. Lond. 14,* 292-315

TOLBERT, B.M., HARKRADER, R.J., JOHNSON, D.O. & JOYCE, B.A. (1976) C-6 oxidation of ascorbic acid: a major metabolic process in animals. *Biochem. Biophys. Res. Commun. 71,* 1004-1009

The search for new cancerostatic agents

GABOR FODOR, RATNAKER MUJUMDAR and JOHN BUTTERICK

Regional Laboratory of the National Foundation for Cancer Research at West Virginia University, Morgantown, West Virginia

Abstract Following the lead given by Albert Szent-Györgyi's bioelectronic theory of cancer, work was continued in two major directions: (*i*) designing new electrophilic molecules, related to methylglyoxal, and (*ii*) using L-ascorbic acid as a (non-toxic) carrier for methylglyoxal and its derivatives in the form of its acetals. The vinylogue of methylglyoxal, 4-oxopent-2-enal, was expected to be a most reactive electron acceptor, on the basis of quantum mechanical calculations by J.J. Ladik's group. A new reaction, the formation of the ene-2, 3-diol acetal and hemiacetal-hemiketal, was found to occur with 'conjugated' aldehydes, such as methylglyoxal, glyoxal, phenylglyoxal, malealdehyde and acrylaldehyde; the reaction proceeded very smoothly with 4-oxopent-2-enal. The structural determination of these new types of acetals by ^{1}H and ^{13}C n.m.r. spectroscopy and by chemical methods is discussed.

The biochemistry of methylglyoxal has been brought to the forefront of research by Albert Szent-Györgyi's bioelectronic theory of cancer (Szent-Györgyi 1976). Glyoxalase, however, a ubiquitous enzyme which decomposes methylglyoxal, has made direct biological screening for methylglyoxal exceedingly difficult. In addition, Szent-Györgyi, the discoverer of vitamin C, has recently found a new role for ascorbic acid—namely, the stimulation of charge transfer between methylglyoxal and methylamine (see Szent-Györgyi, this volume, pp. 3-18). For this reason, we have tried to use ascorbic acid as a carrier for methylglyoxal.

The first experiments, in which we attempted to produce an acetal of these two compounds, were done under strictly anhydrous conditions in the presence of a Lewis acid by a method analogous to the preparation of other 5,6-acetals and ketals of vitamin C (Vargha 1932; Bond *et al.* 1972; Chu & Slaunwhite 1968). We used boron trifluoride etherate as a catalyst and obtained a multitude of products. Surprisingly, however, in a nitrogen atmosphere a 10% aqueous solution of ascorbic acid reacted with added methylglyoxal, whether the latter was previously dissolved or not. During the course of the reaction the potential of ascorbic acid to reduce iodine decreased from 100% to 5–10%. Cellulose powder chromatography in ethyl acetate solution eliminated most of the excess (partly polymeric) of methylglyoxal and ascorbic acid.

This monitoring by iodine pointed to the fact that the dienol grouping had been blocked by methylglyoxal, in sharp contrast with all other acetals of vitamin C. Elemental analysis of the vacuum-dried product was congruent with the empirical formula $C_9H_{12}O_8$, the hydrate of methylglyoxalidene-ascorbic acid (structure 1). ^{1}H n.m.r. spectra (both 60 MHz and 250 MHz)

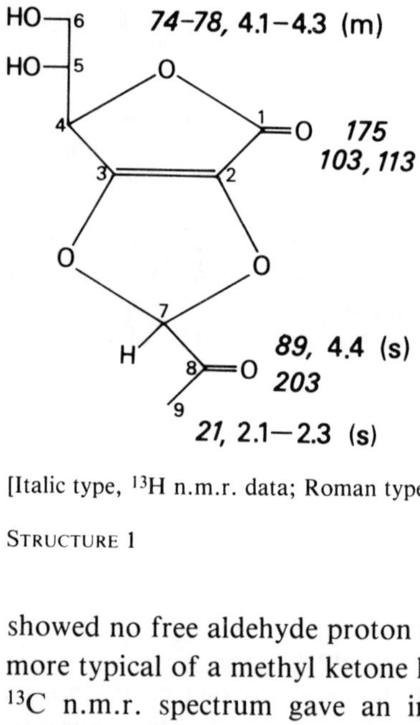

[Italic type, ^{13}H n.m.r. data; Roman type, ^{1}H n.m.r. data.]

STRUCTURE 1

showed no free aldehyde proton and a complex pattern in the methyl region, more typical of a methyl ketone hydrate than of a 'free' methyl ketone. The ^{13}C n.m.r. spectrum gave an indication of the intact, though somewhat shifted, ascorbic acid carbons. It also gave a complex pattern in the methyl region, at higher field than expected from a methyl ketone. Basically, it seemed that besides some polymeric material that formed from methylglyoxal, three sharp methyl signals were present. A broad signal for an acetal carbon was indicative of the character of the compound. The formation of a new chiral centre could account for two diastereomers with their hydrated and non-hydrated forms contributing to a complex methyl pattern in the two n.m.r. spectra. Further drying of a sample over molecular sieves in the ethyl acetate resulted in a sharp rise of the extinction coefficient of the maximum at 243 nm in the u.v. spectrum and the appearance of the deshielded methyl signals predicted for a free methyl ketone. Chemical shifts have been ascribed to the compound.

However, the complexity of the methyl region may cast some doubt on structure (1). A double hemiacetal involving both the aldehyde and ketonic functions of methylglyoxal (2) could not be ruled out. Structure (2), here a

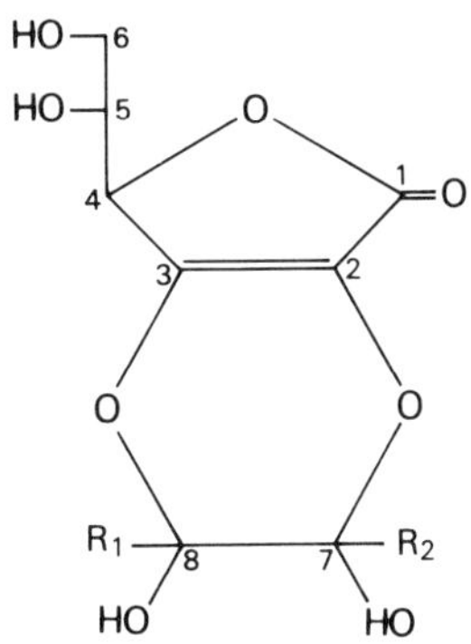

$R_1=CH_3$; $R_2=H$.

$R_1=H$; $R_2=CH_3$.

STRUCTURE 2

hemiacetal–hemiketal, was not expected to be too stable. Besides being a vicinal 5,6 and 7,8-diol it should react with two moles of sodium periodate in strictly neutral conditions, instead of the one mole that it actually consumed, according to an initial experiment by Dr Mujumdar. However, subsequent more systematic experiments by Drs Butterick and Mathelier showed that Dr Mujumdar's crude preparation consumed 3.03 equivalents of sodium periodate whereas the purified ascorbic acid-methylglyoxal product consumed 2.38 equivalents. On the basis of these findings the hemiacetal-hemiketal compounds (structure 2) were shown to predominate. Further purification of the stereoisomeric mixture of these acetals should help to clarify the problem of fine structure, necessary for establishing the relationship between structure and biological activity.

In light of the fact that in all previously described ascorbic acid acetals and ketals (Chu & Slaunwhite 1978; Vargha 1932; Bond *et al.* 1972) the hydroxyl groups at C − 5 and C − 6 were exclusively involved, the spontaneous formation of an ene-2,3-diol acetal, particularly in aqueous solution, is noteworthy.

We have tried to define the limits of the new reaction by reacting a variety of strongly electrophilic—i.e., 'conjugated'—aldehydes with ascorbic acid. 4-Oxopent-2-enal, known in pure form only as the cyclic mixed acetal (Clauson-Kaas & Lumborg 1947), 2,5-dimethoxy-2-methyl-2,5-dihydrofuran

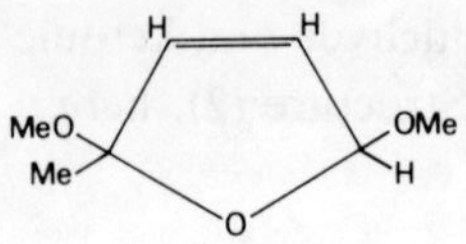

STRUCTURE 3

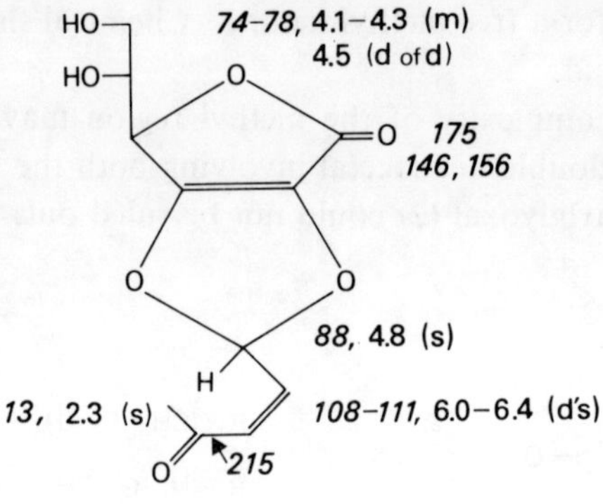

STRUCTURE 4

(3), itself prepared in one step from sylvan (methylfuran), reacted in an analogous way with ascorbic acid and the product (4) free of methoxy groups indicated a successful transacetalization process. In the high-resolution ^{1}H n.m.r. spectrum a sharp singlet appeared in the methyl ketone region, and in the acetal-hydrogen region. The decoupled ^{13}C n.m.r. spectrum clearly showed one acetal and two carbonyl carbons, a sharp singlet for the methyl carbon in the methyl ketone region, and four vinyl carbons, including the residual ascorbic acid carbon atoms. The compound is, however, at present an amorphous solid which does not allow further purification. It causes slight deviations from the calculated C, H and O values in elemental analysis.

Acrylaldehyde in aqueous emulsion also reacted with an aqueous ascorbic acid solution. In a tetrahydrofuran – water mixture in the cold a similar but slow conversion occurred, as indicated by a continuous decrease in the iodine consumption of ascorbic acid. Some polymeric material always accompanied this product, so no perfect separation has been achieved so far.

Malealdehyde cyclic acetal—i.e., 2,5-dimethoxy-2,5-dihydrofuran, the 'nor' derivative of 4-oxopent-2-enal—reacted with ascorbic acid by transacetalization. However, the reaction seemed to stop at the 1:1 product. That only one of the two carbonyls appeared to be involved was rather surprising.

Glyoxal reacted in an analogous manner. Acetalization of the first aldehyde carbonyl breaks down part of the conjugation in malealdehyde and in its entirety in glyoxal, with the electron acceptor property of the compound weakened. It cannot react subsequently with a second mole of ascorbic acid. Interestingly enough, the second aldehyde group, that is not involved in cyclization, did not show any aldehyde proton in the 8.0 p.p.m. region in the ^{1}H n.m.r. spectrum, possibly consistent with its being hydrated. Phenylglyoxal reacted more sluggishly with ascorbic acid than methylglyoxal.

Cinnamic aldehyde, benzaldehyde and acetone did not give any cyclic acetals with ascorbic acid in the same conditions (i.e. in aqueous solution), as indicated by iodine titration.

In summary, ascorbic acid gives a series of new acetals with α-ketoaldehydes and their vinylogues, and some α, β-unsaturated aldehydes, especially with carbonyl derivatives that are strong electron acceptors.

References

BOND, A.D., McCLELLAND, B.W., EINSTEIN, J.R. & FINAMORE, F.J. (1972) Ascorbic acid-2-sulfate of the brine shrimp, *Artemia salina*. *Arch. Biochem. Biophys. 153,* 207

CHU, T.M. & SLAUNWHITE, W.R., JR (1968) The nonenzymic sulfation of androsterone with L-ascorbic acid sulfate. *Steroids 12,* 310

CLAUSON-KAAS, N. & LUMBORG, F. (1947) Nuclear oxidation of sylvan and furfuryl alcohol. *Acta Chem. Scand. 1,* 619

SZENT-GYÖRGYI, A. (1976) *Electronic Biology and Cancer,* Dekker, New York

v. VARGHA, L. (1932) Monoacetone hexuronic acid. *Nature (Lond.) 130,* 847

Discussion

Slater: A propos these interesting new compounds, we have some preliminary data on the biological reactivity of the ascorbate–methylglyoxal adduct (unpublished data of T.F. Slater, M. Ahmed, P. Brunswick, R. Scott, P. Elvin, P. Cadle & J.E. Packer). Methylglyoxal itself is toxic; it has long been known to cause cardiotoxicity (see Takamatsu & Sato 1934). Methylglyoxal complexes have been used as antitumour agents in humans (see Levin *et al.* 1965) but their toxicity is, by and large, not acceptable. The adduct that Professor Fodor described had some interesting biological properties. When we injected it intraperitoneally into rats, for example, their blood pressure dropped transiently. In a less complex system, an isolated perfused rat heart preparation, the complex at low concentrations (10 μmol/l) caused a large decrease in the amplitude of the contractions of the heart and in the frequency of the contractions. Inexplicably, the effect progressively decreased when the dosing was repeated.

In free-radical systems that we have studied (the NADPH/ADP/Fe^{2+} – lipid peroxidation system in liver endoplasmic reticulum, and in pulse radiolysis experiments with experimentally generated radicals) the adduct behaved identically to ascorbic acid itself, on a mole-to-mole basis. This latter point rules out the possibility that the actions are due to small amounts of ascorbic acid being present in the free form in the adduct. In certain cell-culture systems it behaves like methylglyoxal. For example, in 3T3 cells, a mouse cell line, both methylglyoxal and its adduct inhibit incorporation of [^{3}H]thymidine into DNA at 10 μmol/l and higher concentrations (P.A. Riley, personal communication).

However, the effects of the complex on isolated rat hepatocytes are rather different from those produced by methylglyoxal; at 10 μmol/l final concentration both compounds seem to cause no morphological disturbance and appear not to disturb the cells in the resting condition, but when the cells were stimulated to go into division the cells previously exposed to methylglyoxal all died, whereas those treated with the complex remained viable. Thus, at 10 μmol/l the cytotoxicity of methylglyoxal seems to be associated with cell division (P. Foster, P.A. Riley & T.F. Slater, unpublished data).

The adduct worked well when injected intraperitoneally against ascitic tumours (as we might expect) but had no effect on two solid tumours in the mouse. Obviously, many more experiments with tumour-bearing animals are necessary before we can reach an unequivocal conclusion.

Esterbauer: This reaction with ascorbic acid is extremely relevant for those working on the antitumour activity of aldehydes. The acetal structure is particularly important because tumours are more acidic than normal cells and consequently an acetal should liberate a ketoaldehyde more readily in tumour cells than in normal cells.

We have prepared acetylacrylaldehyde (4-oxopent-2-enal) in the free form. Apart from acrylaldehyde this compound is the most active aldehyde in inhibiting DNA synthesis. The incubation concentrations necessary to produce 50% inhibition of DNA synthesis in Ehrlich ascites tumour cells are 5 μmol/l for acrylaldehyde and 13 μmol/l for oxopentenal (Kollaritsch 1973). Thus oxopentenal may have high activity against cancer cells.

Fodor: The formation of the acetal with ascorbic acid was intended to produce a carrier, to prevent the active methylglyoxal from being metabolized on its way to the cancer cell. Other such aldehydes should be equally protected. At the site it can be, and obviously is being, hydrolysed, because it is an acetal and it is in a dilute aqueous solution. We don't have kinetic data for that yet. These results do not contradict Professor Slater's. We were rather surprised that these aldehydes formed the 2,3-acetal rather than the 5,6-acetal.

Williams: Are these adducts mutagens?

Slater: This has not yet been tested.

Williams: Is glyoxal itself a mutagen?

Slater: I don't know about methylglyoxal but the related malonaldehyde has been tested (Mukai & Goldstein 1976) and is mutagenic in an Ames' test, but then almost everything is if you use high enough concentrations and optimal conditions.

Kornberg: The Ames' test has been refined now to form the basis of much industrial and medical screening of the mutagenicity and possible carcinogenicity of new chemicals, to the stage where the number of false positives or false negatives is very small.

Slater: Yes. Flippancy aside, I was thinking of CCl_4 which I believe is positive in the Ames' test but is a very weak carcinogen in normal conditions.

Williams: There is an old saying about carcinogenesis, which is attributed to Sir Alexander Haddow ('Haddow's Paradox'), that those agents which produce cancer cure cancer and *vice versa*. The reason for this is that they are reagents that attack DNA but not proteins. In attacking DNA (under conditions of DNA repair) they cause mutagenesis (the basis of the Ames' test) and probably cancer from the new cell line. At the same time they are anticancer agents because they kill most cells which they attack. That is why we need clever methods of delivery, as Professor Fodor mentioned, into one type of cell as opposed to another. It is all very well to discuss the chemistry of a compound as an attacking group but it has to be delivered *in vivo* specifically to a cancer cell.

Fodor: I should add that all these biological screenings were done with material that was not completely pure; traces of polymeric methylglyoxal remained. The purified material is now being tested.

Slater: The compound we tested, although it contained some polymeric material, as Professor Fodor says, was much less toxic than methylglyoxal itself. The LD_{50} for purified methylglyoxal in mice was 108 mg/kg and for the complex as originally supplied by Professor Fodor was 1500 mg/kg body weight. (See note on p. 174.)

Williams: You suggest that these adducts will act as charge-transfer agents. Charge transfer is generally a one-electron process. But these compounds are highly reactive in two-electron steps. Is their activity at all related to charge transfer or is it associated with simple acid-base chemistry? Consider, for example, a proton on a highly activated carbon atom; the proton could be an extremely good leaving group, and make the compound extremely reactive in acid-base reactions. What makes you suppose that these compounds act in a one-electron sense?

Fodor: When one treats methylglyoxal with an excess of ethylamine, a free radical is formed. The amine reacts with the aldehyde group but cannot form a compound with the keto carbonyl group in a two-electron way; instead, it forms a charge-transfer complex. When one blocks the aldehyde group with ascorbic acid, the same free radical is generated on treatment with ethylamine. Although the two products certainly differ—in colour, in u.v. spectra, in ^{13}C n.m.r. spectra (dramatically)—a charge-transfer complex is formed, as proved by the e.s.r. spectrum. (We still have to isolate the chemical entity.)

Williams: I accept that you see a free radical but what is the proof that this charge transfer has anything to do with the biological activity? Such compounds will probably attack DNA bases in strict acid-base reactions, with no charge transfer. Any extrapolation to protein chemistry is interesting but it has to be proven and set in the context of the other chemical reactions of these reactive species.

Kornberg: The formation of methylglyoxal in microorganisms can be readily demonstrated. In *Escherichia coli* one can show that, for example, when the organisms grow on glucose 6-phosphate as sole carbon source they produce small amounts of methylglyoxal which is excreted into the medium. I discovered some years ago that when I added cyclic AMP to the system the organism stopped growing and that this cessation of growth was associated with the production of massive amounts of methylglyoxal. My colleague Ron Cooper (at Leicester) has investigated the system that is responsible for methylglyoxal formation in *E. coli* and has shown that there is an inducible enzyme that converts dihydroxyacetone phosphate into methylglyoxal which is excreted or is subsequently converted into D-lactate. Now, my point is this; in this system, in which methylglyoxal plays a toxic role, we see an effect on growth but we cannot, on this evidence alone, pinpoint the site at which methylglyoxal acts. It could well be that methylglyoxal reacts not primarily with proteins but perhaps with the amino groups in purines and pyrimidines. What is the reactivity of these ketoaldehydes with purines and pyrimidines?

Fodor: I haven't tried purines or pyrimidines; but your question is very much to the point and worth investigating. We only investigated the most obvious reactions of these new compounds, so far: their formation is reversible; addition of 2,4-dinitrophenylhydrazine hydrochloride gives the methylglyoxal hydrazone almost quantitatively and the ascorbic acid moiety is oxidized by the acryl hydrazine into dehydroascorbic acid which then may undergo further changes.

Slater: We have tested (G.W. White, P. Conroy & T.F. Slater, unpublished data) the complexes formed between certain aldehydes and various compounds containing amino groups against tumour growth in mice. None were active. The compounds tested included malonaldehyde–sulphonamide adducts and 4-hydroxypentenal with polylysine. We have also looked at the reactivity of polyamines such as spermine with aldehydes and, without a doubt, they do react with methylglyoxal (White & Slater 1974). Also, at very low concentrations methylglyoxal is highly reactive towards certain SH-dependent enzymes. D.C.H. McBrien & H. Fraval (unpublished data) have found a strong inhibition by methylglyoxal of the SH-dependent enzyme complex DNA-polymerase III in *E. coli.* I should emphasize that the low (micromolar) concentrations at which these aldehydes act on such SH-dependent enzymes point to specific actions of such aldehydes on metabolism as opposed to *unspecific* inhibitory actions that occur with millimolar concentrations.

Williams: Dr Fodor, when you generate the free radicals in these systems, do you do the reactions with EDTA present?

Fodor: Yes; we did them with and without EDTA.

Williams: So you eliminated the possibility that these reactions are due to metal ions?

Fodor: Yes. Peter Gascoyne obtains the same results using a much more sensitive instrument.

Williams: Were the n.m.r. samples in D_2O or deuteriated solvent?

Fodor: Almost all in organic, anhydrous solvents; that is why there is a discrepancy between the results of HPLC and of the ^1H and ^{13}C n.m.r. spectra. Methylglyoxal gave a sharp two-signal spectrum in chloroform with a 1:3 integration for the aldehyde: methyl protons. But when we added D_2O, the methyl signal broadens (a number of signals appear at 60 MHz around δ 2.6 p.p.m.) and several intermediate signals in addition to that of the aldehyde-hydrate proton appear at δ 4.8 p.p.m. Dissolved in water, methylglyoxal is hydrated (that is why it reacts with ascorbic acid) and the HPLC shows basically the presence of a single compound. The n.m.r. spectra seem to reflect the degree of hydration and not forcibly stereoisomerism.

Williams: So none of the reactive hydrogen atoms are substituted by deuterium?

Fodor: Apart from the OH groups, no. The chiral acetal proton is not substituted.

Williams: In relation to the testing of these compounds, André Lwoff showed a long time ago with a system consisting of a lysogenic strain of bacterium, incorporating a provirus, that the compounds that were cancerostatic or cytostatic (and which were therefore useful in cancer therapy) often were found to cause the production of virus from the lysogenic bacteria (Lwoff 1953). I wonder whether instead of testing for mutagenicity alone one could not devise a bacterial test based on Lwoff's work which could be used for the diagnosis of cancerostatic agents?

Kornberg: That is a most interesting suggestion. A lysogenic bacterium can often be converted into a lytic condition by, for example, interfering with the continued association of the provirus with the host DNA. Agents such as acridines, which intercalate in the DNA and interfere with the normal relationship of one bit of DNA with another, will trigger this conversion. Professor Slater, have you tried to find out whether methylglyoxal or its derivatives have any effect on lysogeny? It would be most valuable to know.

Slater: No. It is a fascinating suggestion that we shall follow up.

Williams: The platinum antitumour agents (*cis*-dichloroplatinum(II)diammine) were developed partly because Dr Reslova knew of Lwoff's work and proved that these compounds did cause a virus to appear in the lysogenic bacteria in *E. coli.* I don't know whether the complexes are entirely safe but certainly this platinum compound is not strongly mutagenic and is extremely effective in these lysogenic tests and against *some* tumours. We see that all the reagents which are linked with cancer have effects in bacteria and we should remember that compounds such as those described by Professor Fodor will not simply attack a protein and form a charge-transfer complex but engage in numerous reactions in a cell. Therefore, although one may observe a biological response and detect chemical evidence such as e.s.r. signals, one should hesitate before claiming a logical connection between the two events.

Note added in proof by Professor Fodor: More recent tests by Professor Slater and his colleagues (unpublished) show the LD_{50} to be greater than 5 g/kg for the acetal, with pure material.

References

KOLLARITSCH, K. (1973) *Über die Wirkung einiger α, β-ungesättigter Carbonylverbindungen und ihrer Serumproteinaddukte auf den Einbau von ^{3}H-Thymidin in die Desoxyribonucleinsäure von Ehrlich-Ascites-Tumorzellen.* Doctoral Thesis, University of Graz

LEVIN, R.H., HENDERSON, E., KARON, M. & FREIREICH, E.J. (1965) Treatment of acute leukemia with methylglyoxal-bis-guanylhydrazone (methyl GAG). *Clin. Pharmacol. Ther. 6,* 31-42

LWOFF, A. (1953) Lysogeny. *Bacteriol. Rev. 17,* 269

MUKAI, F.H. & GOLDSTEIN, B.D. (1976) Mutagenicity of malonaldehyde, a decomposition product of peroxidized polyunsaturated fatty acids. *Science (Wash. D.C.) 191,* 868-869

TAKAMATSU, A. & SATO, A. (1934) *J. Exp. Med. 23,* 506-517

WHITE, G.W. & SLATER, T.F. (1974) *Abstracts 9th FEBS Meeting,* Budapest (abstr. 367)

Thermal copoly(amino acids) as inhibitors of glyoxalase I*

SIDNEY W. FOX, ROBERT M. SYREN and CHARLES RAY WINDSOR

Laboratory of the National Foundation for Cancer Research at the Institute for Molecular and Cellular Evolution, University of Miami, Coral Gables, Florida

Abstract A number of copoly(α-amino acids) have been prepared thermally; some have been found to function as inhibitors of glyoxalase I, an enzyme which occupies a central position in Szent-Györgyi's theory of tumour genesis. These polymers are also of interest in the search for synthetic peptides having carcinostatic activity, since many natural peptides are active. The way in which the inhibitory activity varies with composition of the synthetic polymers has been investigated. Various properties (hydrophobicity, molecular weight, u.v. absorption, kinetic type) have been examined in a search for correlates of inhibitory activity.

The relationship to the origin of enzyme control mechanisms is discussed.

As Szent-Györgyi has emphasized, the understanding, cure, and prevention of cancer are waiting for a better understanding of the living state. The living state may, in turn, be better defined by increased understanding of primitive life and its proliferation (Szent-Györgyi 1977).

In the search for a natural control mechanism for cancer, Szent-Györgyi (1976) has related the enzyme glyoxalase I (*S*-lactoyl-glutathione methylglyoxal-lyase, EC 4.4.1.5) to bioelectronics, biochemical pathways, cellular proliferation, and cancer. The continuing studies of this laboratory are germane to these topics at several interfaces.

Our studies broadly pertain to a theory of a self-ordering origin of life (Fox 1978). These studies are yielding in general the kind of information that relates to an understanding of primitive life (Szent-Györgyi 1977). As a specific example, glyoxal, which is a general inhibitor of cellular proliferation (Schauenstein *et al.* 1977), and is also a substrate for glyoxalase, would have

*Professor Fox was prevented by illness from attending the symposium. His paper was presented by Professor T.F. Slater.

been a prominent substance on the protobiotic Earth (Groth & Suess 1938).

Secondly, the simulated protocells produced in our laboratory serve as simple models for cellular communication (Hsu *et al.* 1971), a process which is impaired in some cancer cells (Azarnia & Loewenstein 1977). These models invite investigation in helping us to understand some of the peptidic factors in various kinds of cell-to-cell interaction.

A third approach has the practical goal of finding therapeutically active copoly(amino acids) inhibitory for enzymes. In this last context, this paper focuses on a class of thermal copoly(α-amino acids) that have been found to function as inhibitors for glyoxalase I. From the variations in these inhibitors, we hope to gain understanding of the enzyme through its inhibitors, and to identify candidate compounds for therapeutic exploration.

THERMAL COPOLY(AMINO ACIDS) AS INHIBITORS OF ENZYMES

The search for inhibitors of glyoxalase I among the thermal copoly(amino acids) was begun because a number of the latter had already been found, in unpublished studies, to be enzyme inhibitors. Polymers of this class have been found to inhibit enzymes: two sperm hyaluronidases (C.B. Metz, R. Gerard, C.R. Windsor & S.W. Fox, unpublished results 1976), pork ATPase (Fox & Adachi, unpublished results 1977), luciferase (S.W. Fox & T. Adachi, unpublished 1977), and α-chymotrypsin (P. Ghosh, personal communication 1977). Copolymers of aspartic acid, glutamic acid, and a third α-amino acid were produced by procedures described in Fox & Dose (1977). The set of polymers tested fall clearly into a non-inhibitory group and a smaller inhibitory set, with the exception of copoly(Asp,Glu,Phe) which was intermediate in this activity (Table 1).

The tritonic* tryptophan copolymer was slightly more inhibitory than the others. Accordingly, the necessity for both aspartic acid and glutamic acid in the polymer was examined. The copoly(Glu,Trp) was found to be somewhat more inhibitory than copoly(Asp,Glu,Trp).

The need for a polymeric structure was shown most clearly by comparison of the activity of a tryptophan homopolymer with that of free DL-tryptophan. Tryptophan was homopolymerized at 265–270 °C for one hour. Of the original weight, 0.4% was isolated as a soluble polymer by liquid NH_3 treatment and Millipore filtration. This polymer yielded a predominant

---

*Tritonic signifies three types of monomeric α-amino acid.

TABLE 1

Tritonic thermal copoly(amino acids) that inhibit and do not inhibit glyoxalase I

Inhibitory polymers (0.33 mg/ml)	*Inhibition[a] (%)*	*Non-inhibitory polymers (0.33 mg/ml)*
Copoly(Asp,Glu,CyS_2)	31	Copoly(Asp,Glu,Ala)
Copoly(Asp,Glu,CySH)	35	Copoly(Asp,Glu,Arg)
Copoly(Asp,Glu,Thr)	42	Copoly(Asp,Glu,Gly)
Copoly(Asp,Glu,Ser)	46	Copoly(Asp,Glu,Iso)
Colopy(Asp,Glu,Trp)	48	Copoly(Asp,Glu,Leu)
		Copoly(Asp,Glu,Lys)
Semi-inhibitory polymer		Copoly(Asp,Glu,Met)
Copoly(Asp,Glu,Phe)	25	Copoly(Asp,Glu,Pro)
		Copoly(Asp,Glu,Tyr)
		Copoly(Asp,Glu,Val)

Mixtures of amino acids were heated for 6 h in an oilbath at 188 °C under nitrogen. After some washing with water and drying, all polymers were dissolved by treatment with liquid NH_3, NH_4OH or 0.1 N-NaOH. Mostly, liquid NH_3 was used. All polymers in this paper were made from equimolar mixtures of DL-amino acids unless otherwise indicated, as in Table 3.

[a]All inhibitions are calculated from averages of two readings. Activities of polymers within pairs agreed closely. Monitored by method of Racker (1951).

peak by high pressure liquid chromatography by separation on a C_{18} column (Waters Associates) of low polarity. After hydrolysis with concentrated HCl, the product gave a single amino acid, tryptophan, on the amino acid analyser. The unhydrolysed polymer showed absorption peaks in potassium phosphate buffer (pH 6.6) in the Beckman recording spectrophotometer at 219 nm and 275–280 nm. These peaks correspond to peptide bond and tryptophan residue respectively.

When DL-tryptophan and poly(Trp) were each tested for inhibition of glyoxalase I, the results of Table 2 were obtained. These results clearly indicate that the tryptophan must be in polymeric form in order to be inhibitory.

TABLE 2

Comparison of activities of monomeric and polymeric tryptophan against glyoxalase I

Substance	*Inhibition (%)*
None (control)	0
DL-Tryptophan[a]	0
Poly(Trp)[a]	100

[a]Saturated in potassium phosphate buffer (pH 6.6; 0.05 M).

All polymers were treated with liquid ammonia, aqueous ammonia or 0.1 N-NaOH in order to solubilize them. One polymer (1:3:1:1 Asp,Glu,Trp, CySH) was dissolved by the detergent Triton X-100 (1%) and also found to be inhibitory.

EFFECT OF VARIATION IN RATIOS OF MONOMERS

The effect of variation in the ratio of the reactants for copoly(Asp,Glu, Trp) is illustrated in Table 3. The fact that the ratios in thermal poly(amino acids) are non-random relative to the reactant mixtures (and by other criteria) has been recognized by many laboratories (cf. Dose & Rauchfuss 1972; Fox 1974; Melius 1977). Similarly, the fact that nearly stoichiometric ratios of amino acids often occur in the polymers has been recognized (Fox & Nakashima 1967; Melius & Sheng 1975; Nakashima *et al.* 1977).

Some differences in activity against glyoxalase I have been observed between these polymers. In a standard test, the 1:2:1 polymer is 50% inhibitory at a concentration of 0.33 mg/ml, whereas the 0:1:1 is 70% inhibitory. The least inhibitory of the series in Table 3 are the 2:2:1 or 2:1:1 polymers, each at 19%.

TABLE 3

Ratios of aspartic acid: glutamic acid: tryptophan in some thermal polymers inhibiting glyoxalase I

Ratios of reactants (Moles Asp:Glu:Trp)	*Ratio in polymers (Moles Asp:Glu:Trp)*
1:1:2	1:2:3
2:1:1	5:2:3
1:2:1	1:1:1
2:2:1	2:1:1
1:1:1	7:5:8
0:1:1	0:2:3

The polymers were prepared by heating for 30 min in a fluidized sandbath at 255 °C. The compositions of the polymers were determined by hydrolysis and automated analysis after the polymers were twice washed with six parts of water, and then solubilized by standing in solution in excess liquid ammonia, which was then allowed to evaporate.

POLYMERS OF FOUR OR FIVE AMINO ACIDS

In the search for increasingly active polymers, four, instead of two or three, amino acids were next polymerized. In the first instance, two amino acids from the active group of Table 1 were included in the same polymer. These were cysteine and tryptophan. This tetratonic polymer proved to be more active than any of the tritonic polymers, showing the maximal inhibition recordable under standard test conditions. It then became of interest to study the effect of a fifth amino acid. The fifth amino acids included in such polymers were those found not to contribute to activity in Table 1. In these results it was possible to observe quantitatively modifying effects from individual amino acids (Table 4).

To understand the inhibitors more thoroughly, it would be helpful to know the conformations of the whole macromolecules, and perhaps the amino acid sequences. That kind of information is, however, accumulated relatively slowly. By thermal polymerization, we can learn relatively rapidly about composition, which is basically related to conformation due to interactions of amino acid side-chains.

The polymer containing tryptophan is the most active polymer in the set of Table 1, although the degree of difference in activity between these five inhibitory polymers is small. The contribution of tryptophan suggests that hydrophobicity (Tanford 1973) can be a fundamental aspect of such inhibitors.

Comparison of the activies of the polymers in Table 4 indicates that inclusion of basic amino acids attenuates the activity of the tetratonic polymer the most. For an effective inhibitor, the basis of one prescription is thus that of a *hydrophobic polyanion.*

TABLE 4

Inhibitory activities of tritonic, tetratonic and pentatonic polymers

Polymer (0.33 mg/ml)	*Inhibition (%)*
None (Standard)	0
Poly(Asp,Glu,Trp)	87
Poly(Asp,Glu,Trp,CySH)I[a]	79
Poly(Asp,Glu,Trp,CySH)II[b]	100
Poly(Asp,Glu,Trp,CySH,Gly)	87
Poly(Asp,Glu, Trp,CySH,Pro)	68
Poly(Asp,Glu,Trp,CySH,Ala)	63
Poly(Asp,Glu,Trp,CySH,Thr)	58
Poly(Asp,Glu,Trp,CySH,His)	53
Poly(Asp,Glu,Trp,CySH,Lys)	47

[a]Made at 210 °C, 6 h in small tube.
[b]Made at 210 °C, 6 h in large flask.

WHAT IS THE EFFECT OF ASCORBIC ACID?

A small difference is found between the activities of the polymer containing cysteine and that containing cystine (Table 1). The quantitative necessity for reducing groups thus cannot be inferred from this comparison.

A further approach to this question is through the use of ascorbic acid with the polymers, because of its known interactions with SH (Szent-Györgyi 1976). In control experiments, the testing of ascorbic acid alone has given the striking result that ascorbic acid is itself an effective inhibitor of glyoxalase I (Fig. 1).

This result confirms the general finding of Ito *et al.* (1976), who studied reductones acting on glyoxalase. Because of the much higher average molecular weight of the polymers, ascorbic acid is not as strongly inhibitory as some of the poly(amino acids) on a molar basis. The combination of polymer and ascorbic acid continues to be of interest in further tests.

The fact that ascorbic acid alone is inhibitory for glyoxalase I is of special interest in a context involving both the Szent-Györgyi theory relative to that enzyme, and Cameron's clinical finding that ascorbic acid prolonged the survival time of terminal cancer patients (Cameron & Pauling 1976). Szent-Györgyi's theory thus receives support from the clinical results observed, when account is taken of the effect of ascorbic acid on this enzyme. Interaction effects of ascorbic acid and poly(amino acid) inhibitors are being studied.

FRACTIONATION OF POLY(GLUTAMIC ACID, TRYPTOPHAN)

Poly(Glu,Trp) was fractionated according to (*a*) molecular weight and (*b*) hydrophobicity.

Fig. 2 shows the distribution brought about in methanolic solution by a μBondagel E-125 cartridge and an E-300 cartridge in series on a Waters Associates Liquid Chromatograph. This limited degree of heterogeneity is common to thermal copoly(amino acids) (Melius 1977). According to estimates of Dr Howard Jordi (Waters Associates), the peaks correspond to molecular weights of $\sim$50 000 (dominant peak), $>$100 000, $<$2000, and $\sim$10 000.

Another preparation of the polymer was fractionated by use of a Sep-Pak C_{18} cartridge. The activities found for the fractions are included in Table 5, which indicates the effect of fractions collected by the use of differing ratios of methanol/water. Progressively more hydrophobic fractions are obtained

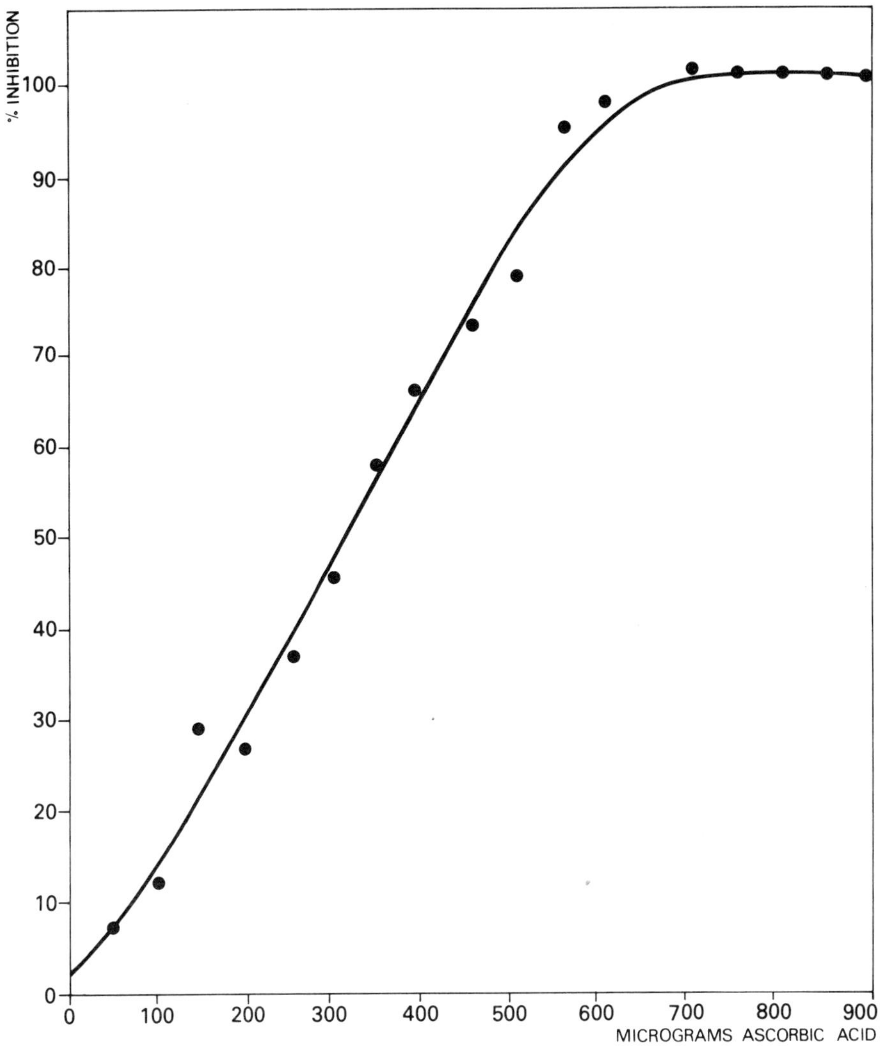

FIG. 1. Inhibition of glyoxalase I by added ascorbic acid in standard tubes.

as the proportion of methanol in the solvent is increased. This is also the direction of increasing inhibitory activity. Such results substantiate the earlier inference (Tables 1 and 3) that hydrophobicity can contribute to inhibitory activity, in at least some polymers.

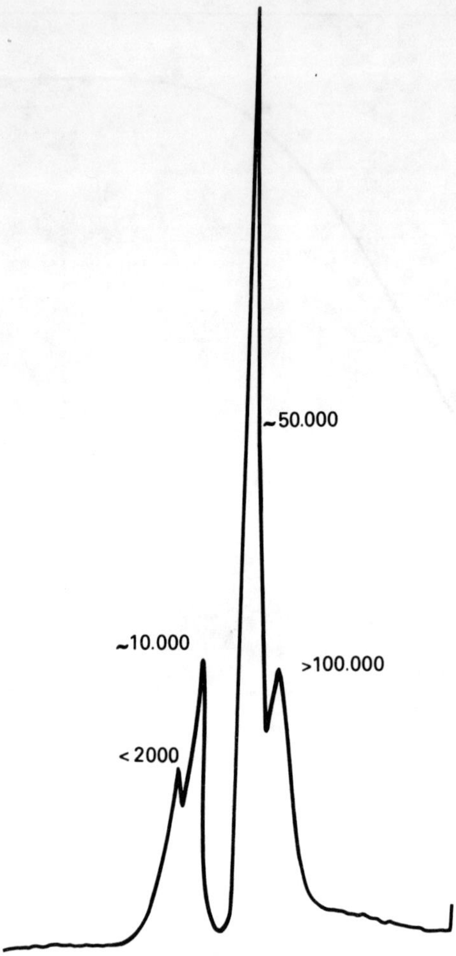

FIG. 2. Fractionation of poly(Glu,Trp) in methanol using μBondagel columns E-125 and E-300 in series. Molecular weights are estimates by Dr H. Jordi of Waters Associates.

TABLE 5

Activities of fractions of poly (glutamic acid, tryptophan) as inhibitors of glyoxalase I

Fraction	*Amount (mg)*	*Relative inhibition (%) at 0.33 mg/ml*
Standard		0
Unfractionated poly(Glu,Trp)	500	55
From 25 MeOH:75 H_2O	270	18
From 50 MeOH:50 H_2O	80	92
From 75 MeOH:25 H_2O	5	100

LIGHT ABSORPTION IN ACTIVE POLYMERS

The relationship of the colour of a compound to its ability to transfer electrons has been discussed by Szent-Györgyi (1976). We were led, therefore, to see if the inhibitory activity of polymers would correlate with their colour. Since the most active of the tritonic series, the tryptophan-containing polymers, are the closest to colourless of the entire group studied, a correlation could not be immediately inferred.

Instead of measuring optical absorption in the visible range, we compared the tritonic polymers at 265 nm. This is a wavelength suitable both for benzenoid amino acids and the pigment that forms with each polymer. A correlation between ultraviolet absorptivity and inhibitory activity was observed (Table 6).

TABLE 6

Absorbance of inhibitory and non-inhibitory tritonic polymers at 265 nm

Inhibitory polymers	*Absorbance, 265 nm*
Poly(Asp,Glu,Trp)	50
Poly(Asp,Glu,Ser)	40
Poly(Asp,Glu,Thr)	35
Poly(Asp,Glu,CySH)	28
Poly(Asp,Glu,CyS_2)	18
Semi-inhibitory polymer	
Poly(Asp,Glu,Phe)	18
Non-inhibitory polymers	
Poly(Asp,Glu,Tyr)	18
Poly(Asp,Glu,Ala)	<10
Poly(Asp,Glu,Arg)	<10
Poly(Asp,Glu,Gly)	<10
Poly(Asp,Glu,Iso)	<10
Poly(Asp,Glu,Leu)	<10
Poly(Asp,Glu,Lys)	<10
Poly(Asp,Glu,Met)	<10
Poly(Asp,Glu,Pro)	<10
Poly(Asp,Glu,Val)	<10

Experimental conditions as for Table 1.

KINETIC ANALYSIS OF INHIBITORS

Some kinetic analysis of two of the inhibitors has been done. A Lineweaver–Burk plot of the effect of copoly(Glu,Trp) suggested non-competitive inhibition over one range of concentration. An Eadie–Hofstee plot of the effect of poly(Asp,Glu,Trp,CySH) indicates competitive inhibition of that polymer with glutathione. The effects appear to be complex; these studies are being extended.

DISCUSSION

In addition to inferences derivable directly from the data, some overviews result.

Structure of inhibitory polymer

The active polymers are polyanionic, according to their composition. Aspartic acid is, however, not an essential monomer. Probably both aspartic acid and glutamic acid serve as trifunctional amino acids, either one of which is essential for the thermal copolymerization of amino acids (Fox & Dose 1977). This is consistent with the concept that biologically active proteins are peptide backbones with activity rooted in large part in the amino acid side-chains at the periphery of the molecule.

A number of the polymers are notably hydrophobic. In the copolytryptophans, and in a homopolytryptophan fraction, inhibitory activity appears to be correlated rather well with tryptophan content. A number of copolymers, however, do not contain tryptophan (Table 1). These need to be examined for hydrophobicity; their compositions suggest little of this property. The inhibitory activity of the polymers containing sulphur is correlated with the presence of sulphur rather than with the sulphur being in the form of SH or S—S. This relationship is being examined more closely, and in conjunction with the effect of added ascorbic acid.

Of special interest is the suggestion of modifying influences, some augmentative, as the number of monomers is increased (see Table 4 especially).

Interaction with glyoxalase I

Hydrophobicity in an inhibitor suggests hydrophobicity in the enzyme. Since virtually all enzyme molecules theoretically must have hydrophobic patches, probably on their exterior, the effectiveness of some copolytryptophans suggests hydrophobe–hydrophobe interactions between enzyme and inhibitor. What kinetic studies have been done suggest that hydrophobicity need not be the only basis for inhibition. Also, hydrophobicity may play a contributory, rather than a dominant, role.

In the light of the recent report claiming that zinc is an essential metal atom in glyoxalase I (Aronsson *et al.* 1978), inhibitors may function, at least in part, as chelators. A mechanism of chelation would be consistent with the polyanionic nature of the inhibitors studied here.

Poly(amino acids) as carcinostats

In addition to providing information relative to glyoxalase I (Szent-Györgyi 1976), poly(amino acids) are also of interest in the more empirical search for carcinostats, most of which may function by some mechanism not involving inhibition of glyoxalase I. A number of natural carcinostats that are peptidic have been recorded in the literature, such as macromomycin (Chimura *et al.* 1968), mitomalcin (McBride *et al.* 1969) and neocarzinostatin (Meienhofer *et al.* 1972). The ease of making large numbers of poly(amino acids) thermally recommends them in a search for synthetic carcinostatic peptides, as does the fact that any active polymer can be modified (Table 3).

The thermal copoly(amino acids) in general are attractive as potential therapeutic agents in that none has been found to cause allergy, despite much testing, and the polymers have not shown toxicity at any of the levels yet tested (Fox & Dose 1977).

Relationship to protobiogenesis

The work described relates in at least two ways to the developing theory of the origin of life (Fox 1978). One of these is the suggestion of the easy spontaneous generation of enzyme inhibitors as parts of primitive enzyme control mechanisms. The work of a number of laboratories has established that thermal copoly(amino acids) on the early Earth, as a class, would have included numerous (proto)enzymic activities (Fox & Dose 1977). The kind of result described here renders it likely that protoenzyme inhibitors would also have arisen at the same time as protoenzymes. The opportunity for the rapid evolution of controlled metabolism would thus have been considerably greater than in the absence of such inhibitors. Poly(amino acid) inhibitors could have been especially beneficial in an era of cellular proliferation preceding the advent of modern repressor molecules (cf. Watson 1970).

In relation to a specific inhibitor for glyoxalase I, one of the active polymers (Table 1) is poly(Asp,Glu,Ser). All three of these amino acids have cosmic credentials, having been found to be releasable from lunar dust (Fox *et al.*

1976) and as products of the reaction of two components of interstellar matter, formaldehyde and ammonia (Fox & Windsor 1970). The possibilities for the origin of threonine, tryptophan, cystine, or cysteine are not as clear.

The production of glyoxal from water and carbon dioxide by ultraviolet light on the early Earth has been suggested by experimental results (Groth & Suess 1938). Both the enzymic and antienzymic control of glyoxal concentrations in the presence of primitive organisms would thus have contributed to the control of the environment and to protometabolism.

The present picture is that the prebiotic Earth had a 'rain forest' of spontaneously synthesized organic molecules, as well as a profusion of protoenzymes and protoenzyme inhibitors, virtually all of the latter being copoly(amino acids). This would have served as a matrix for organisms, in which control mechanisms could begin and then evolve. The organisms then influenced the chemical aspects of the environment (Dubos 1978). New substances, resulting from industrial manufacturing, have intruded into the environment during the present chemical era. These represent a new profusion, and they present a challenge for the production of modern control mechanisms.

ACKNOWLEDGEMENTS

For partial support of this work, we thank the National Aeronautics and Space Administration for Grant No. NGR-10-007-008. This is publication no. 326 of the Institute for Molecular and Cellular Evolution.

References

Aronsson, A.-C., Marmstål, E. & Mannervik, B. (1978) Glyoxalase I, a zinc metalloenzyme of mammals and yeast. *Biochem. Biophys. Res. Commun. 81*, 1235-1240

Azarnia, R. & Loewenstein, W.R. (1977) Intercellular communication and tissue growth: VIII. A genetic analysis of junctional communication and cancerous growth. *J. Membr. Biol. 34*, 1-28

Cameron, E. & Pauling, L. (1976) Supplemental ascorbate in the supportive treatment of cancer: prolongation of survival times in terminal human cancer. *Proc. Natl. Acad. Sci. U.S.A. 73*, 3685-3689

Chimura, H., Ishizuka, M., Hamada, M., Hori, S., Kimura, K., Iwanaga, J., Takeuchi, T. & Umezawa, H. (1968) A new antibiotic, macromomycin, exhibiting antitumor and antimicrobial activity. *J. Antibiot. (Tokyo) 21*, 44-49

Dose, K. & Rauchfuss, H. (1972) On the electrophoretic behavior of thermal polymers of amino acids, in *Molecular Evolution: Prebiological and Biological* (Rohlfing, D.L. & Oparin, A.I., eds.), pp. 199-217, Plenum Press, New York

Dubos, R. (1978) Biological memory, creative associations, and the living Earth, in *The Nature of Life* (Heidcamp, W.H., ed.), pp. 1-21, University Park Press, Baltimore

FOX, S.W. (1974) Origins of biological information and the genetic code. *Molec. Cell. Biochem. 3*, 129-142
FOX, S.W. (1978) The origin and nature of protolife, in *The Nature of Life* (Heidcamp, W.H., ed.), pp. 23-92, University Park Press, Baltimore
FOX, S.W. & DOSE, K. (1977) *Molecular Evolution and Origin of Life*, revised edn., Dekker, New York
FOX, S.W. & NAKASHIMA, T. (1967) Fractionation and characterization of an amidated thermal 1:1:1-proteinoid. *Biochim. Biophys. Acta 140*, 155-167
FOX, S.W. & WINDSOR, C.R. (1970) Synthesis of amino acids by the heating of formaldehyde and ammonia. *Science (Wash. D.C.) 170*, 984-986
FOX, S.W., HARADA, K. & HARE, P.E. (1976) Amino acid precursors in lunar fines: limits to the contributions of jet exhaust. *Geochim. Cosmochim. Acta 40*, 1069-1071
GROTH, W. & SUESS, H. (1938) Bemerkung zur Photochimie der Erdatmosphäre. *Naturwissenschaften 26*, 77
HSU, L.L., BROOKE, S. & FOX, S.W. (1971) Conjugation of proteinoid microspheres: a model of primordial communication. *BioSystems* (ex. *Curr. Mod. Biol.*) *4*, 12-25
ITO, M., OKABE, K. & OMURA, H. (1976) Effect of reductones on glyoxalase I. *J. Nutr. Sci. Vitaminol. 22*, 53-61
MCBRIDE, T.J., AXELROD, M., CULLEN, W.P., MARSH, W.S., SODANO, C.S. & RAO, K.V. (1969) Mitomalcin: a proteinaceous antileukemic antibiotic produced by *Streptomyces malayensis. Proc. Soc. Exp. Biol. Med. 130*, 1188-1190
MEIENHOFER, J., MAEDA, H., GLASER, C.B., CZOMBOS, J. & KUROMIZU, K. (1972) Primary structure of neocarzinostatin, an antitumor protein. *Science (Wash. D.C.) 178*, 875-876
MELIUS, P. (1977) Composition and structure of thermal condensation polymers of amino acids, in *Bioorganic Chemistry* vol. III, *Macro- and Multimolecular Systems* (van Tamelen, E.E., ed.), pp. 123-136, Academic Press, New York
MELIUS, P. & SHENG, J. Y.-P. (1975) Thermal condensation of a mixture of six amino acids. *Bioorg. Chem. 4*, 385-391
NAKASHIMA, T., JUNGCK, J.R., FOX, S.W., LEDERER, E. & DAS, B.C. (1977) A test for randomness in peptides isolated from a thermal polyamino acid. *Int. J. Quantum Chem. Quantum Biol. Symp. 4*, 65-72
RACKER, E. (1951) The mechanism of action of glyoxalase. *J. Biol. Chem. 190*, 685-696
SCHAUENSTEIN, E., ESTERBAUER, H. & ZOLLNER, H. (1977) *Aldehydes in Biological Systems* (Gore, P.H., transl.), pp. 121-126, Pion/Academic, London
SZENT-GYÖRGYI, A. (1976) *Electronic Biology and Cancer,* Dekker, New York
SZENT-GYÖRGYI, A. (1977) The living state and cancer. *Proc. Natl. Acad. Sci. U.S.A. 74*, 2844-2847
TANFORD, C. (1973) *The Hydrophobic Effect*, Wiley, New York
WATSON, J.D. (1970) *Molecular Biology of the Gene*, 2nd edn., Benjamin, New York

Discussion

Crumpton: I find it difficult to evaluate Dr Fox's claims without having more information on the nature of the polymers. Professor M. Sela and his colleagues at the Weizmann Institute have prepared an extensive series of synthetic polypeptides including homo- and heteropolymers with unbranched and branched chains. Although the amino acid sequences of some of the heteropolymers are random and thus their exact structure is unknown, their structures are, nevertheless, much better defined than those of your polymers.

It seems reasonable for you to test some of your hypotheses using these synthetic polypeptides. For example, does polyleucine have any inhibitory action?

Fox: Yes, E. Katchalski and numerous others have made poly(α-amino acids) from *N*-carboxy-α-amino acid anhydrides (Leuchs anhydrides) but we have been the only workers to describe Leuchs copolymers of *all* common proteinous amino acids (Hayakawa *et al.* 1967); we prepared these partly to gain additional comparative experience with the two kinds of heteropoly(α-amino acid). If I were asked whether Leuchs polymers or thermal copoly(amino acids) are more fully characterized, I would say that such a question is pointless because the literature on each is extensive. The Leuchs polymers have been studied for longer but in some aspects, like fractionatable heterogeneity, protobiological and biological properties, I see the thermal polymers as much more fully characterized (for a mainly chemical summary, see Fox & Dose 1977, pp. 154-179).

Advantages of the Leuchs polymers are that L or D configurations of amino acids are retained and homopolymerizations are generally possible. Advantages of the thermal polymers are (*a*) sharply limited heterogeneity, (*b*) the ability to make numerous thermal poly(amino acids) in the time required for a single Leuchs poly(amino acid) (see text), and (*c*) relevance to a comprehensive theory of the origin of protolife, with evolution to L configurations of monomers to have come later.

Crumpton: But do you have any information on the molecular structure of the polymers? Surely such information is an essential prerequisite for a meaningful discussion of any biologically active material?

Fox: The literature on the molecular structure of thermal copoly(amino acids), as studied in other laboratories and ours, is extensive; among the many reviews are those by Oshima (1971), Dose & Rauchfuss (1972), Melius (1977) and Mikes (1977). The most recent compendia that provide entrée to the papers are in Fox & Dose (1977) and Fox (1978). In addition, some of the literature is indexed in biochemical sources under the term proteinoid. Since 1972, each semiannual index of *Chemical Abstracts* has listed many of these polymers under proteins, thermal; these listings deal with significance, characterization, etc.

Crumpton: The tryptophan homopolymer is claimed to be composed solely of tryptophan on the basis of acid hydrolysis and amino acid analysis. Tryptophan in proteins is, however, destroyed under the conditions of acid hydrolysis usually used (i.e. 6M-HCl) and it is normally estimated after alkaline hydrolysis. In this case how can you be so certain of the composition and structure of the polymer?

Neuberger: In the absence of oxygen the tryptophan bound in peptide form is not destroyed by 2M-HCl at 100 °C.

Fox: It is true that one *normally* determines tryptophan *in proteins* by procedures other than acidic hydrolysis. However, proteins contain many kinds of amino acid. It was shown over 30 years ago (Olcott & Fraenkel-Conrat 1947) that decomposition of tryptophan during acidic hydrolysis of proteins is due to reactions with other amino acids liberated from the protein molecules, or with oxygen. Indeed, tryptophan contained in gramicidin, which includes other amino acids but not those that catalyse the decomposition of tryptophan, is, for example, largely recovered after acidic hydrolysis (Olcott 1951). We have seen many results that agree with those.

Accordingly, when the material hydrolysed is not protein and does not contain carbohydrates or aldehydes or aldehydogenic material, and when the hydrolysis is done in sealed tubes containing virtually no air, tryptophan survives hot acidic hydrolysis. We can thus be sure of the composition of our homopolytryptophan, since no other amino acid appears in the analyser chromatogram of the hydrolysate.

We have also hydrolysed the material with 3N-mercaptoethanesulphonic acid, which is known to protect tryptophan. The result was the same.

Eley: It might be worth commenting that colour is a necessary but not sufficient condition for relatively high electronic conduction in solids. For example, metal-free phthalocyanine and manganous phthalocyanine are both highly coloured, but the latter has 10^7-10^{11} times (Day *et al.* 1963; Beales *et al.* 1973) the higher room-temperature conductivity of the two. Bovine plasma albumin is colourless and haemoglobin is coloured, but they have similar conductivities (Eley & Spivey 1960). It is true that all the highly conducting tetracyanoquinodimethane (TCNQ) radical anion complexes have black crystals, so far as I know. But in a series of 17 phenylene polymers, although the black polymers had the lowest room-temperature resistivities of the series, their resistivities were still very high, $\sim 10^{14}\ \Omega$ cm (Eley & Pacini 1968). Presumably the colour reflects ease of *intra*-molecular excitation, and the conductivity then is determined by the permeability of *inter*-molecular barriers.

Williams: I would have presumed that pure polymers with no aromatic amino acids, such as poly(Asp-Glu-Thr), would not absorb at 265 nm, but you are reporting variations in the absorption of peptides that contain no aromatic amino acids (see your Table 6, p. 183). Only sulphur would give a small tail of absorption in that region. Could this phenomenon be due to scattered light differences rather than absorption, with different degrees of polymerization? If so, the result has no meaning.

Fox: The explanation for u.v. absorption in copoly(amino acids) containing serine, threonine or cyst(e)ine is that some intense 265 nm-absorbing pigment is formed from these amino acids; the pigment appears to be part of the polymer. Its constitution is unknown.

Williams: These results reflect on the nature of the enzyme glyoxalase. You referred to the claim that it is a zinc enzyme (Aronsson *et al.* 1978). If it is, the safe way of detecting a zinc enzyme is not just by use of inhibitors (and certainly not the inhibitors you described).

Neuberger: The presence of zinc in the enzyme has been well established for relatively pure preparations from liver and yeast by atomic absorption methods.

Williams: Since these poly(amino acids) prepared in a random way seem to inhibit glyoxalase, do they inhibit other enzymes such as proteinases? To choose glyoxalase alone is deliberately to select experimental information, far away from the hypothesis.

Fox: In any strict sense of the word 'random', the polymers are not produced randomly (Fox & Dose 1977, pp. 155-164; Fox 1978, pp. 27-38). Some have been found to inhibit glyoxalase I; our statements do not apply to the whole glyoxalase complex. As I said in my paper, they also inhibit a proteinase (p. 176). If I understand the last part of this question, our deliberately selected objective is of course the inhibition of glyoxalase I, because it is the one enzyme that figures in Szent-Györgyi's hypothesis (p. 175). Perhaps the comment on hypothesis refers to our concept relating to the origin of enzyme control mechanisms. As I also said in the paper (p. 176), we have found inhibitory effects on other enzymes: hyaluronidases, ATPases, and a proteinase, as well as glyoxalase I.

Slater: The inhibition of glyoxalase by ascorbate has been reported by Ito *et al.* (1976), as you have said and confirmed. We have also found that ascorbate inhibits glyoxalase in normal rat and mouse liver but has the opposite effect on a mouse tumour sample that we have examined (see Fig. 1).

Neuberger: I wonder whether the usual substrate of glyoxalase is not necessarily methylglyoxal or glyoxal but a product or products of intermediate metabolism? For example, δ-aminolaevulinic acid is transaminated by an enzyme to yield a substituted glyoxal with a carboxy side-chain. This product is converted by glyoxalase into hydroxyglutaric acid, just as methylglyoxal is converted into lactic acid. It is conceivable that this may be the prime function of glyoxalase, since methylglyoxal is not found in any of the known pathways of intermediate metabolism whereas the dicarbonyl compound derived from δ-aminolaevulinic acid might easily occur. Glyoxalase thus may be specific simply for α-ketoaldehydes.

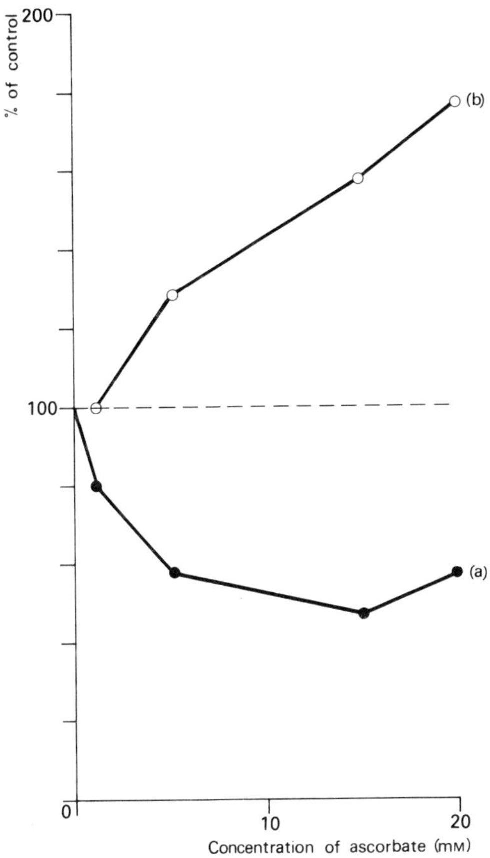

FIG. 1 (Slater). Effects of ascorbate on glyoxalase *(a)* in normal mouse liver supernatant and *(b)* in a mouse lymphosarcoma supernatant. The enzyme was assayed by the method of Alexander & Boyer (1971); incubation was for 5 min at 37 °C. (Data are unpublished studies by S.M. Ahmed & T.F. Slater.)

Williams: Are there standard inhibitors for these enzymes?

Neuberger: I don't know of any inhibitor which is specific for glyoxalase.

Conroy: Several metabolically stable derivatives of glutathione inhibit glyoxalase (Vince *et al.* 1973). I should emphasize that glyoxalase is two enzymes: glyoxalase I and II. Another consideration is that most of the inhibitors are very active *in vitro*: they can increase the cytotoxic effect of methylglyoxal by as much as 18 times (Vince & Daluge 1971). But, on injection into animals they are rapidly broken down by a hepatic glutathionase system (Vince *et al.* 1971; Kester *et al.* 1974) and, by-and-large, they do not significantly increase the antitumour potential of methylglyoxal *in vivo*.

I can support Professor Neuberger's comment: the enzyme system is specific for α-ketoaldehydes and not just for methylglyoxal.

Brandt: Although α-ketoaldehydes are substrates for the enzyme, there is a specific methylglyoxal synthetase found in *Escherichia coli* (Cooper & Anderson 1970). Several reports mention the excretion of methylglyoxal by humans (Platt & Lu 1939) or rats (Salem 1954). There is no doubt that D-lactate is the product of methylglyoxal in humans. Some of the interest in methylglyoxal stems from the fact that it strongly inhibits growth. Also, since methylglyoxal is the simplest α-ketoaldehyde, it makes sense to study it first.

Williams: Do other members of the class inhibit growth?

Brandt: Szent-Györgyi *et al.* (1967) have shown that several of these compounds inhibit growth.

Williams: It occurs to me that the action of glyoxalase is simply a hydration of an aldehyde. Carbonic acid dehydratase hydrates CO_2, which is really the same reaction. Zinc is the catalytic atom of both enzymes. Also, carbonic acid dehydratase is a good catalyst for the hydration of aldehydes. Does it hydrate glyoxal? If it does, you may be looking at the wrong enzyme! Is what is measured as a glyoxalase activity really a carbonic acid dehydratase activity? Furthermore, will methylglyoxal block carbonic acid dehydratase?

Neuberger: The two enzymes may have certain sequences in common. That would be interesting to know.

Williams: The active-site zinc might be bound in identical sequences.

References

ALEXANDER, N.M. & BOYER, J.L. (1971) *Anal. Biochem. 41*, 29-38

ARONSSON, A.-C., MARMSTÅL, E. & MANNERVIK, B. (1978) Glyoxalase I, a zinc metalloenzyme of mammals and yeast. *Biochem. Biophys. Res. Commun. 81*, 1235-1240

BEALES, K.J., ELEY, D.D., HAZELDINE, D.J. & PALMER, T.F. (1973) *Metal Phthalocyanines of the First Transition Period, Katalyse and Phthalocyanines (Symp. Hamburg 1972)* (Kropf, H. & Steinbach, F., eds.), pp. 1-29, G. Thieme Verlag, Stuttgart

COOPER, R.A. & ANDERSON, A. (1970) The formation and catabolism of methylglyoxal during glycolysis in *E. coli. FEBS (Fed. Eur. Biochem. Soc.) Lett. 11*, 273-276

DAY, P., SCREGG, G. & WILLIAMS, R.J.P. (1963) Photoconductivity of manganous phthalocyanine. *Nature (Lond.) 197*, 589-590

DOSE, K. & RAUCHFUSS, H. (1972) On the electrophoretic behaviour of thermal polymers of amino acids, in *Molecular Evolution: Prebiological and Biological* (Rohlfing, D.L. & Oparin, A.I., eds.), pp. 199-217, Plenum Press, New York

ELEY, D.D. & PACINI, B.M. (1968) Semiconduction in phenylene polymers. *Polymer 9*, 159-172

ELEY, D.D. & SPIVEY, D.I. (1960) Semiconductivity of organic substances. Part 6. A range of proteins. *Trans. Faraday Soc. 56*, 1432-1442

FOX, S.W. (1978) The origin and nature of protolife, in *The Nature of Life* (Heidcamp, W.H., ed.), 23-92, University Park Press, Baltimore

FOX, S.W. & DOSE, K. (1977) *Molecular Evolution and the Origin of Life*, revised edn., Dekker, New York

HAYAKAWA, T., WINDSOR, C.R. & FOX, S.W. (1967) Copolymerization of the Leuchs anhydrides of the eighteen amino acids common to protein. *Arch. Biochem. Biophys. 118*, 265-272

ITO, M., OKABE, K. & OMURA, H. (1976) Effect of reductones on glyoxalase I. *J. Nutr. Sci. Vitaminol. 22*, 53-61

KESTER, M.V., REESE, J.A. & NORTON, S.J. (1974) Mouse liver glyoxalase I. Inhibition by S-substituted glutathiones. *J. Med. Chem. 17*, 413-416

MELIUS, P. (1977) Composition and structure of thermal condensation polymers of amino acids, in *Bioorganic Chemistry*, vol. III, *Macro- and Multimolecular Systems* (van Tamelen, E.E., ed.), pp. 123-136, Academic Press, New York

MIKES, O. (1977) Vývoj proteolytických enzymů. *Chem. Listy 71*, 1154-1196

OLCOTT, H.S. (1951) Methods for the determination of amino acids, in *Amino Acids and Proteins* (Greenberg, D.M., ed.), p. 90, Thomas, Springfield, Ill.

OLCOTT, H.S. & FRAENKEL-CONRAT, H. (1947) Formation and loss of cysteine during acid hydrolysis of proteins. Role of tryptophan. *J. Biol. Chem. 171*, 583-594

OSHIMA, T. (1971) Catalytic activities of synthetic polyamino acids: models of primitive enzymes. *Viva Origino 1*, 35-43

PLATT, B.S. & LU, G.D. (1939) The accumulation of pyruvic acid and other carbonyl compounds in beri-beri and the effect of vitamin B_1. *Biochem. J. 33*, 1525-1537

SALEM, H.M. (1954) Glyoxalase and methylglyoxal in thiamine-deficient rats. *Biochem. J. 57*, 227-230

SZENT-GYÖRGYI, A., EGYUD, L.G. & MCLAUGHLIN, J.A. (1967) Ketoaldehydes and cell division. *Science (Wash. D.C.) 155*, 539-541

VINCE, R. & DALUGE, S. (1971) Glyoxalase inhibitors. A possible approach to anticancer agents. *J. Med. Chem. 14*, 35-37

VINCE, R., DALUGE, S. & WADD, W.S. (1971) Studies on the inhibition of glyoxalase I by *S*-substituted glutathiones. *J. Med. Chem. 14*, 402-404

VINCE, R., WOLF, M. & SANFORD, C. (1973) Glutaryl-*S*-(*p*-bromobenzyl)-L-cysteinyl-glycine. A metabolically stable inhibitor of glyoxalase I. *J. Med. Chem. 16*, 951-953

Interactions of methylglyoxal with methylamine

SUHEIL F. ABDULNUR

National Foundation for Cancer Research and Chemistry Department, American University, Washington, D.C.

Abstract *Ab initio* quantum mechanical calculations are used to study the interactions of the aldehydic group of methylglyoxal with the NH_2 groups of protein side-chains, using methylamine as a representative molecule. The hydrogen-bonding interaction, $C=O...H-N$, results in an electronic charge transfer from methylglyoxal to methylamine in both the ground and first excited triplet states. In this latter state a slight possibility is found for the H atom in the hydrogen bond to tunnel from methylglyoxal to methylamine, leading to the possible formation of two free radical fragments. The approach of methylamine to methylglyoxal in the stacked conformation $\begin{matrix}C\\ \vdots\\ N\end{matrix}$ to form a hemiacetal, associated with electron charge transfer from methylamine to methylglyoxal, is energetically unfavourable in vacuum. The concomitant tunnelling of a proton from a proton-donating solvent molecule to the aldehydic oxygen of methylglyoxal is shown to make this approach favourable. The relative stability of the keto and enol forms of methylglyoxal is also investigated, the keto form being found the more stable in vacuum.

Methylglyoxal, and dicarbonyls in general, play an important role as electron acceptors in the submolecular electronic theory of cancer of Dr Albert Szent-Györgyi (1976). Recent investigations (Szent-Györgyi 1978; Gascoyne & Pethig 1978) indicate that methylglyoxal interacts covalently with the NH_2 groups in the lysine (and possibly arginine) side-chains of proteins. Such an interaction is very sensitive to the nature of the solvent, being much more pronounced in some non-aqueous solvents, such as methanol, than in aqueous solvents (Szent-Györgyi 1978; Pethig & Szent-Györgyi 1977). We have been interested in investigating the nature and mechanism of this interaction from a theoretical point of view (Abdulnur 1977). In order to simplify the calculations we have studied the interaction of methylglyoxal with the simplest primary amine, methylamine. This simplified system has been shown by Szent-Györgyi (1976) to reproduce the main features of the interaction of methylglyoxal with proteins.

We have concentrated our attention on the interaction of methylamine with the more reactive aldehydic carbonyl group of methylglyoxal. This leads to the formation of the Schiff base (II) via the intermediate hemiacetal (I) as shown in the equation:

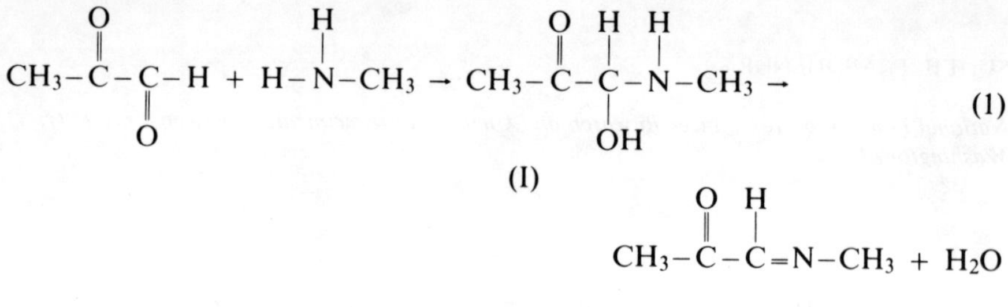

(1)

(I) (II)

In this paper we study the relative stabilities of the keto and enol forms of methylglyoxal and investigate two possible interactions of methylamine with the aldehydic group of methylglyoxal. The first is the hydrogen-bonding interaction with the carbonyl oxygen (p. 198). The second is the stacking interaction in which the N-lone electron pair approaches the electrophilic C atom from a direction perpendicular to the plane of methylglyoxal (p. 199). In order to overcome the large energy barrier encountered in this latter approach, the possibility of a proton transfer from a solvent molecule (H_3O^+) to the carbonyl oxygen is investigated (p. 201). The calculations are done using the *ab initio* Gaussian-70 series of programs (Hehre *et al.* 1973) in the STO-3G basis (Slater-type orbitals, each expanded as a sum of three Gaussian orbitals).

RELATIVE STABILITY OF THE KETO AND ENOL FORMS OF METHYLGLYOXAL

From a study of the pH dependence of the u.v. absorption spectrum of methylglyoxal, Schauenstein *et al.* (1977) have concluded that a keto–enol equilibrium exists in the region of neutral pH as shown in the equation:

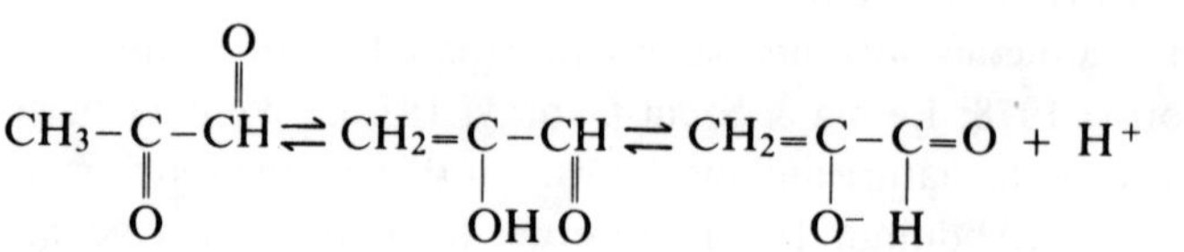

(2)

(III)

The presence of the enol form (III) in the region of neutrality is of interest because it has the α, β-unsaturated aldehyde structure. As discussed at length by Schauenstein *et al.* (1977), α,β-unsaturated aldehydes react very favourably at the (C=C) double bond with thiol and amino groups by an addition mechanism. Such an interaction with methylamine would yield $CH_3-NH-CH_2-CH(OH)-HC=O$ which could compete with or supplement the hemiacetal (I) and Schiff base (II) formation at the aldehydic group, shown in equation (1). When combined, these two reactions would make methylglyoxal a strong cross-linking agent connecting protein side-chains that contain NH_2 or SH groups, thus enhancing the stability of the structured β-state of normal cells in Szent-Györgyi's theory of cancer (1977).

In view of these considerations we investigated the relative stability of the keto and enol forms of methylglyoxal in vacuum by energy-minimizing some of the bond-lengths and angles of each, using the STO-3G minimized values of analogous molecules as a starting point (Del Bene *et al.* 1975). These are shown in Fig. 1. The keto form was found to be the more stable by 17.0 kcal/mol (17.6 kcal/mol, before minimization) in vacuum. The keto – enol equilibrium, predicted by Schauenstein *et al.* (1977) to exist in the region of

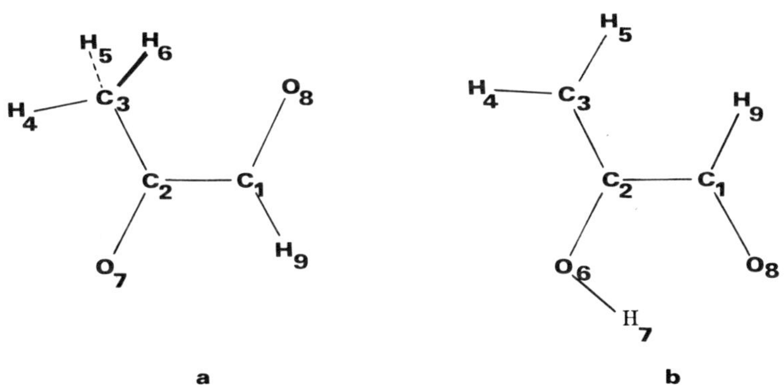

FIG. 1 (*a*) Keto form of methylglyoxal. Minimized in present study: C_1-C_2 = 1.548 Å, C_2-C_3 = 1.541 Å, $<C_1-C_2-C_3$ = 114.6°. STO-3G minimized values of glyoxal (Del Bene *et al.* 1975): C_1-O_8 = C_2-O_7 = 1.22 Å, C_1-H_9 = 1.102 Å, $<C_2-C_1-O_8$ = $<C_1-C_2-O_7$ = 122.4°, $<O_8-C_1-H_9$ = 122.8°. STO-3G minimized values of acetone (Del Bene *et al.* 1975): C_3-H_4 = C_3-H_5 = C_3-H_6 = 1.086 Å, $<C_2-C_3-(H_4,H_5,H_6)$ = 109.9°.
(*b*) Enol form of methylglyoxal. Minimized in present study: C_1-C_2 = 1.517 Å, C_2-C_3 = 1.316 Å, C_2-O_6 = 1.396 Å, O_6-H_7 = 0.991 Å, $<C_3-C_2-O_6$ = 122°, $<C_2-O_6-H_7$ = 103.9°. STO-3G minimized values of acrolein (Del Bene *et al.* 1975): C_1-O_8 = 1.221 Å, C_1-H_9 = 1.104 Å, $C_3-(H_4,H_5)$ = 1.083 Å, $<C_2-C_3-(H_4,H_5)$ = 122.0°, $<C_1-C_2-C_3$ = 122.7°, $<C_2-C_1-O_8$ = 123.9°, $<O_8-C_1-H_9$ = 121.8°.

neutral pH, will thus have to arise from a preferential solute–water interaction of the enol form compared to the keto form. We are now investigating these solvent effects.

INTERACTIONS OF METHYLGLYOXAL WITH METHYLAMINE IN VACUUM

Hydrogen-bonded interaction

Methylamine can hydrogen-bond to the aldehydic oxygen of methylglyoxal to give 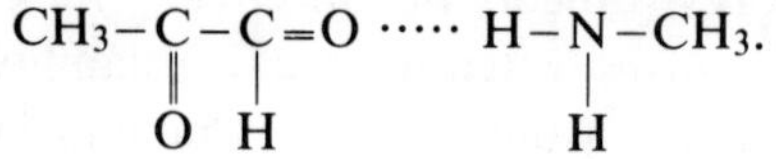 We have investigated this type of interaction for the simplified system methylglyoxal ··· H – NH_2 by the present *ab initio* technique in STO-3G basis. Using the minimized geometry of methylglyoxal given in the previous section, the STO-3G minimized geometry of NH_3 (Newton *et al.* 1970), an O ··· N distance of 3.04 Å (the minimum distance found in our related study of the system formaldehyde ···· methylamine), and 120° for <C–O–N, we found the interaction energy to be – 1.8 kcal/mol; it was also found that 0.018 e^- was transferred from methylglyoxal to NH_3 in this H-bonded complex.

We have also investigated the behaviour of this system in the lowest triplet state using the unrestricted Hartree–Fock(UHF)method. The interaction energy in the above geometry increased slightly to – 2.0 kcal/mol (relative to isolated methylglyoxal in triplet state and NH_3 in ground state); however, the charge transfer from methylglyoxal to NH_3 remained the same as in the ground state. In this study we observed a very interesting phenomenon. The H atom on H–NH_2 has a very small probability of tunnelling from the NH_2 to the oxygen of methylglyoxal in the lowest triplet state, across a barrier height that reaches 77.8 kcal/mol above the separated molecule minimum at an O ···· H distance of 1.5 Å. Beyond the barrier for an O–H distance of 1.0 Å, an unpaired spin of 0.991 e^- is found to have migrated from methylglyoxal (originally with two unpaired electrons) to the NH_2 fragment which is almost neutral, having a net positive charge of only + 0.020. In this geometry the energy of the system is 15.5 kcal/mol above the separated molecule minimum, a difference that is expected to decrease upon geometry minimization of the two free radical fragments thus generated. (A similar situation was found in the system glyoxal ··· methylamine using the semi-empirical INDO technique; S.F. Abdulnur, unpublished data).

The above proposed possible formation of the two free radicals, •NH_2 and

MGH•, in the lowest triplet state must be regarded with caution at present, pending further investigation, as it may be a peculiarity of the UHF method and/or the fact that correlation effects are not fully accounted for. We have previously shown that the UHF method yields a first triplet state energy that is much lower than those for the analogous molecules, glyoxal and dimethylglyoxal (Abdulnur 1978), both of which are experimentally known to have absorption spectra that lie in the visible part of the spectrum (McGlynn *et al.* 1969). The excitation in methylglyoxal was shown to occur mainly in the ketonic group, which acquires a much larger amount of unpaired spin in the triplet state than the aldehydic group. The charge redistribution among the various atomic orbitals upon excitation in this ketonic group was found to be very similar to that in formaldehyde and acetaldehyde. Both of these molecules exhibit similar lower calculated first triplet state energies than are experimentally observed if correlation effects are not taken into account, which led us to ascribe the lower calculated value of methylglyoxal to such effects also. Subsequent work (Nitzche & Davidson 1978) has shown that on inclusion of correlation effects the triplet state energy of methylglyoxal becomes comparable to that of glyoxal.

It is interesting to note that since, as mentioned above, the lowest singlet and associated triplet states of the dicarbonyls glyoxal, dimethylglyoxal and methylglyoxal lie in the visible spectrum, the radical formation described here (if proved correct) could take place on exposure to daylight, and may lead to the initiation of polymerization processes.

Stacking interaction

We have investigated the formation of the hemiacetal (I) in equation (1) through a stacking interaction in which the lone electron pair on the N atom of methylamine would approach the electrophilic C atom on the aldehydic group directly from below, in a direction perpendicular to the plane of the heavy atoms (non-hydrogen atoms) in methylglyoxal. To reduce the computation time involved, the calculation was done for the system:

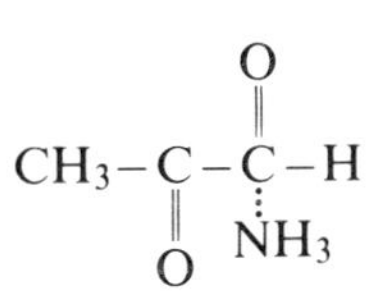

The interaction energies of methylglyoxal and NH_3, together with the electronic charge transferred from NH_3 to methylglyoxal, are given in Table 1 (system I) for various $\mathrm{C}\cdots\mathrm{N}$ distances, R. An analogous calculation for the enol form of methylglyoxal yielded an interaction energy of – 1.5 kcal/ mol and a charge transfer of – 0.0028 to methylglyoxal at the lowest-energy R of 3.0 Å, very comparable to the keto form case.

These results present us with the following dilemma: how can the hydrogen be transferred from NH_3 to the aldehydic oxygen to form the hemiacetal (I) in this reaction path, if the approach of the NH_3 has such a strongly repulsive character? Assuming the present path is the correct path of approach, the calculations in Table 1 suffer from the fact that the bond lengths and angles of methylglyoxal and NH_3 are kept constant, as in the separated molecules, during the approach. Scheiner *et al.* (1976) have included such a relaxation of the aldehydic C=O bond and the puckering of the angle at the C atom (from sp^2 to sp^3) for various R in their analogous study of the system:

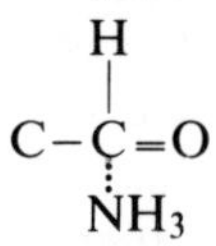

Their work indicates such variation to become important beyond R = 2.0 Å, at which R there is still a strong repulsion, as shown in Table 1. That this formaldehyde–ammonia system bears a close analogy to the methylglyoxal–NH_3 system may be seen from Table 1 also, where the interaction energies and charge transfer to formaldehyde at the same R values are shown. Demoulin *et al.* (1978) have also shown such a

TABLE 1

STO-3G interaction energy (ΔE) (in kcal/mol) and charge transfer in the stacked orientation at various C ··· N distances, R, for the systems (1) methylglyoxal–NH_3 and (II) formaldehyde–NH_3[a]

R (Å)	3.5	3.0	2.5	2.0
ΔE_I[b]	– 1.3	– 1.6	+ 1.9	+ 27.8
Charge transfer to methylglyoxal[c]	– 0.0003	– 0.0030	– 0.0189	– 0.1006
ΔE_{II}[b]	– 0.7	– 0.8	+ 2.7	+ 27.0
Charge transfer to formaldehyde[c]	– 0.0003	– 0.0026	– 0.0174	– 0.0955

[a]The geometries of methylglyoxal and NH_3, in system I, and of formaldehyde and NH_3, in system II, are the STO-3G minimized ones. The H's of NH_3 are maintained in a staggered position relative to the aldehydic carbon bonds.

[b]The interaction energy, ΔE_I, for system I is calculated as the difference between the energy of the composite system methylglyoxal–NH_3 and the sum of the energies of the two isolated molecules methylglyoxal and NH_3. ΔE_{II} is calculated similarly for system II.

[c]The net negative charge transferred from NH_3 to methylglyoxal in system I, and to formaldehyde in system II, as obtained from Mulliken population analysis.

high energy barrier to occur in the similar approach of glyoxal to the NH_2 group of guanine. In the next section the possible effects of a proton transfer from a proton-donating solvent molecule to the aldehydic oxygen of methylglyoxal on the energetics of this process are described.

POSSIBLE INVOLVEMENT OF PROTON TRANSFER FROM SOLVENT IN THE INTERACTION OF METHYLGLYOXAL WITH METHYLAMINE

For simplicity we consider the system $\begin{matrix} F \cdots H - \overset{+}{O}H_2 \\ \vdots \\ NH_3 \end{matrix}$ shown in Fig. 2 (F, formaldehyde) where we have chosen the simplest proton-donating solvent molecule, H_3O^+. The following two possibilities arise: (*a*) An H^+ may be transferred from H_3O^+ to formaldehyde before the approach of NH_3; or

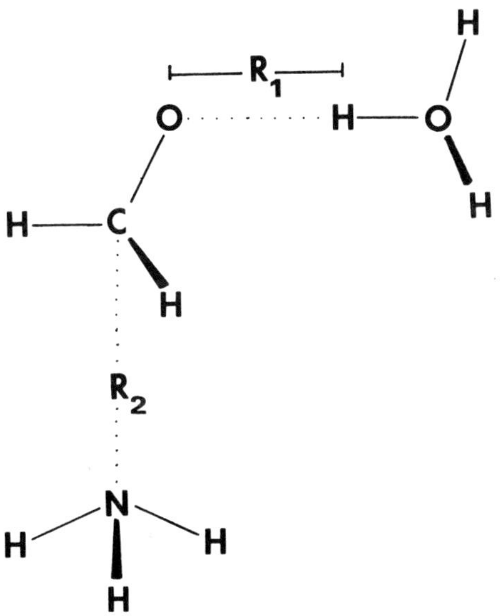

FIG. 2. The system $NH_3 \cdots F \cdots H_3O^+$. The STO-3G minimized geometries of NH_3 and F (formaldehyde) are used. For H_3O^+, an O–H distance of 0.991 Å is used (the same as that found for the O–H distance in the enol form of methylglyoxal, p. 197), and a tetrahedral geometry is assumed. The F, the H_3O^+ oxygen and two of its hydrogens are in the same plane. The NH_3 is below this plane, directly under the C atom of F, with its hydrogens in a staggered conformation relative to the formaldehyde bonds. The $C{=}O \cdots H$ angle is taken as 117°, and the $O \cdots O$ distance as 2.88 Å.

(*b*) an H^+ is transferred from H_3O^+ to formaldehyde simultaneously with the approach of NH_3. Both cases will have to be followed at a later stage by a loss of H^+ from NH_3 on the way to forming the hemiacetal. We report here on the results we have obtained so far regarding these two possibilities.

The calculations have been done for three positions, R_1, of the H in the $F \cdots H_3O^+$ hydrogen bond: when it is in H_3O^+ (R_1 = 1.889 Å), when half-way between the two oxygens (R_1 = 1.44 Å), and when transferred to O=(R_1 = 1.00 Å). The energy at R_1 = 1.44 Å gives us an estimate of the energy barrier involved in the tunnelling of the proton from H_3O^+ to F. In the absence of NH_3, the energies of the system $F \cdots H{-}\overset{+}{O}H_2$ at the above three R_1 values (relative to the sum of the energies of the two isolated F and H_3O^+) are respectively (in kcal/mol): −21.4, −6.7, −12.8. For a neutral H_2O, the corresponding energies for R_1 = 1.889 Å and 1.44 Å are −3.4 and +70.7 (kcal/mol), illustrating the much more favoured proton tunnelling in the protonated system $F \cdots H_3O^+$.

Fig. 3 illustrates the variation in the interaction energy of the three-molecule system as NH_3 approaches formaldehyde for the three R_1 distances given above. The following observations are made:

(1) For a large $F \cdots NH_3$ separation, up to R_2 = 3.0 Å, the most stable position of the H atom in the hydrogen bond is the one in which it is at R_1 = 1.889 Å; that is, still on H_3O^+. The minimum energy of the system for this position occurs around R_2 = 3.0 Å.

(2) The decrease in R_2 is associated with a lowering of the energy barrier for tunnelling (compare curves *b* and *c*) and a much more drastic lowering for the R_1 = 1.00 Å position in which the H is on formaldehyde (curve *d*). Indeed, at R_2 = 2.9 Å, the H^+ on formaldehyde becomes more stable than on H_3O^+, reaching a relatively deep minimum around R_2 = 2.00 Å. The energy barrier is also seen to disappear beyond R_2 = 2.1 Å.

From the above we conclude that as NH_3 approaches formaldehyde the tunnelling of the H^+ from H_3O^+ to formaldehyde increases with the H^+ being completely transferred at around R_2 = 2.1 Å, and the system reaching an energy minimum around R_2 = 2.0 Å. The tunnelling of the H^+ thus makes the approach of NH_3 to formaldehyde a downhill one in energy terms until R_2 = 2.0 Å, which solves the problem posed by the vacuum calculations (p. 200 above). We anticipate the continued approach of NH_3 to formaldehyde beyond R = 2.0 Å to be made more favourable as the geometries of FH^+ and NH_3 are optimized both to change the hybridization of the aldehydic carbon atom from sp^2 to sp^3, and to eliminate H^+ from NH_3. We are currently studying these variations.

The present analysis makes the formation of the hemiacetal dependent on

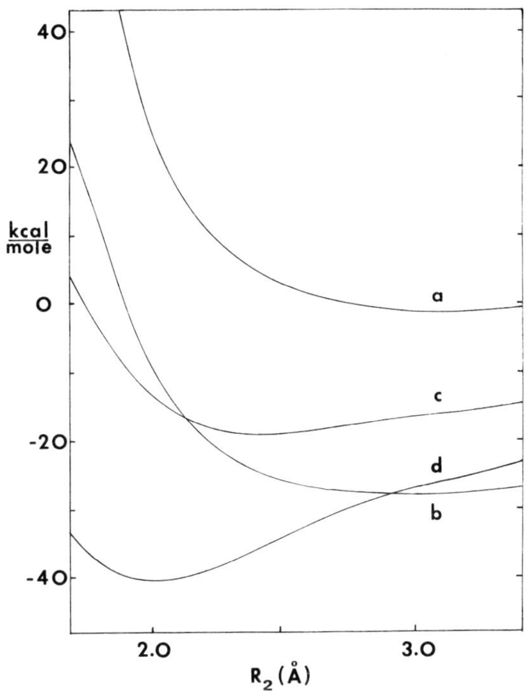

FIG. 3. The variation of the interaction energy of the system of Fig. 2 (energy of system minus the sum of the energies of the isolated molecules) with the H_3N ... F(formaldehyde) distance, R_2, for various R_1 values. Curve (*a*): F + NH_3 only. Curves (*b*), (*c*) and (*d*): F + NH_3 + H_3O^+ for R_1 = 1.889 Å, 1.44 Å and 1.00 Å respectively.

the pH of the solution. In an aqueous solution the pH must be high enough for methylamine to exist in an unprotonated state, but must still be of such a magnitude as to allow a high-enough concentration of H^+ in order to catalyse the reaction. It also predicts that in pure non-aqueous solvents that do not have a proton-donating species, such as dioxane, acetone, formamide, dimethylformamide and dimethyl sulphoxide, the formation of the hemiacetal will not be favoured. On the other hand, solvents like alcohols that could form $H_2O^+{-}R$ species may favour the reaction.

Such predictions are based, however, on one particular solvent effect. In order to account fully for the extent of reaction (1) in various solvents, other important properties of the solvents, such as their dielectric constant, must be taken into account (Szent-Györgyi 1978; Szent-Györgyi & McLaughlin 1978).

ACKNOWLEDGEMENTS

I thank Dr Albert Szent-Györgyi for introducing me to the very interesting reactions for dicarbonyls, for his guidance and for his constant encouragement. I also thank the National Foundation for Cancer Research for its financial support, which made this work possible. The free computer time made available for this project by the Computer Center of the American University is greatly appreciated.

References

ABDULNUR, S.F. (1977) The interaction of methylglyoxal with methylamine and the triplet state. *Int. J. Quantum Chem. Quantum Biol. Symp. 4,* 217-223

ABDULNUR, S. (1978) An unrestricted Hartree–Fock study of the lowest triplet state of methylglyoxal and other carbonyls. *J. Am. Chem. Soc. 100,* 6341-6343

DEL BENE, J., WORTH, G.T., MARCHESE, F.T. & CONRAD, M.E. (1975) An *ab initio* molecular orbital study of substituted carbonyl compounds. *Theor. Chim. Acta (Berl.) 36,* 195-206

DEMOULIN, D., ARMBRUSTER, A.M. & PULLMAN, B. (1978) A quantum-mechanical study of the interaction of glyoxal with guanine. *Theor. Chim. Acta (Berl.) 48,* 143-153

GASCOYNE, P.R.C. & PETHIG, R. (1978) Electronic and dielectric properties of protein–methylglyoxal complexes. *Biol. Bull. 155,* 438

HEHRE, W.J., LATHAN, W.A., DITCHFIELD, R., NEWTON, M.D. & POPLE, J.A. (1973) Gaussian 70. *Program No. 236, Quantum Chemistry Program Exchange,* University of Indiana, Bloomington

MCGLYNN, S.P., AZUMI, T. & KINOSHITA, M. (1969) *Molecular Spectroscopy of the Triplet State,* Prentice-Hall, New York

NEWTON, M.D., LATHAN, W.A., HEHRE, W.J. & POPLE, J.A. (1970) Self-consistent molecular orbital methods V. *J. Chem. Phys. 52,* 4064-4072

NITZCHE, L.E. & DAVIDSON, E.R. (1978) *Ab initio* studies of the $^3n\pi^*$ states of glyoxal and methylglyoxal. *Chem. Phys. Lett. 58,* 171-174

PETHIG, R. & SZENT-GYÖRGYI, A. (1977) Electronic properties of the casein–methylglyoxal complex. *Proc. Natl. Acad. Sci. U.S.A. 74,* 226-228

SCHAUENSTEIN, E., ESTERBAUER, H. & ZOLLNER, H. (1977) *Aldehydes in Biological Systems,* (Gore, P.H., transl.), Pion/Academic, London

SCHEINER, S., LIPSCOMB, W.N. & KLEIER, D.A. (1976) Molecular orbital studies of enzyme activity. 2. *J. Am. Chem. Soc. 98,* 4770-4777

SZENT-GYÖRGYI, A. (1976) *Electronic Biology and Cancer,* Dekker, New York

SZENT-GYÖRGYI, A. (1977) The living state and cancer. *Proc. Natl. Acad. Sci. U.S.A. 74,* 2844-2847

SZENT-GYÖRGYI, A. (1978) *The Living State and Cancer,* Dekker, New York

SZENT-GYÖRGYI, A. & MCLAUGHLIN, J.A. (1978) Charge transfer and dielectric constant. *Biol. Bull. 155,* 468

Discussion

Ladik: You mentioned the triplet state of methylglyoxal. You use a closed-shell Hartree-Fock wave function for the ground state and different orbitals for different spin wave functions in the excited state. That means that you had some correlation in the excited state but none in the ground state. In that way the excitation energy is too small, because you 'push down' the excited state energy. Why were the energies of the first triplet excited states of dimethylglyoxal and glyoxal similarly not too low?

Abdulnur: I attribute this to the fact that the excitation in methylglyoxal occurs on the ketonic group, as in acetaldehyde and formaldehyde, whereas in dimethylglyoxal and glyoxal it occurs on the two symmetrical carbonyl groups.

Neuberger: Many aldehydes carrying electronegative substituents (e.g. trichloroacetaldehyde) exist solely in the hydrated carbonyl form in water. What is the equilibrium for glyoxal between the aldehyde form and hydrated form?

Abdulnur: Using a simpler method (the semi-empirical INDO technique) we calculated that the hydrated form was quite stable. The present calculations are for the reactive unhydrated aldehyde form of methylglyoxal. The nucleophilic attack rate of the amine is assumed to be slow enough for methylglyoxal dehydration to be adequate, and thus not rate-limiting.

Neuberger: You postulate that the hydronium ion H_3O^+ reacts with the non-protonated amine of lysine. In your conditions there must be little of the anionic form of lysine; the pK of the ϵ-amino group of lysine is about 10.4 and, therefore, at pH 7.4 only about 0.1% of lysine will be unprotonated and there will be a low concentration of hydronium ion.

Abdulnur: The model adopted in our calculations for the interaction of methylglyoxal with methylamine favours a high enough pH at which a significant amount of methylamine is present in the unprotonated form, but still not too high to prevent some H^+ being present.

Williams: Below the pK value no help will come from the proton because every bit of help in the catalytic rate constant is lost in the lowering of the reactant concentration term. In water the proton has no effect if protonation of the substrate removes it from the possibility of reaction.

Esterbauer: Theoretical calculations are very important if we are to explain the chemical reaction but we also need to know the equilibrium constants for, say, methylamine with methylglyoxal and the pH dependence of those equilibrium constants. Are there any measurements besides the polarographic one of Königstein and his colleagues (Königstein & Fedoronko 1970)? For reactions of α-dicarbonyl compounds including methylglyoxal with various amines they derived values (dissociation constants) of about 10 mmol/l, which means that the reaction products of the ketoaldehydes and the amines are not very stable.

We also find that proton transfer is very important in the reaction of α,β-unsaturated aldehydes with SH compounds. Here, too, the proton transfer may be the rate-limiting step. For instance, in a phosphate-buffered solution the rate constant may be a hundred times higher than in a buffer-free solution at the same pH. The second-order rate constant for the reaction of 4-hydroxypentenal with thioglycolic acid ethyl ester at pH 7.0 is, for example, 0.10 $l\ mol^{-1}\ s^{-1}$ without buffer and 34 $l\ mol^{-1}\ s^{-1}$ in the presence of 0.2 M-phosphate buffer. This increase is simply due to $H_2PO_4^-$ and HPO_4^{2-}, which act as proton donors (Esterbauer 1970). I am sure that in the reaction of α-ketoaldehydes with primary amines proton transfer reactions are very important too.

Williams: So, you kept the pH constant and added buffer anions?

Esterbauer: Yes; but the ionic strength is not critical.

Williams: It could be that you change the proton transfer rate—that is not the same as a catalytic requirement based on changes of proton concentration.

Pethig: The reaction of methylglyoxal with proteins as a function of pH may be instructive: the reaction takes longer to go to completion, the lower the pH. This would seem to indicate a proton dependence.

Ladik: Dr Abdulnur, how will you take into account the effect of the dielectric constant?

Abdulnur: We are now studying this problem.

Phillips: Dr Ladik, what do you mean by dielectric constant—a bulk concept—when you are considering molecular interactions?

Ladik: This is a problem. Either one approaches it by calculating all the molecules, in which case no dielectric constant is needed, or one treats only a few molecules explicitly and describes the effect of the rest with the aid of a dielectric constant.

Pethig: One could imagine that dielectric screening would reduce the repulsion forces mentioned by Dr Abdulnur in his paper; however, one cannot readily use the concept of the dielectric constant associated with the solvent when the atoms under consideration are, say, 0.2 nm apart.

Abdulnur: Although the two interesting molecules are separated by very short distances, they are nevertheless embedded in the solvent. By affecting the wave functions, and hence charge densities etc., of the two molecules, the surrounding solvent thus indirectly affects their interaction energy. Kirkwood & Westheimer (1938) took account of this by introducing an effective dielectric constant inside an imaginary cavity in which the interacting species reside.

Pethig: But that is a local field problem.

Eley: Using a modern computer the microscopic dielectric constant in the neighbourhood of an ion or molecule might be derived in principle from a potential energy calculation of the configuration and orientation of, say, the first 20 or so water molecules, followed by a calculation of their librational motions. It would be prohibitively expensive, I fear. Clementi is very active in the area (cf. Kistenmacher *et al.* 1974).

Ladik: We can calculate a few important configurations and run a CPA (coherent potential approximation) calculation on the system containing different configurations.

March: In special conditions we use dielectric theory to calculate interatomic or intermolecular interactions, but one needs not a constant but a dielectric function which depends on wavelength or wave number. The theory will only work if the interaction is mediated by the solvent. If we want to use the dielectric function, the interaction has to be weak—it's a linear response theory.

Williams: By discussing models (polylysine plus methylglyoxal systems) we may be missing a vital feature of real proteins. In an α-helical model for a protein, a lysine side-chain will protrude from the main protein backbone. In a 'real' protein some lysines will extend out from the surface but others will be buried in the interior of the protein. For over five years now, phenylglyoxal has become a standard reagent for highly reactive lysine and, to a greater extent, arginine residues in real proteins. This means there are proteins which will pick up glyoxals either on arginine or on lysine and glyoxals will then be powerful specific inhibitors of some enzyme action. The reason for this is simple; the normal pK_a of the 'external' lysine is about 10.4 (in aqueous solution) but that of the 'internal' groups may be totally different. The internal ones can be extremely reactive (because they have a lower pK which would give a difference in rate). Such lysines are known to be specific reagents for enzyme-attacking groups. The glyoxals then block specifically certain enzymes—usually phosphate-handling enzymes.

Fodor: The specificity for arginine is understandable; it will react to give an imidazole.

Neuberger: In many cases one assumes (and the evidence is not clearcut) that the formation of a Schiff base is followed by a reaction with an SH group which may lie near the amino group of lysine to give a stable thiazolidine—it is more stable than the Schiff base. The reaction that you described, that with phenylglyoxal, and the reaction between an amine group and a thiol, are all possibilities. At present we lack the facts.

Williams: Has the fate of radiolabelled glyoxal in cells been investigated? The reaction of polylysine and glyoxal may be fascinating but the use of the tracers at the cellular level will give us sound information.

Conroy: The inhibitory effects of methylglyoxal on enzymes containing thiol groups have been studied (reviewed by Schauenstein *et al.* 1977). Kun (1950) has shown that methylglyoxal deactivates functional SH group enzymes which participate in energy metabolism. A number of enzymes have been reported to be inhibited by methylglyoxal *in vitro:* hexokinase, aldolase, and glyceraldehyde-3-phosphate, glutamate, malate and succinate dehydrogenases. Spolter *et al.* (1965) examined in detail the inhibition of aldolase. It is not certain whether methylglyoxal reacts with some or all of the thiol groups thought to be involved in enzyme catalytic activity, since Miyagi & Tsuiki (1971) reported that the compound strongly inhibited L-glutamine: D-fructose-6-phosphate aminotransferase but failed to inhibit 11 other enzymes involved in carbohydrate metabolism and known to possess thiol groups critical to catalytic activity. In principle, α-ketoaldehydes can react in proteins with the guanidine group of arginine, the ϵ-amino group of lysine, and the SH group of cysteine. It is not certain whether any specificity is shown by methylglyoxal towards these groups or indeed if the concentrations required would be attainable within the cell. The adducts between methylglyoxal and purines have been studied (Shapiro & Hachmann 1966; Litt 1969; Shapiro *et al.* 1969). The glyoxal moiety can be removed by the addition of an excess of a thiol (Litt & Hancock 1967). ^{14}C-labelled methylglyoxal has been used to study its binding to nucleic acids (Krymkiewicz 1973). [^{14}C]methylglyoxal binds to tRNA and DNA. A much greater quantity of [^{14}C]methylglyoxal binds to heat-denaturated DNA than to native DNA. The quantities of [^{14}C]methylglyoxal that bound to polyadenylic and polycytidylic acids were 7% and 3%, respectively, of the quantity binding to polyguanylic acid. Relatively little information is available on the cellular and tissue distribution of [^{14}C]methylglyoxal in mammalian systems.

Slater: We are now using labelled methylglyoxal and the methylglyoxal–vitamin C complex with double labelling to determine the tissue distributions after single intraperitoneal injections. Thereafter we have to study the cellular and intracellular distributions, and the kinetics of excretion.

References

ESTERBAUER, H. (1970) Kinetik der Reaktion von Sulfhydrylverbindungen und α,β-ungesättigten Aldehyden in wässrigem System. *Monatsh. Chem. 101,* 782-810

KIRKWOOD, J.G. & WESTHEIMER, F.H. (1938) The electrostatic influence of substituents on the dissociation constants of organic acids. *J. Chem. Phys. 6,* 506-512

KISTENMACHER, H., POPKIE, H. & CLEMENTI, E. (1974) Study of the structure of molecular complexes. VIII Small clusters of water molecules surrounding Li^+, Na^+, K^+, F^- and C^- ions. *J. Chem. Phys. 61,* 799-815

KÖNIGSTEIN, J. & FEDORONKO, M. (1970) Condensation products of some carbonyl compounds with primary amines and their polarographic estimation. *Proc. Anal. Conf. 3rd, 2,* 113-122

KRYMKIEWICZ, N. (1973) Reactions of methylglyoxal with nucleic acids. *FEBS (Fed. Eur. Biochem. Soc.) Lett. 29,* 51-54

KUN, E. (1950) Inhibition of succinic dehydrogenase by methylglyoxal. *J. Biol. Chem. 187,* 289-297

LITT, M. (1969) Structural studies on transfer ribonucleic acid. 1. Labelling of exposed guanine sites in yeast phenylalanine transfer ribonucleic acid with kethoxal. *Biochemistry 8,* 3249-3253

LITT, M. & HANCOCK, H. (1967) Kethoxal—a potentially useful reagent for the determination of nucleotide sequences in single stranded regions of transfer ribonucleic acid. *Biochemistry 6,* 1848-1854

MIYAGI, T. & TSUIKI, S. (1971) Effect of phosphoglucose isomerase and glucose-6-phosphate on UDP-*N*-acetylglucosamine inhibition of L-glutamine: D-fructose 6-phosphate aminotransferase. *Biochim. Biophys. Acta 250,* 51-62

SCHAUENSTEIN, E., ESTERBAUER, H. & ZOLLNER, H. (1977) *Aldehydes in Biological Systems* (Gore, P.H., transl.), Pion/Academic, London

SHAPIRO, R. & HACHMANN, J. (1966) The reaction of guanine derivatives with 1,2-dicarbonyl compounds. *Biochemistry 5,* 2799-2807

SHAPIRO, R., COHEN, B.I., SHIUEY, S.J. & MAURER, H. (1969) On the reaction of guanine with glyoxal pyruvaldehyde and kethoxal, and the structure of the acylguanines. A new synthesis of N^2-alkylguanines. *Biochemistry 8,* 238-245

SPOLTER, P.D., ADELMAN, R.C. & WEINHOUSE, S. (1965) Distinctive properties of native and carboxypeptidase treated aldolases of rabbit muscle and liver. *J. Biol. Chem. 240,* 1327-1337

Methylglyoxal production in human blood

RICHARD B. BRANDT and STEPHEN A. SIEGEL

Department of Biochemistry and the MCV/VCU Cancer Center, Medical College of Virginia, Virginia Commonwealth University, Richmond, Virginia

Abstract The research of Szent-Györgyi and others has suggested that the three-carbon ketoaldehyde methylglyoxal has a potential role in the control of cell growth. Its metabolism to D-lactate (not the L-lactate of glycolysis) is catalysed by the mammalian enzymes glyoxalase I (*S*-lactoyl-glutathione methylglyoxal-lyase, isomerizing; EC 4.4.1.5) and glyoxalase II (*S*-2-hydroxyacylglutathione hydrolase; 3.1.2.6), with glutathione as a coenzyme. Direct determination of methylglyoxal in biological tissues is difficult because of the active glyoxalase system. However, the product of the glyoxalase system, D-lactate, should indicate formed or added methylglyoxal. A stereospecific assay was used to measure D-lactate in human plasma; it involved the spectrophotometric analysis of NADH at 340 nm catalysed by D-lactate dehydrogenase (D-lactate:NAD$^+$ oxidoreductase; EC 1.1.1.28) from *Lactobacillus leichmannii*. Blood collected by venepuncture was used for the determination of the plasma concentration of D-lactate. The mean concentration for seven normal subjects was 0.023 mM ± 0.002 S.E.M. When the glycolytic pathway in whole blood was inhibited *in vitro* with fluoride, a significant increase in D-lactate was found (about 0.15 mM/hour at 37 °C). Added methylglyoxal also produced an increase in D-lactate formation. Some specific precursors of L-lactate (dihydroxyacetone phosphate, for example) added to whole blood produced an increased concentration of D-lactate, even when glycolysis was not inhibited. This finding indicates that catabolites of glucose lead to methylglyoxal synthesis and suggests a control function for the glyoxalase enzyme system in glycolysis that could be exploited for cancer therapy.

In the 1960s, Szent-Györgyi and his co-workers (Szent-Györgyi *et al.* 1962, 1967; Egyud & Szent-Györgyi 1968) reported the inhibition of tumour growth with extracts from normal tissues. Characterization using derivatization and spectral analysis demonstrated that the inhibitor was an α,β-dicarbonyl (Egyud 1965; Fodor *et al.* 1967; Szent-Györgyi 1967). Methylglyoxal, an α,β-dicarbonyl, is the lowest class member of the ketoaldehydes. A biological system which catalyses the formation of lactate from methylglyoxal was reported early in this century (Dakin & Dudley 1913; Neuberg 1913). For a long time methylglyoxal was thought to be the source of lactic acid formed in animal tissues. It had been reported that D-lactate was formed from glucose in the blood of dogs (Dakin & Dudley 1913) and that methylglyoxal was formed from hexose diphosphates (Neuberg & Kobel 1929). The discovery in the 1930s that glutathione was a coenzyme for the glyoxalase system (Lohmann 1932) but was not required for glycolysis removed methylglyoxal from consideration as an active participant in glycolysis. Racker (1950) was in part responsible for the discovery that the glyoxalase enzyme system consists of two distinct enzymes: glyoxalase I and glyoxalase II. Fig. 1 shows the formation of D-lactate (not the L-lactate of glycolysis) catalysed by the mammalian enzymes, glyoxalase I (*S*-lactoyl-glutathione methylglyoxal-lyase, isomerizing; EC 4.4.1.5) and glyoxalase II (*S*-2-hydroxyacylglutathione hydrolase; EC 3.1.2.6), with glutathione as a coenzyme.

Methylglyoxal inhibits tumour growth in humans and animals—it inhibits protein synthesis (Szent-Györgyi & Egyud 1966), possibly by reaction with tRNA (transfer RNA) (Litt 1969)—and its derivatives have been used in the clinical treatment of cancer (Regelson & Holland 1963). Thus information on its synthesis in humans would seem to be of vital importance. There have been a number of reports of the synthesis of methylglyoxal in tissues (Needham & Lehmann 1937) by polyvalent cations (Riddle & Lorenz 1968) and by lysine (Bonsignore *et al.* 1972); however, these syntheses were non-enzymic in nature. The isolation of a methylglyoxal synthetase from *Escherichia coli* (Cooper & Anderson 1970) suggested a different approach to

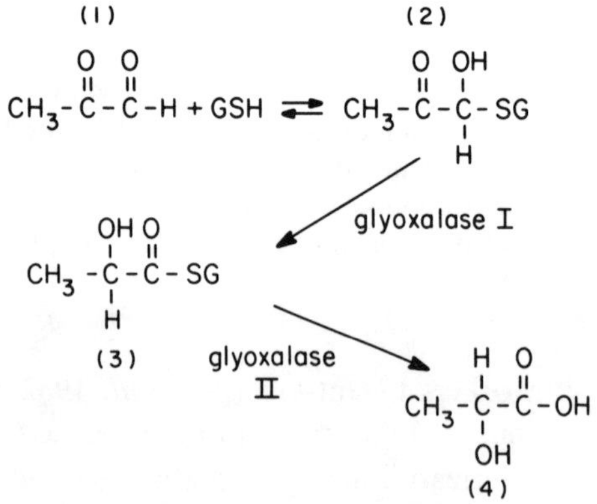

FIG. 1. The glyoxalase enzyme system. (1) Methylglyoxal; (2) the spontaneously formed adduct (hemimercaptal) between methylglyoxal and glutathione (GSH); (3) *S*-lactoylglutathione; and (4) D-lactic acid.

isolating a methylglyoxal synthetase from mammalian tissue, similar to that used in bacteria. When rat liver homogenates were studied in conditions where glyoxalase I activity was inhibited, the addition of dihydroxyacetone phosphate led to the formation of methylglyoxal, as measured spectrophotometrically (Gilbert & Brandt 1975). However, when conditions were used that would be expected to inhibit methylglyoxal synthetase, the catalytic activity for methylglyoxal formation was not decreased (Siegel & Brandt 1978*a*). Great difficulty was encountered in attempting to study the synthesis of methylglyoxal in tissue extracts, since the glyoxalase enzyme rapidly catabolizes it to D-lactate. Indeed, the difficulty of directly measuring methylglyoxal in biological systems, which results from its reactivity with SH groups, amines or the glyoxalase system, suggests that the indirect measurement of D-lactic acid formation by stereospecific means presents a better approach to measuring methylglyoxal than previously used methods.

It was found when using human blood in conditions in which glycolysis was inhibited that significant amounts of D-lactate were synthesized when the blood was incubated at 37 °C for various times. Direct addition of methylglyoxal to the blood showed that an active glyoxalase system was present by the rapid formation of D-lactate (Siegel & Brandt 1978*b*). In this paper D-lactate formation is examined during the incubation of whole human blood with and without inhibition of glycolysis and with the addition of various glycolytic intermediates.

EXPERIMENTAL

Apparatus

A Gilford Recording Spectrophotometer equipped with a constant temperature compartment, automatic cell changer and strip chart recorder was used for absorbance measurements using quartz cells, path length 1 cm.

Reagents and chemicals

All chemicals used in this study were reagent grade and used as supplied. All biochemicals were from Sigma Chemical. Methylglyoxal was purified and the concentration determined as previously described (Gilbert & Brandt 1975).

Procedure

Blood was collected from human volunteers by venepuncture and transferred rapidly into tubes containing heparin or oxalate–fluoride; it was then used immediately in the experiments described. Plasma was assayed for L-lactic acid by Sigma kit 826-3 and D-lactic acid by the method of Siegel & Brandt (1977). The D-lactate assay used deproteinized plasma added to glycine–hydrazine buffer at alkaline pH with NAD^+ in the buffer. Addition of D-lactate dehydrogenase (EC 1.1.1.28) from *Lactobacillus leichmannii* catalyses the formation of NADH which is then spectrophotometrically determined at 340 nm. The D-lactate concentration in plasma was determined from standard curve linear regression data and appropriate dilution factors. Corrections for the effect of L-lactate on the D-lactate assay were determined (R.B. Brandt *et al.*, unpublished 1978); they had a significant effect on the absolute values of D-lactate only when the D-lactate concentration was about 0.05 mM or lower.

The three types of experiments reported here are: *(a)* estimations in normal control subjects where the blood was collected in heparinized tubes and the plasma immediately removed for assay; *(b)* incubation of whole blood in tubes containing fluoride or heparin at 37 °C in a shaker bath from 0 to 3 hours, the blood being removed at various time intervals and the plasma assayed; and *(c)* the effect of glycolytic intermediates: various substrates of known concentration were added to whole blood in tubes containing fluoride or heparin; the tubes were mixed and half the blood was removed for zero-time plasma lactate determination. The remainder of the blood was incubated for two hours at 37 °C before determination of plasma lactate.

RESULTS AND DISCUSSION

Table 1 shows the L- and D-lactate concentration in plasma from seven normal human males (ages 22–29 years). The corrected mean plasma concentration of D-lactate for all subjects was 0.023 mM, which is 1.5% of the L-lactate value (1.51 ± 0.22 S.E.M., $n = 7$). No significant linear correlation between D- and L-lactate was found. The low but consistent (range 0.017–0.028 mM) values in normal plasma for D-lactate suggest a constant turnover of methylglyoxal to D-lactate. Changes in D-lactate levels may reflect alterations in the glyoxalase enzyme system or in methylglyoxal synthesis.

When whole blood is incubated in a system in which the glycolytic pathway is inhibited at the enolase step by fluoride (inhibition of formation of phosphoenolpyruvate from 2-phosphoglycerate), the potential exists for increasing the amounts of the glycolytic substrates prior to the inhibition step, since glucose, as a source of catabolites, is present (normally about 5 mM). The red blood cell is unique in that it cannot synthesize proteins, so that any increase

TABLE 1

D- and L-lactate concentrations in normal human plasma

Subject	[L-Lactate][a] (mM)	[D-Lactate][b] (mM)	[D-Lactate] corrected[c] (mM)
JW	1.48	0.032	0.020
MW	0.74	0.030	0.020
OG	1.29	0.036	0.024
PF	1.47	0.040	0.028
MB	1.06	0.037	0.027
MG	2.40	0.032	0.017
FV	2.12	0.039	0.024

[a]Mean of four determinations, S.E.M. ± 0.01 or less, for all subjects.
[b]Mean of four determinations, S.E.M. ± 0.001 or less, for all subjects.
[c]Mean of four determinations, S.E.M. ± 0.001 or less, for all subjects. Mean for D-lactate for all subjects: 0.023 mM ± 0.002 (S.E.M., $n = 7$), corrected for a small effect of L-lactate on the D-lactate determination.

in the amount of D-lactate formed must be due to increased formation of methylglyoxal and not an effect on increased glyoxalase synthesis. Fig. 2 shows the increase in plasma D-lactate concentration when whole blood was incubated in tubes containing fluoride. The actual values are shown in Table 2. Note that as expected the blood in the tubes containing fluoride showed little change in L-lactate concentration, while the D-lactate concentration increased by 0.478 mM (3700% increase). Heparinized tubes of blood showed a large increase in L-lactate and essentially no change in D-lactate. When fluoride was added to the heparinized tubes the results were indistinguishable from those of the oxalate–fluoride tubes.

TABLE 2

Mean concentration of D- and L-lactate in whole human blood incubated at 37°C

Time (hours)	[D-Lactate] (mM)[a]		[L-Lactate] (mM)[b]	
	Heparin	Fluoride	Heparin	Fluoride
0	0.026 (0.0013)	0.013 (0.0005)	0.70 (0.027)	0.90 (0.019)
1	0.040 (0.0012)	0.146 (0.0014)	3.61 (0.030)	1.16 (0.023)
2	0.046 (0.0006)	0.330 (0.0011)	5.69 (0.051)	1.22 (0.007)
3	0.058 (0.0008)	0.491 (0.0059)	7.22 (0.28)	1.12 (0.007)
Change in 3 hours.	0.032	0.478	6.52	0.22

[a]Values in parentheses are ± S.E.M. for four determinations, uncorrected for small effect of L-lactate on the D-lactate determination.
[b]Values in parentheses are ± S.E.M.

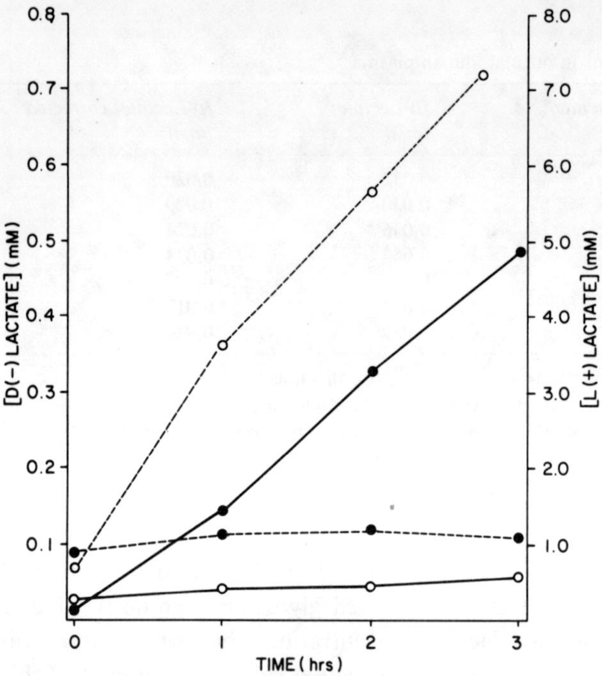

FIG. 2. Plasma concentration of D- and L-lactate (mM) after incubation of whole blood at 37 °C in tubes containing heparin or fluoride. Each point is the mean value of four determinations. ○---○, L-lactate with heparin; •---•, L-lactate with fluoride; ○——○, D-lactate with heparin; •——•, D-lactate with fluoride.

When methylglyoxal was added to blood an increase in D-lactate was found on incubation that was proportional to the concentration of methylglyoxal. From the data shown in Table 3, where the initial concentration of methylglyoxal was 0.25 mM, the two-hour concentration of D-lactate was 0.520 mM. Subtraction of the amount of D-lactate formed from the control value (fluoride-treated blood) shows that a 102% recovery of D-lactate via methylglyoxal had occurred. When a 50 mM initial concentration of methylglyoxal was used, both the zero-time and two-hour samples showed concentrations of greater than 1.7 mM; because of the analytical limits, differences could not be determined, although this rapid increase in D-lactate suggests that blood has a large capacity for utilizing methylglyoxal. For the glycolytic substrates used, a positive change in the percentage of D-lactate from the control indicates a net formation of D-lactate. Large concentrations of substrates were used by comparison with the steady-state levels in red blood cells (Minakami &

TABLE 3

Effect of glycolytic and other substrates on the formation of D- and L-lactate in whole blood with fluoride

Substrate[a] added	Concentration Normal (mM)	Experimental (mM)	Change[b] in [D-lactate] (mM)	Change in [L-lactate] (mM)	% increase[c] D-lactate	% increase L-lactate
None	–	–	0.264	0.273	–	–
Glucose	5.0	50	0.284	0.251	+8	–
G-6-P	0.08	25	0.111	0.040	–	–
F-1,6diP	0.03	17	0.294	0.023	+11	–
DHAP	0.14	10	0.905	0.050	+243	–
3-PG	0.12	25	0.230	0.114	–	–
2-PG	0.03	5	0.240	0.037	–	–
PEP	0.02	14	0.095	0.580	–	+112
Pyruvate	0.05	25	0.177	1.450	–	+431
L-Lactate	2.9	48	0.06	–	–	–
Methylglyoxal	–	0.25	0.520	0.187	+97	–

[a]Glucose-6-phosphate (G-6-P), fructose 1,6-diphosphate (F-1,6diP), dihydroxyacetone phosphate (DHAP), 3-phosphoglycerate (3-PG), 2-phosphoglycerate (2-PG), phosphoenolpyruvate (PEP).
[b]Mean of four determinations for change in D- or L-lactate plasma concentration between zero time incubation and 2 h at 37 °C.
[c]% increase in D- or L-lactate is: (concentration of lactate found with substrate)−(concentration of lactate found when no substrate added) ÷ (concentration of lactate found when no substrate added) × 100.

TABLE 4

Effect of glycolytic and other substrates on the formation of D-lactate in whole blood with heparin

Substrate[a]	Change[b] in [D-lactate] (mM)	% Increase[c] D-lactate
None	0.017	–
Glucose	0.024	+41
G-6-P	0.005	–
F-1, 6diP	0.052	+205
DHAP	1.00	+5800
3-PG	0.013	–
2-PG	0.026	+53
PEP	0.047	+176
Pyruvate	0.200	+1080
L-Lactate	0.013	–
Methylglyoxal	0.308	+1700

[a]Substrate concentrations added as in Table 3. (See Table 3 for abbreviations.)
[b]Mean of four determinations for change in D-lactate plasma concentration over two hours incubation at 37 °C.
[c]See footnote *c* in Table 3.

Yoshikawa 1965), in order to overcome transport barriers into the red blood cell. The largest increase in D-lactate (243%) occurred with dihydroxyacetone phosphate, which agrees with our previous data for liver (Siegel & Brandt 1978*a*). Fructose 1,6-diphosphate and glucose (most likely as precursors of dihydroxyacetone phosphate) caused increased formation of D-lactate, presumably from methylglyoxal. Perhaps, even with the high concentrations of substrates used, a significant barrier to their transport into the red blood cell exists and a better experiment would be one where the cells were disrupted. However, the increase in L-lactate that occurred when either phosphoenolpyruvate or pyruvate was used (after the blocked step in glycolysis) indicates that these substrates entered the red cell, or that significant lysis allowed the reaction to occur in the plasma. Table 4 contains data for heparinized blood (no fluoride). Dihydroxyacetone phosphate is the substrate that gives the largest amount of D-lactate. The large percentage increase for some of the substrates used reflects the normal low control value of 0.017mM-D-lactate. The increase in D-lactate concentration after the addition of pyruvate, phosphoenolpyruvate and 2-phosphoglycerate may indicate the formation of D-lactate directly from pyruvate, catalysed by a D-lactate dehydrogenase in blood. Methylglyoxal added to the heparinized blood also leads to increased synthesis of D-lactate.

CONCLUSION

The experiments reported here showed the formation of D-lactate in human blood incubated at 37 °C when methylglyoxal was added, most likely

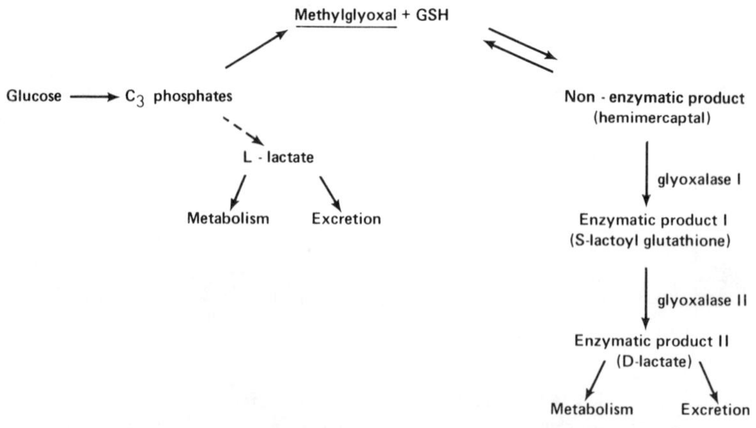

FIG. 3. Formation and catabolism of methylglyoxal from glucose.

catalysed by the glyoxalase enzyme system. When either glycolysis was inhibited, or large concentrations of glycolytic substrates were added, there was a marked increase in D-lactate formation. It is not yet known if the formation of methylglyoxal from dihydroxyacetone phosphate is enzymic, but the existence of this pathway (Fig. 3) provides a potential for growth control, when glycolysis is impaired, by control through the glyoxalase system.

ACKNOWLEDGEMENTS

This research was supported in part by funds from the National Foundation for Cancer Research. The cheerful and competent help of Mike Waters and of the first-year medical students, Mike Bloch, Ozzie Gutterman, Mike Getter, Perry Fine, Frank Viverette and John Wimbush, is gratefully acknowledged. I thank my Co-regional Director for the National Foundation for Cancer Research, Professor Bill Regelson, and my colleague Associate Professor of Biochemistry, Ed Kline, for encouragement and many helpful suggestions.

References

BONSIGNORE, A., LEONCINI, G., SIRI, A. & RICCI, D.(1972) Aminic compounds tested as catalysts for glyceraldehyde conversion to methylglyoxal. *Ital. J. Biochem. 21*, 169-178

COOPER, R.A. & ANDERSON, A. (1970) The formation and catabolism of methylglyoxal during glycolysis in *E. coli. FEBS (Fed. Eur. Biochem. Soc.) Lett. 11*, 273-276

DAKIN, H.D. & DUDLEY, H.W. (1913) On glyoxalase. *J. Biol. Chem. 14*, 423-431

EGYUD, L.G. (1965) Studies on autobiotics: chemical nature of retine. *Proc. Natl. Acad. Sci. U.S.A. 54*, 200-202

EGYUD, L.G. & SZENT-GYÖRGYI, A. (1968) Cancerostatic action of methylglyoxal. *Science (Wash. D.C.) 160*, 1140

FODOR, G., SACHETTO, J.P., SZENT-GYÖRGYI, A. & EGYUD, L.G. (1967) Ketone aldehydes in animal tissues. *Proc. Natl. Acad. Sci. U.S.A. 57,* 1644-1650

GILBERT, R.P. & BRANDT, R.B. (1975) Spectrophotometric determination of methylglyoxal with 2,4-dinitrophenylhydrazine. *Anal. Chem. 47*, 2418-2422

LITT, M. (1969) Structural studies on transfer ribonucleic acid. I. Labeling of exposed guanine sites in yeast phenylalanine transfer ribonucleic acid with kethoxal. *Biochemistry 8*, 3249-3253

LOHMANN, K. (1932) Behavior of the enzymatic attack on methylglyoxal for the synthesis of lactic acid. *Biochem. Z. 254*, 332-354

MINAKAMI, S. & YOSHIKAWA, H. (1965) Thermodynamic considerations of erythrocyte glycolysis. *Biochem. Biophys. Res. Commun. 18*, 345-349

NEEDHAM, J. & LEHMANN, H. (1937) Intermediary metabolism in embryonic life. *Biochem. J. 31*, 1913-1925

NEUBERG, C. (1913) Further investigations concerning the biochemical transformation of methylglyoxal into lactic acid. *Biochem. Z. 51*, 484-508

NEUBERG, C. & KOBEL, M. (1929) Continuation on the catabolic reactions for the formation of methylglyoxal with yeast. *Biochem. Z. 210*, 466-488

RACKER, E. (1950) The mechanism of action of glyoxalase. *J. Biol. Chem. 190*, 685-696

REGELSON, W. & HOLLAND, J.F. (1963) Clinical experience with methylglyoxal bis(guanylhydrazone) dihydrochloride: a new agent with clinical activity in acute myelocytic leukemia and the lymphomas. *Cancer Chemother. Rep. 27*, 15-26

RIDDLE, V. & LORENZ, F.W. (1968) Nonenzymatic polyvalent anion-catalyzed formation of methylglyoxal as an explanation of its presence in physiological solutions. *J. Biol. Chem. 243*, 2708-2724

SIEGEL, S.A. & BRANDT, R.B. (1977) Assay system for D-lactate in plasma or serum: preliminary results in cancer patients. *Virginia Med. 104*, 571

SIEGEL, S.A. & BRANDT, R.B. (1978*a*) Non-enzymatic formation of methylglyoxal from dihydroxyacetonephosphate in rat liver extract. *Virginia J. Sci.*, in press

SIEGEL, S.A. & BRANDT, R.B. (1978*b*) The synthesis of D-lactate in human blood. *Virginia Med. 105*, 645

SZENT-GYÖRGYI, A. (1967) On retine. *Proc. Natl. Acad. Sci. U.S.A. 57*, 1642-1643

SZENT-GYÖRGYI, A. & EGYUD, L.G. (1966) On the regulation of cell division. *Proc. Natl. Acad. Sci. U.S.A. 56*, 203-207

SZENT-GYÖRGYI, A., HEGYELI, A. & MCLAUGHLIN, J.A. (1962) Constituents of the thymus gland and their relation to growth, fertility, muscle and cancer. *Proc. Natl. Acad. Sci. U.S.A. 48*, 1432-1439

SZENT-GYÖRGYI, A., EGYUD, L.G. & MCLAUGHLIN, J.A. (1967) Ketoaldehydes and cell division. *Science (Wash. D.C.) 155*, 539-541

Discussion

Williams: You quote the mean concentration of D-lactate in human plasma as 0.023 mmol/l. If that were in equilibrium with methylglyoxal and a possible reductase system, how much methylglyoxal would there be?

Brandt: Making several assumptions I calculate that the highest concentration of methylglyoxal, both bound and unbound, will be about 0.02 mmol/l. Our technique for measuring D-lactate in plasma uses perchloric acid for deproteinization and, therefore, this figure represents not only free circulating but plasma protein-bound forms. The steady-state concentration of free methylglyoxal in a tissue would probably be considerably less, maybe by a factor of 10 or even 100. When the concentrations of glycolytic intermediates were raised (see Table 4, p. 217) the D-lactate concentration increased considerably; this suggests that there would be a strong elevation of methylglyoxal if carbohydrate metabolism were impaired.

Williams: What steady-state concentration does methylglyoxal reach when it is added to blood (with the blood enzymes as they are)? Also, at what concentration does it become inhibitory of any enzyme in blood?

Brandt: A concentration of 1mM-methylglyoxal is sufficient to kill mice (R.B. Brandt, unpublished work 1975). In bacterial studies (Egyud & Szent-Györgyi 1966) concentrations lower than 1 mmol/l inhibit growth. In cell culture, Greeg (1968) found that 0.1mM-methylglyoxal is not an unusual concentration to inhibit growth.

Williams: That is a factor of about 10 between any biochemically known action in this sort of system and the concentration you suspect is present in blood.

Brandt: Yes and no! On the one hand, in the microenvironment around an enzyme the local concentration may be much higher whereas, on the other hand, the blood concentration may be low because of the high glyoxalase activity in red blood cells.

Slater: With some cell systems in culture (e.g. the incorporation of [^{3}H]-thymidine into 3T3 cells, mentioned earlier, p. 169) or with isolated enzyme systems (e.g. stimulation of lipid peroxidation in rat liver microsomes) one can see strong effects (i.e. 50% inhibition or stimulation) produced by methylglyoxal at concentrations of 1–10 μmol/l—that is, the same order of concentration as Dr Brandt quoted .

Dianzani: In a reconstructed protein synthetic system we find inhibition at methylglyoxal concentrations of about 0.5 mmol/l. For other aldehydes, however, we find inhibition at lower concentrations. With malealdehyde, for instance, inhibition starts at 0.1 mmol/l. With other types of cellular targets, methylglyoxal causes damage at lower concentrations. For instance, the binding of bromosulphophthalein to ligandin (or protein Y) from liver cells is decreased by 2.3 μmol/l methylglyoxal.

Bondy: As many tumours seem to have an active anaerobic glycolytic system, it seems, from your data on blood, as though inhibitory concentrations of methylglyoxal might be found in tumours. Have you checked this?

Brandt: We have considered that and I plan to repeat some of the old Warburg studies, using the respirometer, to determine whether the lactate that was claimed to be formed is D-lactate. In a symposium on lactic acid, Brin (1965) showed by optical rotation that it is not D-lactate but L-lactate which is produced from anaerobic glycolysis of Jensen sarcoma and Ehrlich carcinomas.

Bondy: When one adds a precursor of L-lactate to the blood system, some D-lactate is formed. Some of this must 'leak over' into methylglyoxal. If that is so, does it happen in tumours?

Brandt: I don't know. We only recently finished developing the assay. Looking forward, I would expect to find elevated concentrations of D-lactate in diabetes or other diseases involving abnormal carbohydrate concentrations. There is a genetic disease, D-lactaturia, in which the patient urinates D-lactate but seems to have no other problems (Duran *et al.* 1977).

Neuberger: This is nice work, Dr Brandt, but I have to make the point that the evidence that D-lactate is derived from methylglyoxal is indirect. At present we don't know any other precursor of D-lactic acid but the possibility cannot be dismissed that there is an enzyme which acts on some precursor from the glycolytic cycle to produce this acid. To judge from the effect of fluoride and from the fact that dihydroxyacetone phosphate is a precursor, I conclude that it is likely to be connected with the glycolytic cycle. That methylglyoxal is an intermediate is a reasonable assumption but it is not proved. Before one accepts the idea one would like to have evidence (which I realize is difficult to obtain) of an increase in the concentration of methylglyoxal in the blood.

Brandt: I accept your reservations. We tried to substantiate our theory by inhibiting the glyoxalase I in liver both by removing glutathione and by using a strong inhibitor of glutathione (Siegel & Brandt 1978). It did not occur to us during that study that the blood system was much simpler, in that the red cell does not have a nucleus, it does not synthesize proteins and thus does not make more glyoxalase I. Our next step will be to lyse red cells, purify the system by removing all substrate, and then to add radioactive precursors. In such a system we shall not have to worry about glutathione and glyoxalase catalysing the formed methylglyoxal to D-lactate.

Williams: Is there a path back from pyruvate in red cells?

Brandt: That is possible. The step from phosphoenolpyruvate to pyruvate is usually written with a unidirectional arrow because pyruvate kinase (EC 2.7.1.40) normally (in conditions that obtain in biological tissues) keeps the equilibrium strictly in favour of pyruvate. The conversion of pyruvate into phosphoenolpyruvate goes through other steps that need mitochondrial enzymes. However, the addition of a large excess of pyruvate *in vitro* (see Table 4, p. 217) did result in some D-lactate being formed when the enolase step was not inhibited. That implies that the step from phosphoenolpyruvate is slightly reversible and therefore leads to synthesis of methylglyoxal.

Williams: But racemic lactate should be formed non-enzymically in your system.

Brandt: With a large excess of pyruvate (25 mmol/1) Table 3, p. 217) the amount of L-lactate increased (with fluoride in the system) because even though glycolysis was blocked, pyruvate can be reduced to L-lactate. The amount of D-lactate did not increase, showing that D-lactate is not formed by direct reduction of pyruvate or from L-lactate.

Neuberger: Methylglyoxalase may have another function not connected with methylglyoxal. In green plants the synthesis of porphyrins does not proceed along the normal δ-aminolaevulinic acid pathway from glycine and succinate. Instead, 2-oxoglutaric acid is a direct precursor. The most reasonable assumption is that the 1-carboxy group of 2-oxoglutaric acid is reduced, so forming a glyoxal ($CHO.CO.CH_2.CH_2COOH$). This glyoxal is transaminated to δ-aminolaevulinic acid. The plant would then need to protect itself against the toxic effect of this glyoxal intermediate, in which case glyoxalase would have an extremely important function. I wonder whether glyoxalase did not develop when this pathway for the synthesis of haem and porphyrin was in operation. This is an alternative answer to the question of what is the function of glyoxalase.

Brandt: Again, that is possible. We may be able to elucidate the enzyme's function with some of the stereospecific enzymes and radioactive tracers that are now available. We cannot disregard the possibility that it prevents damage to cells instead of or as well as being involved in growth promotion.

References

BRIN, M. (1965) The synthesis and metabolism of lactic acid isomers. *Ann. N.Y. Acad. Sci. 119*, 942-956

DURAN, M., VAN BIERVLIET, J.P., KAMERLING, J.P. & WADMAN, S.K. (1977) D-Lactic aciduria, an inborn error of metabolism? *Clin. Chim. Acta 74*, 297-300

EGYUD, L.G. & SZENT-GYÖRGYI, A. (1966) Cell division, SH, ketoaldehydes and cancer. *Proc. Natl. Acad. Sci. U.S.A. 55*, 388-393

GREEG, C.H. (1968) Inhibition of mammalian cell division by glyoxals. *Exp. Cell Res. 50*, 65-72

SIEGEL, S.A. & BRANDT, R.B. (1978) Non-enzymatic formation of methylglyoxal from dihydroxyacetonephosphate in rat liver extract. *Virginia J. Sci.*, in press

Formation and properties of reactive aldehydes

E. SCHAUENSTEIN and H. ESTERBAUER

Department of Biochemistry, University of Graz, Austria

Abstract 4-Hydroxyalkenals react easily with SH groups in enzymes and structural proteins to give stable adducts in which the aldehyde is bound through a thioether linkage to the protein. This reaction represents the mechanism by which these compounds block certain metabolic processes (DNA, RNA and protein synthesis, respiration, glycolysis) and exert their antitumour activity.

A highly significant inverse correlation exists between the number of hydroxypentenal-sensitive SH groups in the soluble proteins and the doubling time of 12 transplantable animal tumours.

Hydroxyalkenals applied peritumourally completely prevent or significantly delay growth of the solid forms of four animal tumours. The almost non-toxic adducts of 4-hydroxypentenal–cysteine 1:1 and crotonal–cysteine 1:2 applied intraperitoneally prevent or inhibit growth of Ehrlich ascites tumour in mice.

The possible formation of unsaturated aldehydes *in vivo* from lipids, carbohydrates and diamines is discussed.

During our work on lipid peroxidation in aqueous systems and the effect of lipid oxidation products on tumour cells a very interesting aldehyde was discovered among the neutral, water-soluble oxidation products. This aldehyde was identified as *trans*-4-hydroxyoct-2-enal and proved to be a compound of considerable biochemical and biological interest because of its highly specific reactivity towards SH groups of proteins and enzymes involved in various metabolic processes (DNA, RNA and protein synthesis, glycolysis, respiration), and because of its antitumour properties. Subsequent search for other hydroxyalkenals with similar or improved antitumour activity led to the synthesis of a number of analogous 4-hydroxyalkenals with various chain lengths and additional functional groups. Among the aldehydes synthesized, *trans*-4-hydroxypent-2-enal (HPE) (Fig. 1) appeared to be the most promising model compound for a general extensive investigation of the chemical, bio-

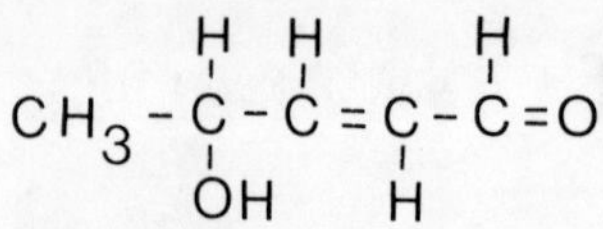

FIG. 1. Formula of *trans*-4-hydroxypent-2-enal.

chemical, physiological and antitumour properties of aldehydes of the *trans*-4-hydroxy-2-alkenal series. Consequently, most of the work done in this field by our laboratory deals with HPE and its effects on normal and tumour cells. Other α,β-unsaturated aldehydes were included in these studies in order to understand better the special features of the 4-hydroxyalkenals (for review see Schauenstein 1967; Schauenstein *et al.* 1977).

REACTION OF HPE AND OTHER 2-ALKENALS WITH SH GROUPS OF LOW MOLECULAR WEIGHT COMPOUNDS AND PROTEINS

HPE is a highly reactive electrophilic agent which reacts easily with the SH group in glutathione and other low molecular weight thiols such as cysteine, cysteamine, penicillamine, coenzyme A, mercaptoacetic acid and mercapto-acetic acid ethyl ester. The reactive nucleophilic species in fact is the RS^- ion formed by dissociation of the SH group. All reactions between HPE and RSH are therefore strongly catalysed by increasing the pH. The reaction products initially formed are saturated aldehydes in which the SH compound is bound at C-3 by a thioether linkage. The free aldehyde group of the primary product then reacts with the hydroxy group to give a cyclic hemiacetal.

Figure 2 shows as an example the reaction of HPE with glutathione. 2-Alkenals without a hydroxy group at C-4 (i.e. acrolein [propenal, acrylalde-hyde], crotonal), react in the same manner except that no cyclic hemiacetal is formed.

The investigation (Esterbauer *et al.* 1976) of the reaction between HPE and cysteine revealed that up to a mixing ratio of aldehyde:CySH = 1, a monoadduct with cyclic hemiacetal structure is formed (Fig. 3). If cysteine is present in excess the free aldehyde group present in the oxo-cyclo equilibrium mixture condenses with a second mole of cysteine to give a diadduct which was identified as a thiazolidine derivative. As the second cysteine molecule reacts approximately 25 times more slowly than the first one, both the monoadduct 1 and the diadduct 2 can be prepared as solid substances. In principle, a similar reaction takes place between acrolein or crotonal and

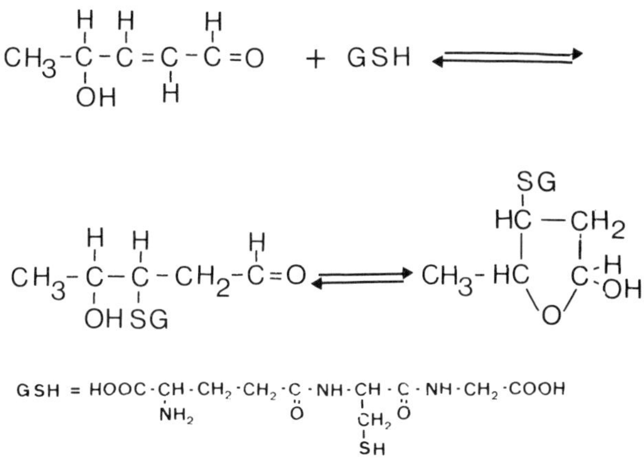

FIG. 2. Reaction of 4-hydroxypentenal with glutathione (GSH).

cysteine (Fig. 4). Kinetic measurements and the results of structural analysis, however, indicate that the rate-limiting step is the addition of cysteine to the C=C double bond of the aldehyde leading to the monoadduct which immediately reacts with additional cysteine yielding the diadduct, which is therefore the only product formed.

The reaction of HPE with SH groups of proteins and enzymes proceeds in the same way as described for glutathione. The final product is a modified protein in which the aldehyde is bound via a thioether linkage to cysteine residues of the polypeptide chain (Fig. 5). While the SH groups of all low molecular weight thiols investigated so far react completely with HPE, the reactivity of protein SH groups towards HPE may vary considerably. HPE in fact by no means blocks all the SH groups present in a protein but only a certain number, which we call the HPE-sensitive SH groups.

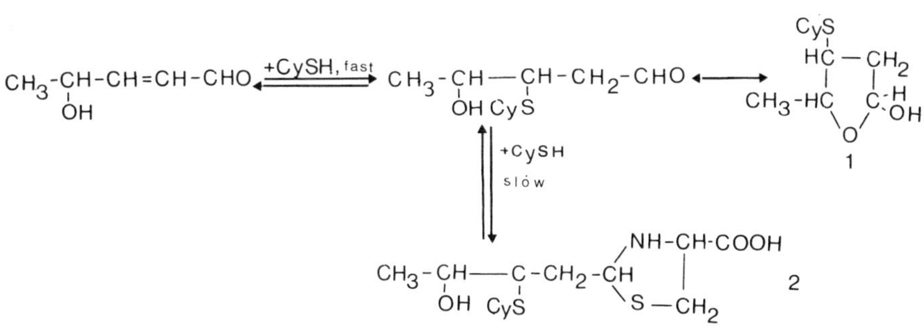

FIG. 3. Reaction of 4-hydroxypentenal with cysteine.

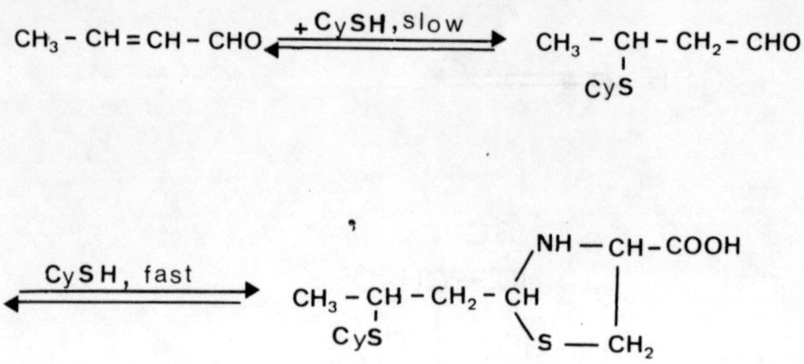

FIG. 4. Reaction of crotonal with cysteine.

The number of protein-bound aldehyde molecules can be estimated by reacting the modified protein with 2,4-dinitrophenylhydrazine (2,4-DNPH) and measuring the absorbance of the protein-bound 2,4-DNPH derivative (Schauenstein *et al.* 1971). Fig. 6 shows, for example, the u.v. absorption spectrum of glyceraldehydephosphate dehydrogenase after treatment with HPE and 2,4-DNPH. The absorption maximum at 366 nm is indicative of a

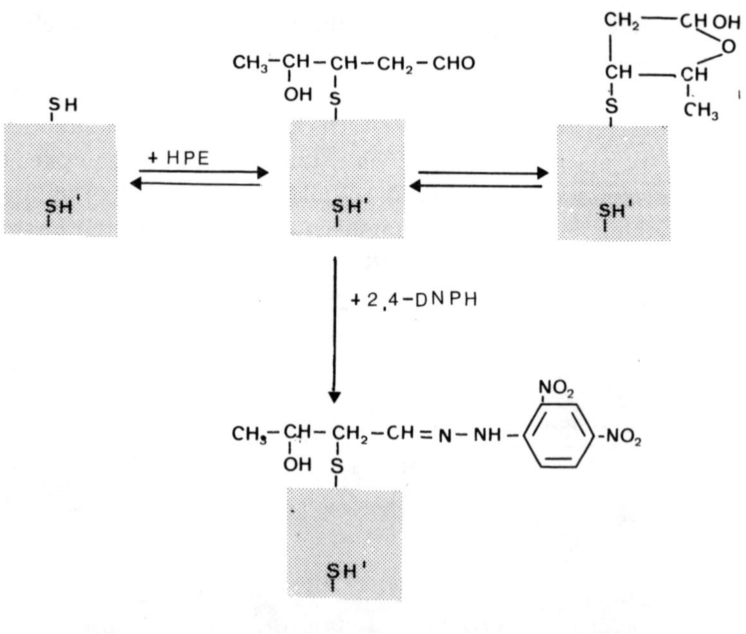

FIG. 5. Reaction of 4-hydroxypentenal (HPE) with SH groups of proteins and detection of the protein-bound aldehyde by 2,4-dinitrophenylhydrazine. SH, HPE-sensitive SH group. SH′, HPE-insensitive SH group.

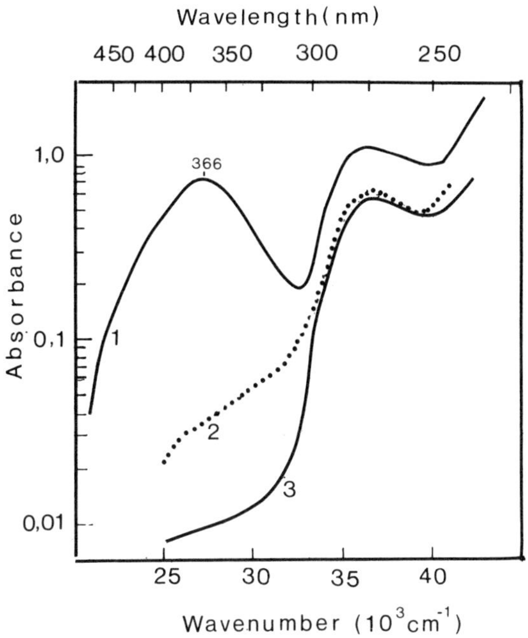

FIG. 6. Ultraviolet absorption spectrum of glyceraldehydephosphate dehydrogenase: (1) after preincubation with HPE and subsequent treatment with 2,4-dinitrophenylhydrazine (2,4-DNPH); (2) after treatment with 2,4-DNPH; (3) without treatment.

hydrazone of a saturated aldehyde, which is expected to be formed according to the reaction shown in Fig. 5.

The quantitative assessment of the intensity of the hydrazone maximum gives about four hydrazone residues, indicating that four HPE molecules are bound per enzyme molecule. This corresponds to the number of SH groups of glyceraldehydephosphate dehydrogenase responsible for enzyme activity. Hence it was also found that enzyme activity completely disappeared on binding with four molecules of HPE. The remaining 12 SH groups of this enzyme do not react with HPE even under extreme conditions. Similarly, in other enzymes and proteins, HPE blocks only certain SH groups, whereas others do not react at all. Ovalbumin, for instance, possesses four SH groups but none of these interacts with HPE. Aldolase possesses 28 SH groups, of which only one reacts with HPE. Although the reasons for the observed selectivity of HPE are not completely apparent, it seems plausible that some SH groups of a protein are much more reactive than others because they are in a chemical environment which favours the interaction with the aldehyde.

It is well known that SH groups play a significant role in nucleic acid and protein biosynthesis, respiration, glycolysis and many other reactions carried out by living cells. Consequently, SH-blocking agents may inhibit such processes. When Ehrlich ascites tumour cells are incubated *in vitro* with HPE the aldehyde enters the cells and then interacts in a complex way with the various SH compounds present, inducing an inhibition of vital processes controlled by SH groups.

As shown in Fig. 7, with increasing concentrations of HPE ranging from 0.1 mM to 5 mM the non-protein SH groups (mainly glutathione) are first consumed, whereas the bulk of the protein SH groups are not yet attacked to any great extent. The latter enter into the reaction only when the level of glutathione in the cell is reduced nearly to zero. Whereas cellular glutathione can be completely bound by HPE, only about half the SH groups in the soluble proteins are HPE-sensitive and can thus be blocked by the aldehyde. From the total protein SH groups (soluble and insoluble proteins), only 31% are HPE-sensitive as shown by cytospectrometric measurements. The relation between the decrease in both glutathione and protein SH groups and the inhibitory effect of HPE on metabolism appears highly interesting to us, as it offers an explanation for the mechanism by which HPE exerts antitumour activity.

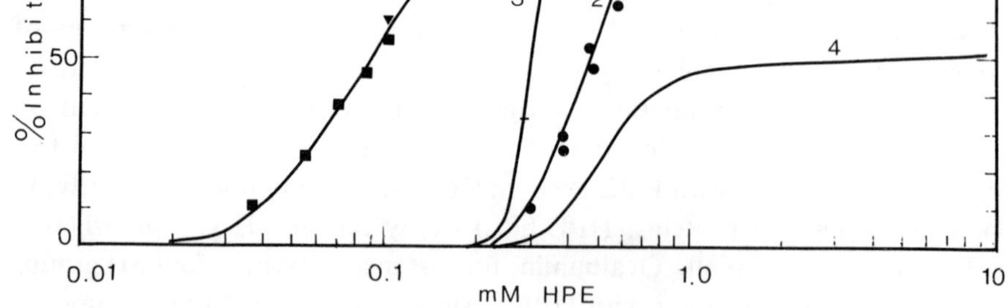

FIG. 7. Effects of 4-hydroxypentenal on Ehrlich ascites tumour cells. (1) inhibition of incorporation of [8-^{14}C]adenine (▼) and [^{3}H]thymidine (■) into DNA. (2) inhibition of O_2 uptake. (3) decrease (= inhibition) of glutathione. (4) decrease (= inhibition) of the SH groups of the soluble proteins. The arrows indicate where protein synthesis, RNA synthesis and glycolysis are inhibited to 50%.

As shown in Fig. 7, HPE primarily attacks DNA synthesis, indicating that the SH groups involved in this biosynthetic process are the most 'HPE-sensitive' ones. After preincubation for 30 minutes, half-maximal inhibition takes place at an incubation concentration of 0.09 mM-HPE. The fact that this biosynthetic process is already inhibited in conditions where the glutathione level is not decreased substantially indicates that the SH groups involved in DNA synthesis are not protected by glutathione against attack by HPE. On the other hand, the groups essential for respiration appear to be protected by glutathione, as O_2 uptake can be inhibited to a considerable extent only when the glutathione of the cell is almost completely consumed. Protein synthesis, RNA synthesis and glycolysis are inhibited to 50% at concentrations of 0.14, 0.32 and 0.25 mM-HPE, respectively.

Of importance in relation to the therapeutic use of HPE as antitumour agent is the fact that the metabolism of cells of normal tissues is not at all or considerably less affected by HPE. As shown in Table 1, 5 mM-HPE does not

TABLE 1

Effect of 4-hydroxypentenal on O_2 uptake and SH groups of soluble cell proteins of several normal organs of rat and of transplanted tumours

	Loss of SH (nanomol/mg protein)	*% inhibition of O_2 uptake*
Normal tissue	0-23	0-30
Slowly growing tumours	18-27	30-50
Rapidly growing tumours	50-60	100

Normal tissue: liver, kidney, jejunum, diaphragm.
Slowly or moderately growing tumours: Ehrlich ascites tumour-cells solid, Nemeth-Kellner lymphosarcoma solid, Sarcoma-180 solid, Yoshida sarcoma.
Rapidly growing tumours: Ehrlich ascites tumour cell line Heidelberg and London, ascites. Nemeth–Kellner lymphosarcoma ascites, Sarcoma-180 ascites.

inhibit the respiration of liver slices, and has only slight effects on other normal tissues. The respiration of slowly dividing tumour cells shows medium inhibitory effects, whereas the respiration of rapidly dividing tumour cells is completely inhibited. This differential sensitivity is also reflected in the finding that in such conditions HPE preferentially attacks the protein SH groups of cancer cells. Furthermore, a highly significant association exists between the doubling time of various animal tumours and their content of HPE-sensitive protein SH groups (Fig. 8) (Schauenstein & Schindler 1976; Schauenstein *et al.* 1978).

Szent-Györgyi (1976) and others (Knock 1967; Harington 1967) have emphasized the importance of the SH group for the metabolism and

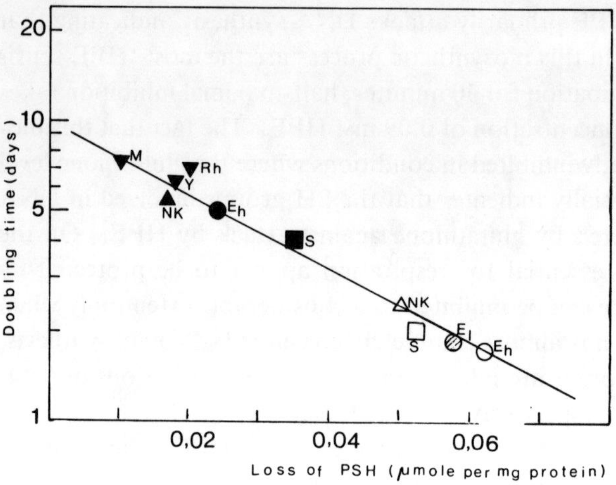

FIG. 8. Relation between 4-hydroxypentenal-sensitive SH groups of soluble proteins and the doubling time of various tumours.
E_h, Ehrlich ascites tumour cells-Heidelberg, ascites (○) and solid (•).
E_l, Ehrlich ascites tumour cells-London ascites (⊘).
S, Sarcoma-180, ascites (□) and solid (■).
NK, Nemeth-Kellner lymphosarcoma, ascites (△) and solid (▲).
Rh, Rhabdomyosarcoma (▼).
Y, Yoshida sarcoma (▼).
M, Harding-Passey melanoma (▼).

regulation of cell life in normal and cancerous tissue. Our results shown in Table 1 and Fig. 8 suggest that the protein SH groups of rapidly dividing cells are in general more reactive towards HPE than those of slowly dividing cells. Possibly, SH groups of the rapidly dividing cells are more accessible to HPE or are protected to a lower degree by glutathione against the action of HPE.

Finally, we want to report results which clearly demonstrate the binding of HPE to the SH groups of the cell proteins. The number of protein SH groups in a single cell can be measured cytospectrometrically after quantitative staining with 2,2′-dihydroxy-6,6′-dinaphthyl disulphide (DDD) and Fast Blue B (Esterbauer *et al.* 1973). A single Ehrlich ascites tumour cell contains on average 1.60×10^{-14} mole protein SH groups. After incubation with 5 mM-HPE, 1.06×10^{-14} were found, indicating a loss of 0.54×10^{-14} mole protein SH/cell. A similar measurement of the aldehyde bound to the cell proteins revealed a value of 0.56×10^{-14} mole/cell. This good agreement between bound aldehyde and loss of SH provides proof that HPE has been chemically bound to the SH groups of the cell proteins (Nöhammer *et al.* 1973).

THERAPEUTIC ACTION OF HPE AND HPE–CYSTEINE AND CROTONAL–CYSTEINE ADDUCTS

A consequence of the high affinity of HPE for SH groups is its extraordinarily short persistence within the organism. On intravenous injection into mice or rats, after only a few minutes practically no HPE is detectable in the blood. When HPE is injected intraperitonally only traces of it can be found in the peritoneal cavity two to three minutes later, and in the venous blood none is found. One should not therefore expect HPE to influence considerably the growth of tumours on intravenous or intraperitoneal injection. Subcutaneous peritumoral application of HPE, however, definitely inhibits the growth of Ehrlich solid tumour, Sarcoma-180, Nemeth–Kellner lymphosarcoma and HP-melanoma (Table 2). (For details see Schauenstein *et al.* 1977, pp. 77–79).

TABLE 2

Therapeutic effects of HPE (subcutaneous peritumoral injection, 1 ml) on transplantable tumours of mice

Tumour	*Therapeutic index*	*Dose (mg/10 g body weight)*
Ehrlich, 3 days	0.23	2 × 0.2 mg on 3rd day
Ehrlich, 8 days	0.23	0.2 mg on both 9th and 12th days
Sarcoma-180, 3 days	0.19	0.2 mg on both 4th and 7th days (Sarcoma-180 and NKLY)
Sarcoma-180, 8 days	0.38	
NKLY[a], 3 days	0.17	
NKLY, 8 days	0.54	
Melanoma, 7 weeks	0.37	1 × 0.2 mg
Melanoma, 7 weeks	0.12	2 × 0.2 mg
Melanoma, 7 weeks	0.09	3 × 0.2 mg

$$\text{Therapeutic index} = \frac{\text{tumour weight of treated animals}}{\text{tumour weight of control animals}}.$$

0.0–0.3, marked or good inhibition.
0.4–0.6, moderate or weak inhibition.
0.7–1.0, no inhibition.
[a]NKLY, Nemeth–Kellner lymphosarcoma.

Local application of HPE (linen cloth steeped in aqueous 5% HPE solution) may also heal human basaliomata, as shown by Musger. Of six cases, three were completely cured (Musger 1968).

A new approach to increasing the therapeutic effectiveness of HPE and other potentially tumour-affecting, α,β-unsaturated aldehydes is their reversible binding with SH compounds. All adducts formed by the reaction of α,β-unsaturated aldehydes with SH compounds can release the aldehyde again, slowly and gradually, to the same extent as it is bound by another SH compound with greater affinity for the aldehyde. As shown in Fig. 9, the rate of the reverse reaction depends on the nature of the adduct. About 5–25 % of the bound aldehyde is released from most adducts within a few hours (Esterbauer *et al.* 1975*a*, 1976).

For a preliminary screening, Ehrlich ascites tumour cells were preincubated (1–2 hours) *in vitro* in solutions (1–10 mM) of the adducts and then reimplanted into healthy mice. Tumour growth was then followed. The following 12 adducts were investigated in this way: HPE–bovine serum albumin, HPE–reduced bovine serum albumin, HPE–reduced β-lactoglobulin, HPE–reduced insulin, HPE–glutathione, HPE–thioglycolate, HPE–cysteine 1:1, HPE–cysteine 1:2, crotonal–cysteine 1:2, acrolein–cysteine 1:2, HPE–penicillamine 1:1, crotonal–penicillamine. Although all these adducts can release the bound aldehyde *in vitro* only two have pronounced carcinostatic activity, namely the HPE–cysteine 1:1 adduct and the crotonal–cysteine 1:2 adduct (Tillian *et al.* 1976, 1978; Conroy *et al.* 1977). The reasons why some adducts are effective *in vivo* while others are not are not completely apparent. Possibly cell permeability and the rates of aldehyde liberation, detoxification and excretion play a significant role *in vivo*.

Table 3 and 4 give the results of an extended study of the therapeutic effects of these two adducts on mice bearing the Ehrlich ascites tumour. If used in

TABLE 3

Therapeutic effect of five doses of 4-hydroxypentenal–cysteine 1:1 adduct, given intraperitoneally to mice 72, 96, 120, 144 and 168 hours after implantation in the peritoneal cavity of 4–5 × 10^5 Ehrlich ascites tumour cells/10 g body weight

Number and sex of mice	*Dose (mg/10 g body weight)*	*% healed*	*% significantly delayed*	*% not significantly delayed*
9 f	0.9	0	0	100
10 f	1.7	40	10	50
10 f	3.5	40	60	0
9 f	7.0	22	78	0
15 f	14.0	0	100	0
10 m	3.5	0	0	100
14 m[a]	12.0	22	78	0
38 f	12.0	30	0	70

[a]Single dose one day after implantation with 4–5 × 10^6 Ehrlich ascites tumour cells /10 g body weight.

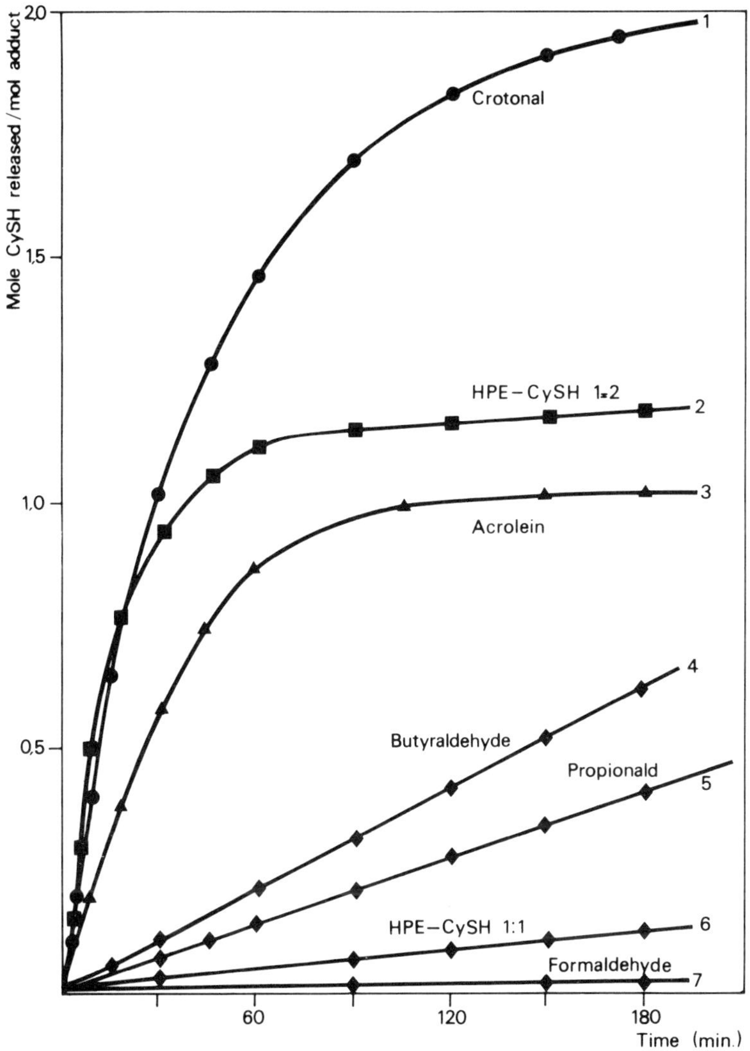

FIG. 9. Dissociation of various aldehyde–cysteine adducts measured by the rate of cysteine release.
(1) Crotonal diadduct (crotonal–CySH 1:2).
(2) 4-Hydroxypentenal (HPE) diadduct (HPE–CySH 1:2).
(3) Acrolein diadduct (acrolein–CySH 1:2).
(4) 2-Butylthiazolidine-4-carboxylic acid (Butyrald.–CySH 1:1).
(5) 2-Propylthiazolidine-4-carboxylic acid (Propionald.–CySH 1:1).
(6) 4-Hydroxypentenal monoadduct (HPE–CySH 1:1).
(7) Thiazolidine-4-carboxylic acid (Formald.–CySH 1:1).

the appropriate dose, both adducts completely prevent or significantly inhibit growth of this tumour in mice. If, for instance, female mice were treated with the HPE adduct, 22–40 % of the animals remained free from ascites and solid tumour for 90 days, while in the remainder of the animals the survival time was extended up to 289 %. Similar results were found for the crotonal adduct. In male animals the adducts seem to be less therapeutically active. Although a significant increase in survival time was observed, no animal was completely free of tumour on the 90th day. So far, however, no reasonable explanation can be given of the factors which produce the observed sex difference.

TABLE 4

Therapeutic effect of five doses of crotonal–cysteine 1:2 adduct, given to mice i.p. 72, 96, 120, 144 and 168 hours after implantation into the peritoneal cavity of $4-5 \times 10^5$ Ehrlich ascites tumour cells/10 g body weight

Number and sex of mice	*Dose (mg/10 g body weight)*	*% healed*	*% significantly delayed*	*% not significantly delayed*
20 f	2.5	30	10	60
20 f	5.0	30	70	0
19 f	10.0	36	64	0
9 f	20.0	22	78	0
10 m	5.0	10	90	0
19 m	10.0	0	100	0

Histochemical methods in combination with cytospectrometric measurements clearly show that the adducts are taken up by the cancer cells, and that within the cell the aldehyde is released and subsequently bound to the SH groups of proteins located mainly in the nuclear area. If, for instance, Ehrlich ascites tumour cells were incubated with HPE–cysteine 1:1 adduct (two hours with 100 mM, 37 °C) a reduction of 0.234×10^{-14} mole per cell in protein thiol was found, whereas 0.247×10^{-14} mole of aldehyde was bound to the cell proteins. We assume that the liberated aldehyde preferentially inhibits SH groups of functional importance for DNA synthesis and thereby stops the proliferation of the tumour cells. *In vitro*, both adducts may inhibit the incorporation of thymidine into DNA. Ehrlich ascites cells (2×10^6/ml) preincubated (two hours at 37 °C) with HPE–cysteine 1:1 adduct (0.05 mM and 0.20 mM) show a 50 % and 100 % inhibition of the incorporation of labelled thymidine into the trichloroacetic acid insoluble fraction. These concentrations are in the range of the doses required for the free aldehyde to produce the same inhibition. However, the preincubation period necessary to establish the inhibition is considerably longer (Fig. 10).

Similar results were obtained with the cysteine adduct of crotonal. Compared to the control incorporation (in the absence of adduct), treatment of Ehrlich ascites cells with 0.03 and 0.20 mM cysteine adduct inhibited the incorporation of thymidine by 50 % and 100 % respectively.

A quantitative comparison of the effect of different adducts on DNA biosynthesis can be made if we compare on a molar basis the doses of aldehyde

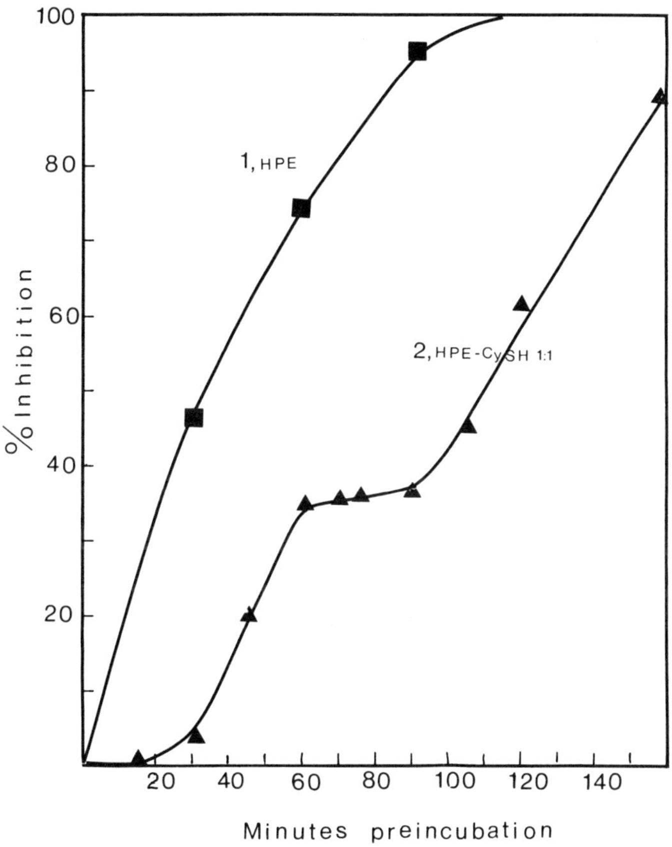

FIG. 10. Time dependence of inhibition of [^{3}H]thymidine incorporation by 4-hydroxypentenal (HPE) (1) and HPE–CySH 1:1 (2) into DNA of Ehrlich ascites tumour cells. The cells (2 × 10^6/ml) were preincubated for the indicated periods and subsequently thymidine incorporation was measured for 1 h. Final concentrations: 0.085 mM for HPE and 0.110 mM for HPE–CySH 1:1.

TABLE 5

Inhibition of [^{3}H]thymidine incorporation into DNA of Ehrlich ascites tumour cells by various aldehyde–RSH adducts. Effective dose required to reduce incorporation to 50 % of the control in μmol of bound aldehyde (ED_{50})

Adduct	*ED_{50} (μM)*
Acrolein–CySH 1:2	28
Crotonal–CySH 1:2	32
HPE–CySH 1:1	52
Acrolein–red. BSA	~ 50
HPE–BSA	100
HPE–red. BSA	280
Propionaldehyde–CySH	360
HPE–CySH 1:2	510
Formaldehyde–CySH	>1500

Four million cells in a final volume of 2 ml were preincubated with various concentrations of the compounds for 2 h, followed by 1 h of [^{3}H]thymidine incorporation. ED_{50} values were obtained from a double logarithmic dose response plot.
BSA, bovine serum albumin; red. BSA, BSA reduced with thioglycolate.

required to produce 50 % inhibition. The data summarized in Table 5 clearly demonstrate that the inhibition of the incorporation of labelled thymidine strongly depends on the structure of the adduct. From these data it is also evident that such studies should only be taken as preliminary experiments as they do not allow us to predict whether an adduct is therapeutically effective or not *in vivo*. The acrolein–cysteine 1:2 adduct, for example, is the most potent inhibitor of DNA synthesis among the adducts tested. *In vivo*, however, this compound produces no noteworthy inhibition of tumour growth. Adducts affecting DNA synthesis only at rather high concentration (i.e. HPE–cysteine 1:2 adduct) are possibly therapeutically inactive because the required high concentrations cannot be reached *in vivo*.

It seems to us that the transformation of potentially tumour-affecting α,β-unsaturated aldehydes into storage forms with proper stability opens up a promising new method of cancer chemotherapy.

ARE α,β-UNSATURATED ALDEHYDES PRESENT IN LIVING MAMMALIAN TISSUE?

Although the presence of α,β-unsaturated aldehydes in living mammalian tissues has not yet been conclusively demonstrated, it is plausible to assume that such aldehydes are continuously formed *in vivo*, at least in low amounts. This assumption is supported by several studies made in different fields.

1. There is no doubt that a great number of different aldehydes are formed

in vitro during lipid peroxidation as breakdown products. Lipid peroxides appear also to be present in small amounts in animal tissue in physiological conditions and their formation may be greatly enhanced in some pathological conditions such as, for example, CCl_4-induced lipid peroxidation (Slater 1972). From our investigations of the mechanism of lipid autoxidation in aqueous systems we would expect, therefore, that endogenous lipid peroxides give rise to a low intracellular level of α,β-unsaturated aldehydes.

2. Heating dilute aqueous solutions of 2-deoxysugars produces α,β-unsaturated hydroxyaldehydes (Esterbauer *et al.* 1975*b*). The resulting product from 2-deoxyribose is a *trans*-4,5-dihydroxypent-2-enal (Fig. 11). It was shown that measurable amounts of this aldehyde are formed, even at 37 °C. Therefore, if deoxysugars are formed *in vivo* during biosynthesis or degradation, possibly such hydroxyalkenals are formed at the same time.

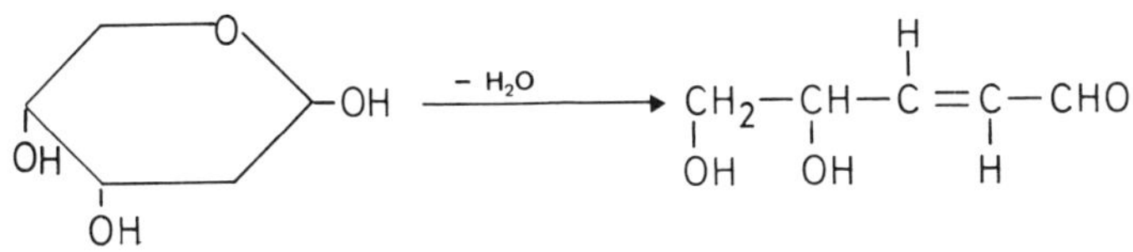

FIG. 11. Formation of 4,5-dihydroxypentenal by dehydration of 2-deoxyribose.

3. According to Quash & Taylor (1970), β-aminopropionaldehyde is a component of human serum. In animal tissue this aldehyde is formed by the enzymic breakdown of spermidine and it may be converted into malonaldehyde which itself exists at physiological pH in the salt form of β-hydroxyacrolein. On the other hand, β-aminopropionaldehyde should easily eliminate NH_3 to give acrolein (acrylaldehyde, propenal), which is a powerful inhibitor of DNA synthesis. This suggestion receives support from the observations of Alarcon (1970) and Kimes & Morris (1971), who found that the aminoaldehydes formed from spermine and spermidine by the action of amine oxidase are unstable compounds and spontaneously form acrolein by β-elimination (Fig. 12).

It appears to us that in studying biogenic aldehydes one has to be extremely careful. The main difficulties in demonstrating the occurrence of α,β-unsaturated and other aldehydes in living tissue are the problems of separating and estimating very small amounts of aldehydes, the artificial formation of aldehydes during the work-up procedure, and the dragging in of aldehydes from reagents and solvents. The first problem seems to be solved by the new technique of high performance liquid chromatography, which allows the rapid separation and quantitative estimation of 2,4-dinitrophenylhydrazine

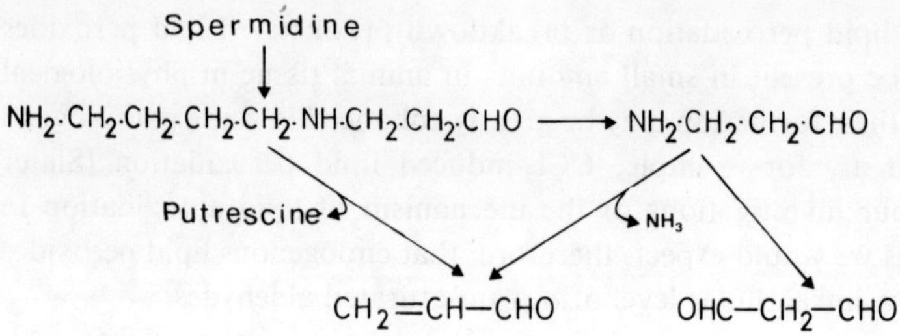

FIG. 12. Formation of acrolein and malonaldehyde from spermidine.

derivatives of aldehydes in the nanogram range. More serious, according to our experience, is the problem of ruling out completely the artificial formation of aldehydes and avoiding contamination from traces of aldehydes present in even highly purified reagents.

We are confident that aldehydes in general and α,β-unsaturated aldehydes in particular will increasingly attract the attention of biochemists and biologists and that collaboration and an exchange of practical knowledge between laboratories working in this field will help to eludicate the occurrence and the biological significance of this group of compounds.

References

ALARCON, R.A. (1970) Evidence for the formation of the cytotoxic aldehyde acrolein from enzymatically oxidized spermine and spermidine. *Arch. Biochem. Biophys. 137*, 365-372

CONROY, P.J., NODES, J.T., SLATER, T.F. & WHITE, G.W. (1977) The inhibitory effects of a 4-hydroxypentenal:cysteine adduct against sarcoma-180 cells in mice. *Eur. J. Cancer 13*, 55-63

ESTERBAUER, H., NÖHAMMER, G., SCHAUENSTEIN, E. & WEBER, P. (1973) Beitrag zum quantitativen histochemischen Nachweis von Sulfhydrylgruppen mit der DDD-Färbung. 3. Quantitative cytospektrometrische Bestimmungen an Ehrlich-Ascites-Tumorzellen. *Acta Histochem. 47*, 106-114

ESTERBAUER, H., ZOLLNER, H. & SCHOLZ, N. (1975*a*) Reaction of glutathione with conjugated carbonyls. *Z. Naturforsch. Sect. C BioSci. 30*, 466-473

ESTERBAUER, H., SANDERS, E.B. & SCHUBERT, J. (1975*b*) Isolation and characterisation of an unsubstituted 2,3-unsaturated sugar, *trans*-2,3-dideoxy-D-glycero-pent-2-enose, produced by thermal dehydration of 2-deoxy-D-*erythro*-pentose. *Carbohydr. Res. 44*, 126-132

ESTERBAUER, II., ERTL, A. & SCHOLZ, N. (1976) The reaction of cysteine with α,β-unsaturated aldehydes. *Tetrahedron 32*, 285-289

HARINGTON, J.S. (1967) The sulfhydryl group and carcinogenesis. *Adv. Cancer Res. 10*, 247-309

KIMES, B.W. & MORRIS, D.R. (1971) Preparation and stability of oxidized polyamines. *Biochim. Biophys. Acta 228*, 223-234

KNOCK, F.E. (1967) *Anticancer Agents*, Thomas, Springfield, Ill.

MUSGER, A. (1968) Behandlungsversuche mit Hydroxypentenal. *Hautarzt 20*, 375-376

NÖHAMMER, G., SCHAUENSTEIN, E. & WEBER, P. (1973) Quantification of selective binding of 4-hydroxypentenal to structural protein thiols in Ehrlich tumor cells. *J. Histochem. Cytochem. 21*, 1082-1089

QUASH, G.A. & TAYLOR, D.R. (1970) Serum β-aminopropionaldehyde: identification and origin. *Clin. Chim. Acta 30*, 17-23

SCHAUENSTEIN, E. (1967) Autoxidation of polyunsaturated esters in water: chemical structure and biological activity of the products. *J. Lipid Res. 8*, 417-428

SCHAUENSTEIN, E. & SCHINDLER, R. (1976) Malignancy and SH-compounds, in *Molecular Base of Malignancy (Selected Papers from Int. Symp.)* (Deutsch, E. *et al.*, eds.), pp. 55-59, Thieme, Stuttgart

SCHAUENSTEIN, E., TAUFER, M., ESTERBAUER, H., KYLIANEK, A. & SEELICH, Th. (1971) Über die Reaktion von Protein-SH-Gruppen mit 4-Hydroxypentenal. *Monatsh. Chem. 102*, 517-529

SCHAUENSTEIN, E., ESTERBAUER, H. & ZOLLNER, H. (1977) *Aldehydes in Biological Systems* (Gore, P.H., transl.), Pion/Academic, London

SCHAUENSTEIN, E., GÖLLES, J., WALTERSDORFER, H. & SCHAUR, R.J. (1978) Association between the doubling time of various cells and tissues, and the SH-content of their soluble proteins. *Z. Naturforsch. Sect. C Biosci. 33*, 79-83

SLATER, T.F. (1972) *Free Radical Mechanisms in Tissue Injury*, Pion/Academic, London

SZENT-GYÖRGYI, A. (1976) *Electronic Biology and Cancer. A New Theory of Cancer*, Dekker, New York

TILLIAN, H.M., SCHAUENSTEIN, E., ERTL, A. & ESTERBAUER, H. (1976) Therapeutic effects of cysteine adducts of α,β-unsaturated aldehydes on Ehrlich ascites tumor of mice. I. *Eur. J. Cancer 12*, 989-993

TILLIAN, H.M., SCHAUENSTEIN, E. & ESTERBAUER, H. (1978) Therapeutic effects of cysteine adducts of α,β-unsaturated aldehydes on Ehrlich ascites tumor of mice. II. *Eur. J. Cancer 14*, 533-536

Discussion

Purcell: Are these adducts new compounds?

Esterbauer: All the 4-hydroxyalkenals and the cysteine adducts of the hydroxyalkenals or 2-alkenals are new compounds and are described by us for the first time. The free 2-alkenals have of course been known for a long time.

Purcell: Did you do toxicity studies on mice? What was the LD_{50}?

Esterbauer: We could not obtain an LD_{50} with the pure adducts—they seem to be completely non-toxic. We had difficulty in injecting more than 100 mg into mice and at that level we observed no acute toxicity. However, six or seven days after injection about half the mice showed some signs of chronic toxicity, but we do not have any details about that yet.

Dianzani: The aldehydes you use are probably metabolized by aldehyde dehydrogenases. The activity of the dehydrogenases towards different aldehydes may, however, be different. Is the rate of breakdown of the aldehydes you use known? If so, does it coincide with the therapeutic index?

Esterbauer: In tumour or in liver?

Dianzani: In liver tumours. In some preliminary experiments we found that mitochondria from liver contain an aldehyde dehydrogenase that metabolizes 4-hydroxypentenal. Its activity is much less in mitochondria from ascites hepatomas. Consequently, the hepatoma mitochondria are much more sensitive to this aldehyde than normal liver mitochondria.

Esterbauer: The two main processes for the detoxification and removal of these aldehydes in the living organism are probably: (1) metabolism by aldehyde dehydrogenases, and (2) the reaction with glutathione and the subsequent formation of mercapturic acids in the liver.

The effectiveness of one or both of these processes may be decreased in tumour tissue. This would offer a reasonable explanation why in general the metabolism of tumour cells (synthesis of DNA, RNA and protein, mitochondrial respiration, glycolysis) is much more sensitive towards aldehydes than the metabolism of normal cells.

Laki: What is the doubling time of the tumour?

Esterbauer: In the ascites form the doubling time is about 1.5 to 2.5 days (average 2.0 days).

Laki: How soon does it make metastases?

Esterbauer: Mice usually die 15 days after implantation; that is probably too short a time for the tumour to metastasize.

Bondy: There are no metastases from such transplanted tumours.

Dianzani: Transplanted ascites tumours rarely give metastases. The general immunological defence in transplanted rats is generally high, so the metastases, if any, are usually seen very late.

Slater: Tissues contain several enzymes that metabolize aldehydes such as the 4-hydroxyalkenals. This rapid catabolism and excretion, when considered together with the known interactions of such aldehydes with thiol and amino groups, I believe explains their only modest antitumour activity when injected into animals at a site distant to the tumour. For example, the activity of 4-hydroxypentenal against solid tumours was much less when it was injected at a distant site than when it was injected intraperitoneally into animals with ascites tumour cells (P. Conroy, P. Elvin & T.F. Slater, unpublished data).

Esterbauer: The free aldehydes are probably not the right ones to use therapeutically as major drugs. Unfortunately, immediately after intravenous injection such aldehydes will bind to the serum proteins and to the glutathione present in large amount in erythrocytes and display no activity. I should like to be able to induce the formation of such aldehydes within the tumour cells, perhaps by injecting adducts which release the aldehyde within the tumour cells or, perhaps, by inducing lipid peroxidation.

Slater: You have reported (Schauenstein *et al.* 1977) that direct application of hydroxypentenal to women with carcinoma of the cervix was beneficial. Have you followed up this observation?

Esterbauer: That was done by Professor Richter and his colleagues

(Ratzenhofer *et al.* 1975). They gave HPE in the form of a sodium alginate gel in a pessary directly to the cervix for several weeks. After resection of the uteri histological examination revealed that in two out of eight cases the tumour was destroyed. In the remaining six cases only partial destruction of the tumour cells was observed. Usually a cervical carcinoma is treated clinically by a common surgical technique and the patient has to stay in hospital for only a few days. The application of hydroxypentenal requires the woman to stay in hospital for three weeks or more. Musger encountered (see p. 233) the same problem in the treatment of basaliomata with HPE; usually doctors treat these by total excision.

Williams: The compounds are optically active at the hydroxy-substituted carbon atom. Do you separate or use the D- and L-forms? One form might be more active.

Esterbauer: The compounds we use are the racemates and we have never separated the D and L forms. I agree that one form might be more active.

Williams: Why did you use the *trans* isomers?

Esterbauer: The *cis* aldehydes are not stable; they readily cyclize to dihydrofuran derivatives.

Fodor: The cyclization is a reversible process; it resembles the formation of a hemiacetal.

Esterbauer: We have no indication that *cis*-4-hydroxyalkenals exist in the open-chain form; it was always a cyclic form.

Fodor: Still, it could react in a biological medium as the free aldehyde.

Esterbauer: Under the conditions of our synthesis the cyclic hemiacetal (dihydrofuran derivative) itself is unstable and easily loses a water molecule to give a furan derivative.

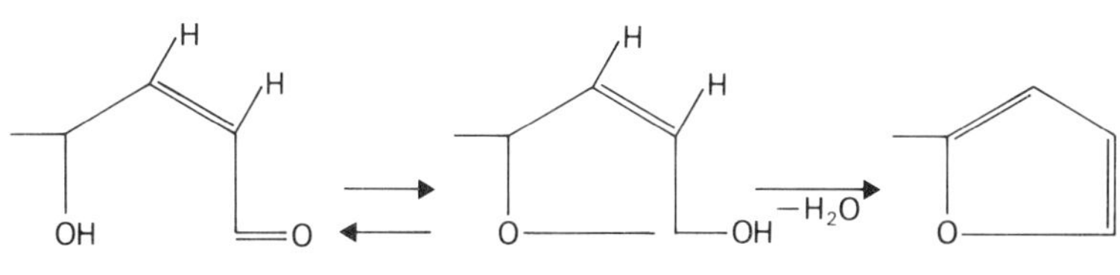

Fodor: And the furan is not rehydrated?

Esterbauer: I do not believe so. We prepare the *trans*-hydroxyalkenals by hydrogenating the triple bond of hydroxyalkinal acetal with $LiAlH_4$ and subsequent saponification of the acetal. When hydrogen and a palladium catalyst were used the *cis* double bond was obtained. When, however, these acetals were saponified (aqueous oxalic or citric acid) only 2-alkyl substituted furans were formed (Esterbauer & Weger 1967; Seyfarth 1968). The furan derivatives we obtained did not form 2,4-dinitrophenyl hydrazones and did not exhibit (in aqueous or alcoholic solution) the u.v. absorbance maximum at 222 nm characteristic of α,β-unsaturated aldehydes. Thus it seems unlikely to me that a rehydration takes place to any considerable extent.

Williams: Do the unconjugated isomers (e.g. 4-hydroxypent-3-enal) have no activity?

Esterbauer: They should show no activity; the addition reactions are the consequence of the conjugated system.

Neuberger: Do the hydroxyenals react with nucleic acids?

Esterbauer: Using [^{14}C]hydroxypentenal we found no radioactivity in the nucleic acids. I admit, however, that the concentrations we used were not high (i.e. less than 10 mmol/l). At low concentrations (50–200 μmol/l), however, which produced an inhibition of DNA and RNA biosynthesis, there was no measurable direct reaction between HPE and the nucleic acids.

Neuberger: One would expect a Michael addition to the DNA bases.

Esterbauer: Michael addition may take place between HPE and the amino group of guanosine residues, for instance, but these adducts are usually not very stable. Dialysis to remove the excess of reagents could reverse the addition. With hydroxyalkenals we see no other reaction besides that with the SH group and we believe that this is one of the major characteristic differences between these aldehydes and other aldehydes or, for example, methylglyoxal.

Williams: The only other group that might be reactive is selenium, namely the selenol (SeH) in glutathione peroxidase. Aldehydes might knock out this enzyme; aldehydes do remove selenium from such enzymes.

References

ESTERBAUER, H. & WEGER, W. (1967) Über die Wirkung von Aldehyden auf gesunde und maligne Zellen. 3. Mitt.: Synthese von homologen 4-Hydroxy-2-alkenalen II. *Monatsh. Chem. 98*, 1994-2000

RATZENHOFER, M., RICHTER, K. & SCHAUENSTEIN, E. (1975) Wirkung von 4-Hydroxypentenal auf die nicht karcinomatös und die karcinomatös veränderte Portio uteri. *Exp. Pathol. (Jena) 11*, 83-106

SCHAUENSTEIN, E., ESTERBAUER, H. & ZOLLNER, H. (1977) *Aldehydes in Biological Systems* (Gore, P.H., transl.), Pion/Academic, London

SEYFARTH, H.E. (1968) Synthese von 4-Hydroxy-octen-(2)-al-(1)-diäthylacetal und Übergang in 2-n-Butylfuran. *Chem. Ber. 101*, 619-622

Biological activity of methylglyoxal and related aldehydes

M.U. DIANZANI

Institute of General Pathology of the University of Torino and National Foundation for Cancer Research, Regional Laboratory of Torino, Italy

Abstract The effect of methylglyoxal and other aldehydes on several biochemical variables has been studied. Aldehydes inhibit amino acid incorporation into proteins, both in reconstituted systems and in isolated hepatocytes. They also decrease the secretion of protein and lipoprotein from hepatocytes into the incubation medium. This inhibition is seen even with prelabelled proteins, which indicates damage to the secretory mechanism itself. This conclusion is strengthened by the fact that aldehydes also decrease the binding of colchicine to liver tubulin. Aldehydes decrease the respiratory rate of mitochondria, as well as mitochondrial swelling induced by phosphate, by Ca^{2+} or by K^+ plus valinomycin. They also partially inhibit cytochrome *P*-450. When injected into normal rats, aldehydes produce a decrease in the mitotic index of bone marrow cells and of the epithelial lining of the small intestine. A decrease in mitotic index and in cellularity is seen after injecting aldehydes into the peritoneal cavity of rats bearing transplanted ascites AH-130 Yoshida hepatoma. Aldehydes also impair the function of liver cell ligandin and potentiate the increase in cell permeability induced by 5-hydroxytryptamine (serotonin). The meaning of these results is discussed with special reference to the pathogenesis of cellular lesions in carbon tetrachloride poisoning.

Interest in the biological activity of aldehydes has greatly increased for two main reasons: (1) the hypothesis by Albert Szent-Györgyi that these substances, and especially methylglyoxal, may have a regulatory influence on the electronic state of proteins, and consequently on cell division; and (2) the demonstration that lipid peroxidation, which leads to the production of considerable amounts of aldehydes, is a basic mechanism in cell pathology.

Our experimental approach to this problem mostly derived from studies of the effects of carbon tetrachloride on mammalian cells. CCl_4 is broken down in the smooth endoplasmic reticulum of liver cells, so producing the free radicals $Cl^{\bullet}$ and $CCl_3^{\bullet}$. These damage liver cell structures in at least two ways:

(1) by covalent binding; and (2) by removal of hydrogen atoms from suitable molecules, especially from polyunsaturated fatty acids contained in microsomal phospholipids. This starts lipid peroxidation, which produces severe morphological and functional damage to membranes. There is good reason to think that the early damage at the level of the smooth endoplasmic reticulum is mostly due to these types of free radical effects. According to Recknagel *et al.* (1975), the major disturbances are especially due to lipid peroxidation, whereas Castro's group (Villaruel *et al.* 1976; de Toranzo *et al.* 1978) think they relate to covalent binding. Both lipid peroxidation and covalent binding seem, however, difficult to accept as direct causes of damage occurring at cell sites far from the smooth endoplasmic reticulum. In fact, both free radicals and lipoperoxides are extremely reactive in general and are hardly likely to be mediators of such 'remote' damage. The substances ultimately responsible for a direct interaction with distant cell sites must show the following essential features: (1) they must be produced in sufficient amounts; (2) they must be diffusible throughout the cytoplasm; and (3) they must be able to react with targets that are widely distributed within the cell.

It seems to me that all these requirements are fulfilled by aldehydes derived from the peroxidative degradation of unsaturated fatty acids. It is well known that malonaldehyde is produced in large amounts by the peroxidation of fatty acids. According to Schauenstein *et al.* (1977), several other aldehydes and ketoaldehydes are produced in peroxidizing biological systems. Among others, aldehydes related to *trans*-4-hydroxypent-2-enal (HPE) seem to be particularly important (Schauenstein 1967). Methylglyoxal has also been found in oxidized fat (Dempsey & Snell 1963).

The diffusibility of aldehydes throughout the cytoplasm is clearly shown by the experiment described in Table 1. CCl_4 was allowed to diffuse from the central well of a closed flask, whose main compartment contained isolated rat hepatocytes. This treatment initiates lipid peroxidation, as shown both by the production of malonaldehyde and by the appearance of a big diene absorption band in microsomal phospholipids. Malonaldehyde accumulates in the extracellular fluid in conditions where the viability of cells, as assessed both by the dye exclusion test and by the measurement of the release of enzymes from the hepatocytes, is unaffected. The accumulating malonaldehyde could not be derived from plasma membrane lipids, since plasma membranes are unable to metabolize CCl_4 (R. Le Page, K. Cheeseman & T.F. Slater, unpublished data). Furthermore, the addition of different concentrations of promethazine does not affect the intracellular and extracellular accumulation of malonaldehyde differently.

This paper describes the results of experiments we did to study the influence

TABLE 1

Malonaldehyde production by rat hepatocytes incubated *in vitro* with CCl_4. Influence of promethazine on intracellular and extracellular accumulation of malonaldehyde

Cell treatment	*Malonaldehyde production*[a]	
	Intracellular	*Extracellular*
None	0.080 ± 0.016	0.099 ± 0.025
5 μM-Promethazine	0.085 ± 0.002	0.089 ± 0.007
2.5 μM-Promethazine	0.078 ± 0.004	0.096 ± 0.011
CCl_4, 7.5 μl/flask	0.124 ± 0.011 (+55%)	0.312 ± 0.036 (+215%)
5 μM-Promethazine + CCl_4	0.105 ± 0.012 (+23%)	0.215 ± 0.013 (+141%)
2.5 μM-Promethazine + CCl_4	0.118 ± 0.016 (+51%)	0.296 ± 0.038 (+208%)

The cell suspension contained 4×10^6 cells. It was centrifuged before trichloroacetic acid (TCA) was added, to separate the cells from the supernatant. The cells were resuspended in 3 ml 5% TCA. The supernatant was brought to 3 ml with water and TCA to a final TCA concentration of 5%. The determinations were done on 1.5 ml samples, so they correspond to the malonaldehyde content of 2×10^6 cells. For other technical details, see Gravela *et al.* (1979).

[a]Absorbance at 543 nm of the thiobarbituric acid-reacting compounds produced/h by 2×10^6 hepatocytes. Values represent mean of two triplicate experiments ± S.D. Figures in parentheses are % increase over the corresponding controls.

of the aldehydes on several biochemical parameters among those known to be affected by CCl_4 poisoning.

METHODS

Male rats of the Wistar strain, weighing 200–300 g, fed *ad libitum* on a semi-synthetic diet (Fratelli Piccioni, Brescia, Italy) and fasted 15–18 h before killing, were used. Homogenates were prepared in a Potter-Elvehjem apparatus. They were submitted, if necessary, to differential centrifugation in a Spinco-Beckman Model L centrifuge according to the procedure of de Duve *et al.* (1955).

Amino acid incorporation into rat liver post-mitochondrial supernatant was studied as described elsewhere (Dianzani & Gravela 1975; Dianzani *et al.* 1976). The amino acid source was in most cases a ^{14}C-labelled protein hydrolysate of *Chlorella* obtained from the Radiochemical Centre, Amersham, having a specific activity of 52 Ci atom carbon. When poly(U) was used instead of endogenous mRNA, [^{14}C]phenylalanine was used as an amino acid source. Details of the method are described elsewhere (Dianzani *et al.* 1976).

Liver mitochondria were isolated according to Feo *et al.* (1973). The polarographic determination of respiratory rate and of the acceptor/control ratio, as well as of the ADP/O ratio, was done according to Feo *et al.* (1973, 1975). A spectrophotometric method (Feo *et al.* 1975) was used to measure mitochondrial swelling.

Lysosome-rich fractions (referred to here as 'M + L' fraction) were prepared according to de Duve *et al.* (1955). Acid phosphatase was measured according to de Duve *et al.* (1955) and β-galactosidase according to Sellinger *et al.* (1960), with the *p*-nitrophenyl derivative of galactose as a substrate. Both 'free' and 'total' activities were determined. They were expressed as units/g wet liver. When activation of enzymes by acid autolysis was studied, samples of the 'M + L' fraction were preincubated for different lengths of time with 50 mM-sodium acetate:acetic acid buffer, pH 5.0, at 37 °C. Cathepsin B_1 activity was measured on homogenates according to Barrett (1972).

The microsomal fraction, isolated according to de Duve *et al.* (1955), was tested for glucose-6-phosphatase activity according to Zuretti *et al.* (1973), for aminopyrine demethylase according to Orrenius (1965), and for cytochrome *P*-450 according to Omura & Sato (1964).

In the experiments on tubulin, rats were killed by bleeding under nembutal anaesthesia after rapid liver perfusion through the abdominal aorta with cold 0.9% NaCl. The livers were then rapidly removed, minced and homogenized with five strokes of a motor-driven teflon pestle of the Potter-Elvehjem homogenizer. The medium consisted of three volumes (w/vol) of a 0.24 M-sucrose solution in buffer containing 1×10^{-2} M-sodium phosphate, pH 7.0, 1×10^{-4} M-GTP and 5×10^{-3} M-$MgCl_2$ (PMG buffer) according to Weisenberg *et al.* (1968). The supernatants obtained after centrifugation for 1 h at 100 000 ***g*** at 4 °C with the rotor 30 of the Spinco-Beckman L centrifuge were diluted to give a protein concentration of 10 mg/ml. 100 μl aliquots of the diluted supernatant (1 mg protein) were preincubated with different concentrations of the tested aldehydes for 15 min at 37 °C. 10 μl of an aqueous solution of ^{3}H-labelled colchicine (0.11 mM, corresponding to 1 800 000 c.p.m.) were then added to each sample and the incubation was continued for a further 90 min. The reaction was stopped by the addition of 1 ml of a 0.3% suspension of charcoal in distilled water at 4 °C. Blanks and standards contained 100 μl of 0.24 M-sucrose in PMG buffer instead of the undiluted supernatant. They were brought to the same final volume as the samples and kept in the same experimental conditions (Sherline *et al.* 1974). Charcoal (Aktivkohle, Merck) was added to the blanks and distilled water in standards. After addition of charcoal, all tubes were mixed, allowed to stand at 4 °C for 10 min and then centrifuged at 2200 ***g*** for 15 min at 4 °C in the IED model centrifuge. 100 μl aliquots of the supernatants were added to 10 ml of Bray scintillation liquid and radioactivity was detected in the Mark 1 type Nuclear

Chicago liquid scintillation counter. In some experiments, the effect of aldehydes was tested on both polymerized and soluble tubulin from liver, prepared according to Pipeleers *et al.* (1977).

The activity of ligandin was studied according to Levy *et al.* (1969). Proteins were separated on a G-75 Sephadex column. Conjugation of bromosulphophthalein to the proteins (0.24 μmol added to 6 mg protein and incubated at room temperature for 15 min) was used to test the binding activity of the proteins.

Capillary permeability was studied by the trypan blue test, as used by Ugazio *et al.* (1976).

Hepatocytes were isolated from normal rat liver by the method of Berry & Friend (1969), with some modifications introduced by Gravela *et al.* (1977). Cell viability was assessed by the dye exclusion test and by following the release of glutamic pyruvic transaminase (alanine aminotransferase) and lactate dehydrogenase from the cells into the incubation fluid. These enzymes were measured by reagent kits from Biochemical Test Combination, Boehringer, Mannheim, Germany. In the experiments shown in Table 1 (p. 247) the cell suspension was centrifuged at 5000 ***g***-min in order to separate the cells from the incubation fluid before the addition of 5% trichloroacetic acid. The cells were then suspended in 3 ml trichloroacetic acid and the supernatant was brought to the same volume with water and trichloroacetic acid (final concentration 5%). Aliquots of 1.5 ml were used for malonaldehyde estimation.

In the experiments on Yoshida hepatoma cells, rats were injected intraperitoneally with approximately 5×10^6 tumour cells. These rats received 50 mg of either methylglyoxal or 4-hydroxypentenal (HPE) per kg intraperitoneally on both Days 4 and 5 after the injection of tumour cells and were killed 24 h later. The ascitic fluid was then collected, its volume recorded and its cellularity estimated by the aid of a Burker counting slide. The mitotic index was determined after the injection of 4 mg colchicine/kg in saline, i.p. The injection was done 6 h before killing.

In order to study the influence of aldehydes on the proliferation of normal cells, 100 mg/kg methylglyoxal or 50 mg/kg HPE were injected intraperitoneally into rats. Control rats received 0.9% NaCl solution. The mitotic index was measured in the bone marrow and in the small intestine epithelial lining, as described above. Smears of femoral bone marrow, as well as of ascitic fluid, were stained by the Giemsa method. Mitotic nuclei in metaphase were counted as a proportion of total nuclei.

The following aldehydes were used: methylglyoxal; *trans*-4-hydroxypent-2-enal; malealdehyde; 2-oxopent-2-enal; malonaldehyde; acetaldehyde. They were kindly supplied by Professor T. F. Slater (Brunel University, Uxbridge).

Only malonaldehyde was prepared directly, in solution by treating commercial 1,2,3,3-tetraethoxypropane with Dowex 50 WX2 (Serva, Heidelberg). Aldehyde samples were used immediately after dilution. In the experiments where aldehydes were incubated *in vitro* with cells or cell particles, they were added to the cell fraction and then kept at 37°C for 10 min before the determination of the activity to be tested. All chemicals used were Sigma or Boehringer products, when not stated otherwise.

RESULTS AND DISCUSSION

Influence of aldehydes on protein synthesis

In a first group of experiments we studied the incorporation of labelled amino acids from *Chlorella* hydrolysate into postmitochondrial supernatant (PMS) proteins. Table 2 shows that all tested aldehydes inhibited amino acid incorporation but only at rather high concentrations.

Some variation of the activity was found among different aldehydes. The least effective was malonaldehyde, which produced only 10% inhibition at 5 mM concentration; 52 and 89% inhibition was obtained with 10 and 20 mM respectively. The most effective aldehyde was malealdehyde, which inhibited

TABLE 2

Effect of different concentrations of aldehydes on amino acid incorporation into post-mitochondrial supernatant (PMS) proteins from normal rat liver. (The figures are percentages of control incorporation)

Aldehyde	*Aldehyde concentration (mM)*								
	0	*0.1*	*0.25*	*0.5*	*1*	*2*	*5*	*10*	*20*
Malealdehyde	100	82	52	15	14	14			
4-Hydroxypentenal	100		96	82	46	26	16	12	
2-Oxopent-2-enal	100					92	75	22	
Malonaldehyde	100					98	90	48	
Methylglyoxal	100		68	31	9				

Amino acid incorporation was studied according to Richardson *et al.* (1971). The reaction mixture (1 ml) contained: 0.2 ml PMS (1 mg protein); 0.1 M-KCl; 0.01 M-Mg acetate; 0.02M-Tris-Cl buffer, pH 7.4; 2 mM-ATP; 10 mM-creatine phosphate; 10 μg creatine kinase; 0.25 mM-GTP; a mixture of amino acids containing 1.25 μM-phenylalanine and other amino acids according to Richardson *et al.* (1971); 0.5 μCi of ^{14}C-labelled *Chlorella* protein hydrolysate. After 15 min at 37 °C the reaction was stopped by adding 1 ml 10% trichloracetic acid. The precipitates were collected on glass fibre filters, purified and counted for radioactivity. For methylglyoxal, poly(U) was used as a messenger and labelled phenylalanine was used as a substrate.

by 18% at 0.1 mM, by 48% at 0.25 mM and by 85% at 0.5 mM. Thiol compounds, such as glutathione, dithiothreitol and cysteine, protected against inhibition, whereas excess amino acids did not protect. This seems to show that the chief target of the aldehydes is the SH group (Guidotti *et al.* 1964; see also Schauenstein & Esterbauer, this volume, p. 225).

With the system used, amino acid incorporation mainly occurs by chain elongation, as only a little reattachment of ribosomes to endogenous mRNA has been shown to take place. With poly(U), however, one measures not only chain elongation but also chain initiation. Aurintricarboxylic acid, a well-known inhibitor of chain initiation, inhibits amino acid incorporation into PMS proteins with endogenous mRNA by only 35%, whereas it is able to block completely poly(U)-coded phenylalanine incorporation into polyphenylalanine (Gravela 1975). We therefore decided to study the effect of aldehydes on a poly(U)-directed system. Table 3 shows in parallel the results obtained with endogenous mRNA and with poly(U). In the first case, uniformly labelled amino acids and labelled phenylalanine were used as substrates. It is clear from Table 3 that poly(U)-coded polyphenylalanine synthesis was much more affected by aldehydes than the endogenous mRNA-coded incorporation of amino acids into PMS proteins. The reason for this is not damage to poly(U). In fact, the extent of inhibition does not decrease when the concentration of poly(U) is increased to 200 μg. It seems probable, therefore, that the aldehydes affect chain initiation more than chain elongation. The same results was obtained with CCl_4 (Dianzani *et al.* 1976). It is

TABLE 3

Effect of different concentrations of malealdehyde and 4-hydroxypentenal on endogenous mRNA-dependent amino acid incorporation into PMS, and on poly(U)-directed polyphenylalanine synthesis. (The figures are percentages of control incorporation)

		Amino acid incorporation with endogenous mRNA	*Phenylalanine incorporation with endogenous mRNA*	*Polyphenylalanine synthesis with poly(U)*
Control		100	100	100
Malealdehyde	0.2 mM	52	58	22
	0.3 mM	39	35	8
	0.5 mM	25	41	4
4-Hydroxypentenal	0.5 mM	72	63	31
	1.0 mM	37	49	13
	2.0 mM	26	44	15

Methods as in Table 2. Polyphenylalanine synthesis was calculated as the difference in specific radioactivity in the presence and in the absence of poly(U) (0.1 mg/ml).

well known that SH groups are required for chain initiation (Bermek *et al.* 1971). This is another argument for thinking that the most important target for aldehydes are SH groups.

In other experiments, amino acid incorporation was studied with isolated hepatocytes. In this case [^{14}C]valine, added to an amino acid pool, was used as a label. Table 4 shows that a remarkable inhibition of the incorporation of

TABLE 4

Effect of methylglyoxal on [^{14}C]valine incorporation into proteins in isolated hepatocytes. Effect on the secretion into the incubation medium of prelabelled proteins. Cell viability is assessed by the release of enzymes into the medium

Methylglyoxal concentration (mM)	*[^{14}C]Valine incorporation into hepatocyte proteins (c.p.m./60 min)*	*Prelabelled protein secretion (c.p.m./30 min)*	*% release of enzymes in 60 min*	
			Glutamic pyruvic transaminase	*Lactate dehydrogenase*
0	14 250	1600	4	7
0.5	13 600	–	4	5
1.0	–	1160	–	–
1.25	9700	–	4	5
2.5	4380	845	5	5
5	1330	1190	6	9
10	250	2610	13	31

Amino acid incorporation methods as in Table 2, with labelled valine in place of phenylalanine.
Prelabelling of intracellular proteins and measurement of protein secretion into the incubation medium according to Gravela *et al.* (1977).

the label into liver cell proteins occurred with methylglyoxal even at a concentration of 1.25 mM. Inhibition was about 26% at 1mM and about 47% at 2.5 mM in the case of the secretion of proteins from the cells into the medium. At higher concentrations of the aldehyde there was an increase in the release of labelled proteins. At the low concentrations mentioned, the tests of viability showed no cell death. The release of glutamic pyruvic transaminase and of lactate dehydrogenase from the cells into the incubation fluid was unchanged in these conditions, whereas it was considerably increased with higher aldehyde concentrations. This shows that the action of methylglyoxal on protein synthesis and on protein secretion by liver cells at low concentrations is not related to cell death, whereas the increased release of proteins seen at high concentrations of aldehyde, in the presence of a major inhibition of protein synthesis and of enzyme release, was mostly due to cell death. In some experiments we prelabelled proteins by incubating the cells in labelled amino acids; we then stopped protein synthesis by adding 0.7 μM-cycloheximide. At this point, the aldehydes (methylglyoxal or HPE) were added. In these conditions we observed a marked inhibition of secretion. This means that the aldehydes act on protein release not only by inhibiting protein synthesis but also by some other mechanism involving the regulation of the intracellular traffic of presynthesized proteins.

Influence of aldehyde on tubulin

According to LeMarchand *et al.* (1973, 1974) and Gravela *et al.* (1977) the microtubular system is involved in the secretion of lipoproteins and proteins from liver cells. Colchicine, the best-known inhibitor of microtubular function, inhibits secretion and at the same time promotes an increase in the triglyceride content of isolated hepatocytes. We have recently described (Poli *et al.* 1979) how CCl_4, added *in vitro* to isolated hepatocytes, blocks the secretion of prelabelled lipoproteins or proteins in conditions where protein synthesis is blocked by cycloheximide. This shows that the onset of triglyceride accumulation in CCl_4-poisoned hepatocytes can be related to cell damage, either in the piling up or in the release of pre-existing macromolecules. As these results were mimicked by using aldehydes instead of CCl_4, we decided to study the influence of the aldehydes on tubulin, the major protein of the

TABLE 5

Percentage decrrase in rat liver tubulin activity after incubation with several aldehydes

Aldehyde	*Aldehyde concentrations (mM)*							
	0.1	*0.5*	*1*	*2*	*5*	*10*	*20*	*50*
2-Oxopent-2-enal			0		0	18.4		
4-Hydroxypentenal	0.5	0.5	10.3		45.0	73.0	89.5	94.0
Malealdehyde	0		1.1	11.1	30.4	60.4	82.2	
Methylglyoxal	0	0	5.8	9.2	16.5	93.4	100	
Acetaldehyde	0.9	3.4	4.6		14.5	17.1	34.0	78.1

microtubular system. Table 5 shows that preincubation with several aldehydes produces a big decrease in colchicine binding to rat liver supernatant tubulin. The block is higher with methylglyoxal and with HPE than with the other aldehydes. Both methylglyoxal and HPE were active at 1 mM concentration, whereas a 2 mM concentration was needed for malonaldehyde and even higher amounts for 2-oxopent-2-enal and acetaldehyde.

In other experiments the effect of HPE on the polymerized and soluble forms of tubulin was studied. The method used was that described by Pipeleers *et al.* (1977). It consists of separating the two forms of tubulin by centrifuging liver homogenates at room temperature in a microtubule-stabilizing buffer containing dimethyl sulphoxide and glycerol. The two forms of tubulin are then quantified by the colchicine-binding assay. The soluble form is measured directly; the polymerized form is measured by difference after depolymerization at 4 °C with a tubulin-depolymerizing buffer. Table 6 shows that 1 mM-HPE produces a 17% decrease in the total tubulin content,

TABLE 6

Percentage decrease of different forms of tubulin induced by different concentrations of 4-hydroxypentenal

4-Hydroxypentenal concentration (mM)	*Total tubulin*	*Polymerized tubulin*	*Soluble tubulin*	*Polymerized tubulin as % of total tubulin*
1	17.1	12.5	18.6	27.7
5	62.5	34.4	72.5	45.9
10	71.7	51.0	79.1	45.4
20	74.9	55.4	81.8	46.7

with a similar decrease in both forms of tubulin. At this HPE concentration about 28% of tubulin occurs in the soluble form, as in the absence of the aldehyde. With 5 mM-HPE, however, there is a larger decrease in total tubulin, mostly due to a decrease in the soluble form. Polymerized tubulin increases to about 46% of the total. This situation does not change after a further increase of aldehyde concentration up to 20 mM. It is clear, therefore, that both forms of tubulin contain sites sensitive to the aldehyde, but soluble tubulin contains more of the sensitive groups. It seems that depolymerization releases some sensitive sites that are masked in the polymerized state. As the damage to tubulin produced by the aldehydes is prevented by dithiothreitol or reduced glutathione, it seems probable that in this case, also, SH groups are the major targets of the aldehydes. It is well known that SH groups are required for tubulin polymerization (see Gabriel *et al.* 1977, for references). The action of the aldehydes seems therefore to mimic to a certain extent that of colchicine, which is known to prevent polymerization. This seems particularly relevant not only to the interpretation of the mechanism of triglyceride accumulation in the liver of CCl_4-treated animals, but also in relation to Albert Szent-Györgyi's views on the inhibitory effects of aldehydes on cell division.

Influence of aldehydes on the mitotic index

Functioning microtubules are required to support the movements of centrioles and chromosomes during mitosis. In order to check whether the aldehydes really exert a colchicine-like activity *in vivo* we studied the mitotic index in tissues of rats injected intraperitoneally with the aldehydes. Two tissues were chosen for this study because of their high rate of cellular proliferation, namely bone marrow and the epithelial lining of the small intestine. Table 7 shows that in both tissues treatment with methylglyoxal or HPE pro-

TABLE 7

Effect of methylglyoxal and 4-hydroxypentenal (HPE) on the mitotic index in bone marrow and small intestine epithelial cells

Treatment	*Bone marrow*	*Small intestine epithelium*	
		Upper part	*Lower part*
Control	13.1 ± 1.44(100%)	11.2 ± 0.36(100%)	15.2 ± 0.67(100%)
Methylglyoxal	11.1 ± 1.52(85%)	8.8 ± 0.77(79%)*	12.8 ± 1.10(84%)
HPE	10.5 ± 1.26(80%)	7.4 ± 0.45(66%)**	11.4 ± 0.93(75%)*

The mitotic index is the number of mitoses/1000 counted nuclei. Three rats per group, killed 24 h after treatment. Means ± S.D.
* Significance of differences ('t' test), $P < 0.05$.
**Significance of differences, $P < 0.01$.

duced a remarkable decrease in the mitotic index. The decrease was greater for HPE than for methylglyoxal. In the small intestine the measurements were made both in the upper part (luminal half of the intestine crypts) and in the lower part of the epithelial lining. The mitotic index was more affected in the upper than in the lower part. This may reflect an impaired migration of cells towards the top of the villi.

With the background of these results on normal tissues, we repeated the experiment on AH-130 Yoshida hepatoma cells growing in the peritoneal cavity of rats. Table 8 shows that the mitotic index greatly decreases in cancer cells after treatment with methylglyoxal. Similar results were obtained with HPE. The cellularity of the ascitic fluid was decreased by 50–60% after treatment with the aldehydes, when given in two separate doses on Days 4 and 5 after hepatoma transplantation. In a pilot experiment rats injected with approximately 5×10^6 hepatoma cells received either HPE (25 or 12.5 mg/kg) or methylglyoxal (50 or 25 mg/kg) intraperitoneally on Days 2,3,4,5 and 6 after hepatoma transplantation. After 18 days, 100% of animals receiving the lower dose of HPE and 80% of those receiving the lower dose of methylglyox-

TABLE 8

Effect of methylglyoxal on the mitotic index (%) of AH-130 Yoshida hepatoma cells

	Controls	*Methylglyoxal treatment*	*% of controls*
Number of rats	4	6	
Ascitic fluid (ml)	38.9 ± 13.1	29.2 ± 7.3	74**
Number of cells/ml (10^6)	66.6 ± 5.65	35.8 ± 9.55	54**
Total number of cells	2.58 ± 0.916	1.03 ± 0.314	39**
Mitotic index (per 100 nuclei)	23.4 ± 3.42	7.7 ± 1.23	34**

Details as in Table 7. Data on the effect of treatment on the cellularity of ascitic fluid are included.

al were alive without any tumour growth. The higher dose, however, led to the death of several animals. Further experiments are in progress on this point.

It seems clear from these experiments that aldehydes have a big influence on cell multiplication. Of course, the inhibition of the growth of tumour cells was certainly influenced by the fact that the aldehydes were injected into the same compartment as the tumour cells, namely the peritoneal cavity. It is remarkable, however, that the aldehydes were even able to produce some decrease in mitotic index in tissues of normal rats, where the substances probably arrived at their target through the blood. As far as the mechanism of the decrease in mitotic index goes, we have no conclusive information. The fact that aldehydes partially block tubulin might suggest the possibility of a colchicine-like mechanism but other interpretations are possible.

The influence of aldehydes on microsomal enzymes

Table 9 reports the results of experiments on glucose-6-phosphatase. It is clear that both malealdehyde and HPE inhibit this enzyme. Malealdehyde inhibited by about 28% at 0.25 mM and by about 90% at 1 mM. HPE inhibited only by 5% at 0.5 mM and by 70% at 2 mM. Reduced glutathione, added to the medium before the aldehydes, gave some protection, despite the fact that it was slightly inhibitory if added alone (20% inhibition by 2 mM-glutathione). 1 mM-malealdehyde inhibited by 85% if added alone but the inhibition was decreased to 38% and 44% respectively when 2 mM-glutathione or 2 mM-dithiothreitol were present. Similar protection was afforded by thiol substances. Preincubation for 5 min of 1 mM-malealdehyde with 2 mM-glycine or 2 mM-cysteine before addition of the microsomes almost completely prevented inhibition. This was expected, as it is known that aldehydes react easily with both NH_2 and SH groups. Preincubation as described above pro-

TABLE 9

Inhibition by methylglyoxal and 4-hydroxypentenal of liver microsomal glucose-6-phosphatase activity

Aldehyde concentration (mM)	*Enzyme activity (U/mg protein) (Mean ± S.D.)*	
	Malealdehyde	*4-Hydroxypentenal*
0	0.188 ± 0.002	0.193 ± 0.014
0.10	0.205 ± 0.002	
0.25	0.137 ± 0.004	
0.50	0.042 ± 0.008	0.181 ± 0.005
1.00	0.021 ± 0.019	0.165 ± 0.005
2.00		0.135 ± 0.008
3.00		0.105 ± 0.009
6.00		0.065 ± 0.004

Activity is expressed as U/mg protein. 1 U = μmoles inorganic orthophosphate set free from the substrate/min. Liver microsomes (0.2 mg protein) were exposed to aldehyde for 10 min at 37 °C, then incubated for a further 30 min for the assay of enzymic activity. The reaction mixture (1 ml) contained: EDTA, 1 mM; 20 mM-imidazole–HCl buffer, pH 6.5; 40 mM-glucose-6-phosphate. The reaction was stopped by adding 5 ml of 8% trichloroacetic acid.

bably resulted in the trapping of aldehydes by the amino acids. When glycine or cysteine, however, were added to microsomes 5 min before the addition of malealdehyde, glycine was no longer protective, whereas cysteine was still protective. This may mean that the aldehyde reacts more rapidly in the system used, with SH groups than with NH_2 groups, thus blocking the SH functional groups of the enzyme. A major involvement of SH groups is confirmed by the fact that the content of SH groups freely reacting with *N*-[^{14}C]ethylmaleimide decreases in the system used from 16.3 to 1.7 nmol/mg protein (Bertone & Dianzani 1977).

The aminopyrine demethylase activity of microsomes was slightly influenced by both methylglyoxal and HPE, whereas cytochrome *P*-450 was depressed to a greater extent. By contrast, the lipoperoxidative intermediate, linolenic acid hydroperoxide, was strongly inhibitory both to aminopyrine demethylase and to cytochrome *P*-450. The fact that aminopyrine demethylase activity was practically unaffected by the aldehydes, whereas cytochrome *P*-450 is affected to a certain extent, might be explained in the sense that residual cytochrome may be sufficient to ensure practically normal enzyme activity.

In other experiments we found that pretreatment of microsomes with 2 mM-malealdehyde or 5 mM-HPE considerably decreases the diene conjugation band usually induced by adding CCl_4 to the system. This shows that the metabolic pathway of CCl_4 is affected by the aldehydes.

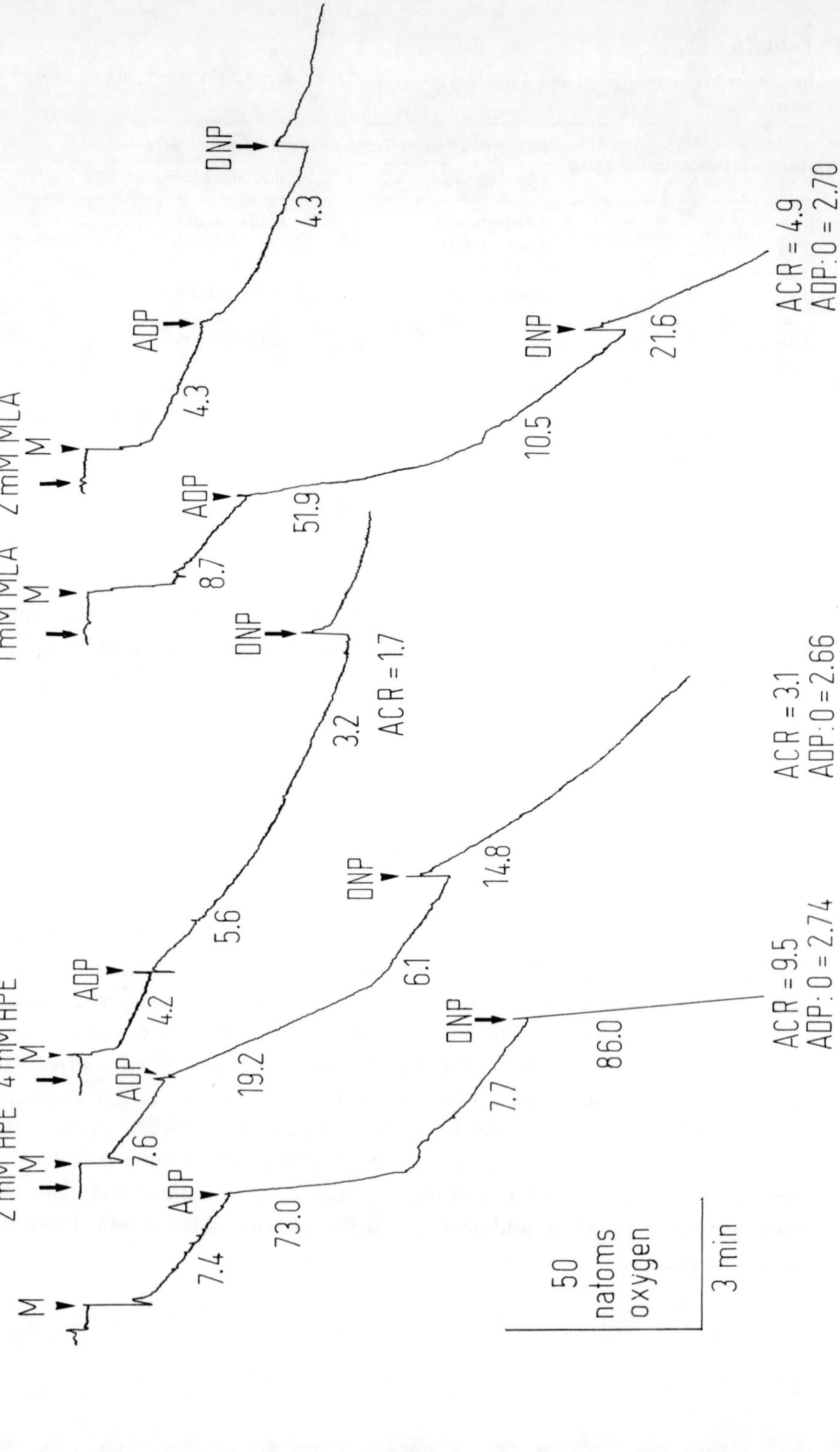
M
ADP
7.4
73.0
2 mM HPE
M
7.6
ADP
19.2
7.7
DNP
86.0
ACR = 9.5
ADP:O = 2.74
6.1
DNP
14.8
ACR = 3.1
ADP:O = 2.66
4 mM HPE
M
4.2
ADP
5.6
3.2
DNP
ACR = 1.7
1 mM MLA
M
8.7
ADP
51.9
10.5
DNP
21.6
ACR = 4.9
ADP:O = 2.70
2 mM MLA
M
4.3
ADP
4.3
DNP
50 natoms oxygen
3 min

Influence of aldehydes on mitochondria

Fig. 1 shows that the respiratory rate of mitochondria with glutamate plus malate as substrates is strongly affected by both malealdehyde and HPE. Control mitochondria show a high State 3 respiration and a high acceptor/control ratio (Feo *et al.* 1973) on addition of ADP. Stimulation of respiration by ADP was decreased, in this experiment, from 73 to about 52 natoms O_2/min per mg protein by 1 mM-malealdehyde, whereas the acceptor/control ratio decreased from 9.5 to 4.9. Stimulation of respiration by dinitrophenol was considerably decreased by the aldehydes. Similar results were obtained with 2 mM-HPE. Inhibition was increased by increasing the aldehyde concentration. It is noteworthy that in each case the ADP/O ratio was unaffected. The effect of the aldehydes were prevented by the presence of 2.5 mM-dithiothreitol.

Addition of aldehydes to mitochondria resulted in an increased rate of swelling. Aldehyde-swollen mitochondria showed no further swelling when phosphate was added. Aldehyde-induced swelling was prevented to some extent by dithiothreitol, which even preserved the sensitivity to phosphate. The aldehydes were also able to decrease Ca^{2+}-induced swelling and K^+ plus valinomycin-induced swelling (Fig. 2). These results were obtained with relatively high concentrations of aldehydes, but it must be noted that mitochondria contain an active aldehyde dehydrogenase which might have influenced the results by rapidly decreasing the local concentration of the added aldehydes.

Influence of aldehydes on lysosomes

Neither 'total' nor 'free' activities of acid phosphatase and β-galactosidase of the 'M + L' fraction were influenced by malealdehyde or HPE at concentrations ranging from 1 mM to 5 mM; 1 mM-malealdehyde did not even in-

←

FIG. 1. Respiratory rate and acceptor/control ratio in liver mitochondria as affected by HPE and malonaldehyde (MLA). The reaction mixture contained 135 mM-KCl, 10 mM-Tris-Cl (pH 7.4), 10 mM-L-glutamate (Tris salt), 5 mM-L-malate (Tris salt) and 5 mM-Tris-phosphate in a final volume of 2 ml. Malonaldehyde and HPE were added at the concentrations indicated. The reaction was started by adding 0.2 ml of mitochondrial suspension (2.0-2.5 mg protein). State 3 respiration was induced by adding 0.2 mM-ADP. Temperature was 23 °C. Dinitrophenol (DNP) concentration was 0.1 μM, when present. The respiratory rate is expressed as natoms oxygen taken up per min per mg protein.

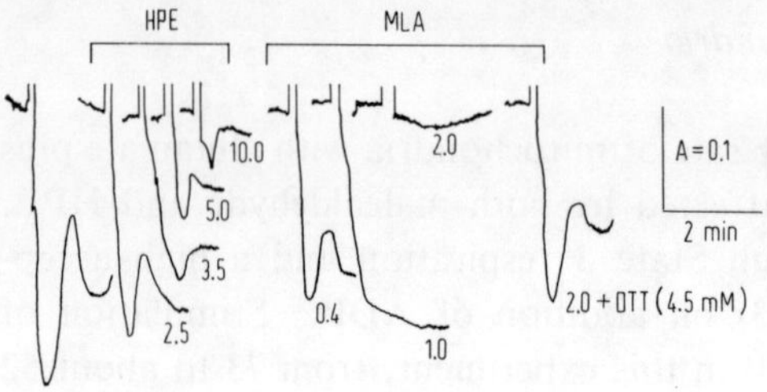

FIG. 2. Mitochondrial swelling as affected by malonaldehyde (MLA) or HPE. The test system contained, in 3 ml, 10 mM-KCl, 2 mM-Tris-Cl buffer (pH 7.4), 2 mM-Tris-phosphate, 3 mM-Tris-glutamate, 1.5 mM-Tris-malate, 280 mM-sucrose and mitochondria containing 1.0-1.6 mg protein/ml. The numbers under the curves refer to the concentration of the aldehydes indicated in the headings. DTT, dithiothreitol. The reaction was started by adding 1×10^{-3}M-valinomycin.

fluence the acid phosphatase activation usually produced by incubation at pH 5.0 at 37 °C. Cathepsin B_1 activity, on the contrary, was inhibited by about 30% by 0.2 mM-HPE. The inhibition increased to about 60% when the HPE concentration was increased to 3 mM. 8 mM-dithiothreitol produced a big activation of the enzymic activity (about 300% activation). Addition of HPE up to 1 mM concentration to the dithiothreitol-stimulated system did not modify the results. A decrease in stimulation was, however, observed with HPE concentrations above 3 mM. Stimulation was abolished only by 40–50 mM-HPE. The subsequent addition of HPE to the dithiothreitol-stimulated system decreased the stimulation, starting from HPE concentrations of 1 mM. The intraperitoneal administration of 50 mg/kg body weight of HPE to hepatoma-bearing rats did not modify the cathepsin B_1 activity of the tumour cells, which was basically much lower than that of normal liver.

Influence of aldehydes on ligandin

Ligandin (protein Y) and protein Z are known to act as intracellular transporters for several substances coming from the blood. Free fatty acids, bilirubin, steroid hormones and bromosulphophthalein are transported especially by protein Y, whereas protein Z is more specific for free fatty acids. The binding of bromosulphophthalein is usually taken as a measure of the activity of ligandin. The results of our experiments to test the influence of aldehydes on the binding activity of ligandin are preliminary, but it is clear that high concentrations of methylglyoxal inactivate ligandin as well as protein Z (Fig. 3 and Table 10).

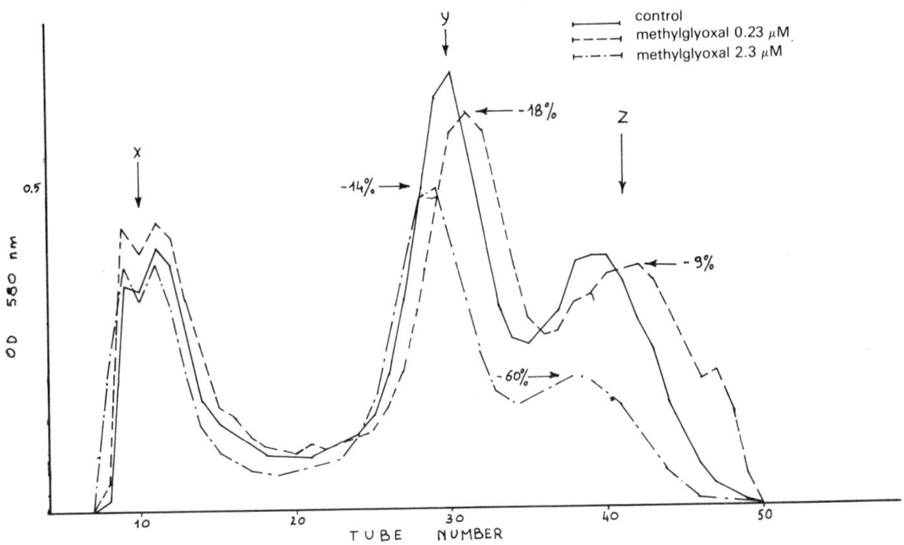

FIG. 3. Binding of bromosulphophthalein (BSP) to X,Y and Z fractions of rat liver supernatant, as influenced by methylglyoxal. The supernatant was obtained by centrifuging the liver homogenate at 110 000 *g*. The values refer to 0.125 g rat liver supernatant, containing about 6 mg protein. Added BSP was 0.24 μmol. Preincubation with methylglyoxal (0.23 or 2.3 μM) was done for 15 min before the addition of BSP. After incubation the proteins were eluted from a G-75 Sephadex column (90 × 60 mm) at a flow rate of 21 ml/h. Collections were 1.7 ml/tube. Optical density was estimated at 580 nm after alkalinization to measure BSP.

Influence of aldehydes on capillary wall permeability

These experiments were done in order to understand the reasons for the terminal circulatory shock occurring in rats given lethal amounts of CCl_4. Ugazio *et al.* (1976) have recently found that a peroxidative derivative of linoleic acid, linoleic acid hydroperoxide, greatly increases capillary permeability. Aldehydes have a similar effect, but at higher doses. The smallest active dose was 7.1 pmol for serotonin (5-hydroxytryptamine), 4 nmol for linoleic acid hydroperoxide, 0.15 μmol for HPE, 1.15 μmol for 2-oxopent-2-enal and 2 μmol for malonaldehyde. If HPE is injected in combination with serotonin there is a much larger increase in capillary permeability. With half the active dose of HPE, the lowest effective dose of serotonin becomes 3.5 μmol. The effect of serotonin on capillary permeability was doubled by injecting 0.075 μmol malealdehyde or 1.5 μmol methylglyoxal (doses that correspond to half the smallest active doses of these aldehydes). These results support the idea that aldehydes derived from CCl_4-induced lipid peroxidation play a part in causing death after CCl_4 administration.

TABLE 10

Inhibition by methylglyoxal of the binding of bromosulphophthalein (BSP) to proteins Y and Z from rat liver

(A) Percentage precipitation of total of liver supernatant by methylglyoxal

	Methylglyoxal concentration (μmol/mg protein)					
	2300	*230*	*23*	*11.5*	*2.3*	*0.23*
Precipitated proteins(%)	37	7	5	7	6	0

(B) Inhibition of BSP binding by low concentrations of methylglyoxal

	Binding to protein Y (μmol/g)	*% inhibition*	*Binding to protein Z (μmol/g)*	*% inhibition*
Control	0.365	0	0.100	0
Methylglyoxal				
0.23 μmol/mg protein	0.306	18	0.092	9
2.3 μmol/mg protein	0.314	14	0.040	60

BSP(2.1 μmol/g)was added *in vitro* to rat liver supernatant in the presence of different final concentrations of methylglyoxal after preincubation with methylglyoxal for 15 min at room temperature. The supernatant contained approximately 48 mg protein.

Binding to proteins Y and Z is expressed as μmol BSP bound by amounts corresponding to 1 g wet liver.

The values were extrapolated from the elution graphs, after measurement of the optical densities of the eluates at 580 nm. (Values are means of 3–4 experiments.)

CONCLUSIONS

It is clear from the experiments described here that aldehydes exert several damaging effects on the biochemical parameters investigeated. The results speak strongly in favour of the idea that these substances contribute to the lesions occurring in cell regions ‘far’ from the CCl_4-metabolizing sites. Of course, there is no proof that aldehydes are the only low molecular weight substances producing such ‘remote’ damage. Experiments by Roders *et al.* (1977, 1978), Willis *et al.* (1977) and Comporti *et al.* (1977; personal communication) have shown, for instance, that low molecular weight substances derived from lipids allowed to peroxidize *in vitro* are able to produce haemolysis when added to red blood cells. The aldehydes used throughout this work are not haemolytic at all. We have therefore to accept the idea that substances other than aldehydes are responsible for this type of cell damage.

Another necessary comment is that the inhibitory effects of aldehydes that we observed took place at rather high concentrations. It seems probable that this is the result of the presence in the homogenates and other enzymic preparations of many substances which trap or metabolize the added aldehydes, so preventing their reaction with the enzymic targets. Among such substances one has to consider SH and NH_2 groups. From the stoichiometric point of view, however, only one molecule of aldehyde is required to react with a single functional SH or NH_2 group in the enzyme molecule. The problem of the action of aldehydes produced *in vivo* during lipid peroxidation therefore becomes mainly that of the probabilities of the recently produced aldehyde molecule reaching the enzymic target, rather than other shielding molecules. It is clear that a continuous flow of aldehydes from peroxidizing lipids might lead to saturation of the shields and consequently to an easier contact with enzymic targets. As aldehydes are destroyed in the cells by aldehyde dehydrogenase, as well as by other enzymes, it is possible that another way of increasing the efficiency of the aldehydes would be by inhibiting this enzyme. We observed in some preliminary experiments that mitochondria from the AH-130 Yoshida hepatoma, which contain less aldehyde dehydrogenase than normal liver cell mitochondria, are more sensitive to added aldehydes than the normal mitochondria. In any case, the ability of aldehydes to diffuse from the production sites throughout the cytoplasm is clearly demonstrated by the experiments of Table 1 (p. 247).

To me, one of the most interesting effects of aldehydes is the effect on tubulin. As it is now clear that the onset of triglyceride accumulation after CCl_4 depends on disruption of the release of lipoproteins synthesized before CCl_4 addition this early block in lipoprotein micelle traffic might at least partially depend on aldehyde-induced loss of function of tubulin. It is possible that even the colchicine-like action of aldehydes on mitosis is explained, at least in part, by tubulin inhibition. Further experiments are, of course, needed, and other interpretations are possible.

References

BARRETT, A.J. (1972) in *Lysosomes, A Laboratory Handbook* (Dingle, J.T., ed.), pp. 46-135, North-Holland, Amsterdam

BERMEK, F., MÖNKEMEYER, H. & BERG, R. (1971) The role of SH-groups of human ribosomal subunits in polypeptide synthesis. *Biochem. Biophys. Res. Commun. 45,* 1294-1299

BERRY, M.N. & FRIEND, D.S. (1969) High-yield preparation of isolated liver parenchymal cells. *J. Cell Biol. 43,* 506-520

BERTONE, G. & DIANZANI, M.U. (1977) Inhibition by aldehydes as a possible further mechanism for glucose-6-phosphatase inactivation during CCl_4^--poisoning. *Chem.-Biol. Interact. 19,* 91-100

DE DUVE, C., PRESSMANN, B.C., GIANETTO, R., WATTIAUX, R. & APPELMANS, F. (1955) Tissue fractionation studies. 6. Intracellular distribution patterns of enzymes in rat-liver tissue. *Biochem. J. 60,* 604-617

DEMPSEY, W.O. & SNELL, E.E. (1963) Pyridoxamine-pyruvate transaminase. II. Characteristics of the enzyme. *Biochemistry 2,* 1414-1419

DE TORANZO, E.G.D., DIAZ GÒMEZ, M.I. & CASTRO, J.A. (1978) Carbon tetrachloride activation, lipid peroxidation and liver necrosis in different strains of mice. *Res. Commun. Chem. Pathol. Pharmacol. 19,* 347-352

DIANZANI, M.U. & GRAVELA, E. (1975) Inhibition of protein synthesis in carbon tetrachloride-induced liver injury, in *Pathogenesis and Mechanisms of Liver Cell Necrosis* (Keppler, D., ed.), pp. 225-238, MTP Press, Lancaster

DIANZANI, M.U., GABRIEL, L., GRAVELA, E. & PARADISI, L. (1976) Interference of CCl_4 metabolites with subcellular structures, in *Recent Advances in Biochemical Pathology. Toxic Liver Injury* (Dianzani, M.U., Ugazio, G. & Sena, L.M., eds.), pp. 29-38, Minerva Medica, Torino, Italy

FEO, F., CANUTO, R.A. & GARCEA, R. (1973) Acceptor control ratio of mitochondria. Factors affecting it in Morris hepatoma 5123 and Yoshida hepatoma AH-130. *Eur. J. Cancer 9,* 203-214

FEO, F., CANUTO, R.A., GARCEA, R. & GABRIEL, L. (1975) Effect of cholesterol content on some physical and functional properties of mitochondria isolated from adult rat liver, fetal liver, cholesterol-enriched liver and hepatomas AH-130, 3924 A and 5123. *Biochim. Biophys. Acta 413,* 116-134

GABRIEL, L., BONELLI, G. & DIANZANI, M.U. (1977) Inhibition of colchicine binding to rat liver tubulin by aldehydes and by linoleic acid hydroperoxide. *Chem.-Biol. Interact. 19,* 101-109

GRAVELA, E. (1975) *In vitro* inhibition of polyphenylalanine synthesis by bromotrichloromethane. Evidence for an oxidative mechanism. *Res. Commun. Chem. Pathol. Pharmacol. 11,* 615-625

GRAVELA, E., POLI, G., ALBANO, E. & DIANZANI, M.U. (1977) Studies on fatty liver with isolated hepatocytes. I. The action of colchicine, phalloidin, cytochalasin B and cycloheximide on protein and triglyceride synthesis and secretion. *Exp. Mol. Pathol. 27,* 339-352

GRAVELA, E., ALBANO, E., DIANZANI, M.U., POLI, G. & SLATER, T.F. (1979) Effects of CCl_4 in isolated hepatocytes: inhibition of protein and lipoprotein secretion. *Biochem. J. 178,* 509-512

GUIDOTTI, G., FONNESU, A. & CIARANFI, E. (1964) Inhibition of amino acid incorporation into protein of Yoshida ascites hepatoma cells by glyceraldehyde. *Cancer Res. 24,* 900-905

LEMARCHAND, Y., SINGH, A., ASSIMACOPOULOS-JEANNET, F., ORCI, L., ROUILLER, C. & JEANRENAUD, B. (1973) A role for the microtubular system in the release of very low density lipoproteins by perfused mouse livers. *J. Biol. Chem. 248,* 6862-6870

LEMARCHAND, Y., PATZELT, C., ASSIMACOPOULOS-JEANNET, F., LOTEN, E.J. & JEANRENAUD, B. (1974) Evidence for a role of the microtubular system in the secretion of newly synthesized albumin and other proteins by the liver. *J. Clin. Invest. 53,* 1512-1517

LEVY, A.J., GATMAITAN, Z. & ARIAS, I.M. (1969) Two hepatic cytoplasmic protein fractions Y and Z, and their possible role in the hepatic uptake of bilirubin, sulfobromophthalein and other anions. *J. Clin. Invest. 48,* 2156-2167

OMURA, T. & SATO, R. (1964) The carbon-monoxide-binding pigment of liver microsomes. I. Evidence for its hemoprotein nature. *J. Biol. Chem. 239,* 2370-2378

ORRENIUS, S. (1965) On the mechanism of drug hydroxylation in rat liver microsomes. *J. Cell Biol. 26,* 713-723

PIPELEERS, D.G., PIPELEERS-MARCHAL, M.A., SHERLINE, P. & KIPNIS, D.M. (1977) A sensitive method for measuring polymerized and depolymerized forms of tubulin in tissues. *J. Cell Biol. 74,* 341-350

POLI, G., GRAVELA, E., ALBANO, E. & DIANZANI, M.U. (1979) *Exp. Mol. Pathol. 30,* 116-127

RECKNAGEL, R.O., HRUSZKEWICZ, A.M. & GLENDE, E.A. (1975) Absolute-dependence of CCl_4 induced loss of glucose-6-phosphatase and cytochrome P-450 on lipid peroxidation, in *Recent Advances in Biochemical Pathology. Toxic Liver Injury* (Dianzani, M.U., Ugazio, G. & Sena, L.M., eds.), pp. 93-98, Minerva Medica, Torino, Italy

RICHARDSON, A., MCGOWN, E., HENDERSON, L.M. & PATRICIA, B. (1971) *In vitro* amino acid incorporation by the postmitochondrial supernatant from rat liver. *Biochim. Biophys. Acta 254,* 468-477

RODERS, M.K., GLENDE, E.A. & RECKNAGEL, R.O. (1977) Prelytic damage of red cells in filtrates from peroxidizing microsomes. *Science (Wash. D.C.) 196,* 1221-1222

RODERS, M.K., GLENDE, E.A. & RECKNAGEL, R.O. (1978) NADPH-dependent microsomal lipid peroxidation and the problem of pathological action at a distance. New data on induction of red cell damage. *Biochem. Pharmacol. 27,* 437-443

SCHAUENSTEIN, E. (1967) Autoxidation of polyunsaturated esters in water: chemical structure and biological activity of the products. *J. Lipid Res. 8,* 417-428

SCHAUENSTEIN, E., ESTERBAUER, H. & ZOLLNER, H. (1977) *Aldehydes in Biological Systems* (Gore, P.H., transl.), pp. 1-205, Pion/Academic, London

SELLINGER, O.Z., BEAUFAY, H., JACQUES, P., DOYEN, A. & DE DUVE, C. (1960) Tissue fractionation studies. 15. Intracellular distribution and properties of β-*N*-acetylglucosaminidase and β-galactosidase in rat liver. *Biochem. J. 74,* 450-456

SHERLINE, P., BODWIN, C.K. & KIPNIS, D.M. (1974) A new colchicine binding assay for tubulin. *Anal. Biochem. 62,* 400-407

UGAZIO, G., TORRIELLI, M.V., BURDINO, E., SAWYER, B.C. & SLATER, T.F. (1976) Long-range effects of products of carbon tetrachloride-stimulated lipid peroxidation. *Biochem. Soc. Trans. 4,* 353-356

VILLARUEL, M.C., DE TORANZO, E.G.D. & CASTRO, J.A. (1976) CCl_4 administration to strain A/J mice or rats and the arachidonic acid content of their liver microsomal phospholipids. *Res. Commun. Chem. Pathol. Pharmacol. 14,* 193-196

WEISENBERG, R.C., BORISY, G.G. & TAYLOR, E.W. (1968) The colchicine-binding protein of mammalian brain and its relation to microtubules. *Biochemistry 7,* 4466-4477

WILLIS, R.J., PESH-IMAM, M. & RECKNAGEL, R.O. (1977) Hemolytic activity of a lipid material obtained from peroxidized microsomes. *Res. Commun. Chem. Pathol. Pharmacol. 18,* 433-437

ZURETTI, M.F., PANI, P. & PERNIGOTTI, L. (1973) Studio della glucoso-6-fosfatasi nella intossicazione con CCl_4. *Boll. Soc. Ital. Biol. Sper. 49,* 513-517

Discussion

Slater: A part of Albert Szent-Györgyi's general speculation on cell growth control is that aldehydes such as methylglyoxal tend to inhibit cell division and, therefore, may be useful in the treatment of established tumours. What Professor Dianzani has shown, however, is that the general biochemical reactivities and toxicities of these aldehydes are such that the chances of being able to treat a tumour in the whole animal by parenteral administration of aldehydes like methylglyoxal or even hydroxypentenal, as such, are remote, because of the high doses required and the general toxicity of these compounds. This emphasizes Professor Esterbauer's earlier comments that if these aldehydes are to be made effective against tumours we must try to generate them intracellularly rather than infuse them into the whole animal or, alternatively, to transport them to the tumour in the form of a dissociable, less toxic adduct.

Dianzani: This is an important point. However, it seems clear from our experiments that thiols in the incubation medium prevent most of the damage. So, I think that what has to be demonstrated is whether the rate of intracellular release of active aldehyde from the injected adduct is sufficient to guarantee the results. Thiols released from the adducts might also protect against the damage that the aldehydes are expected to provoke at intracellular targets.

Esterbauer: Free 4-hydroxypentenal (HPE) is toxic. The monoadduct and the diadduct of HPE and cysteine, however, are not toxic. We have measured *in vitro* the rate of aldehyde release from several aldehyde–cysteine adducts and found that the rates differ considerably. Certainly the speed with which a reversible equilibrium is set up, as well as the equilibrium state itself, is of great importance for the therapeutic effectiveness of an adduct. Both parameters depend on the structure of the adduct, i.e. the parent aldehyde (Esterbauer *et al.* 1967). In therapeutic tests it was found, for example, that the very stable acrolein adduct is more or less inactive therapeutically while the somewhat less stable crotonal or HPE adducts are therapeutically effective (Tillian *et al.* 1976).

Brandt: Some approaches to increasing the production of methylglyoxal or other aldehydes within cells have in effect been described, for example when thiamine deficiency has been induced in animals so that carbohydrate metabolism is impaired, leading to the piling-up of three-carbon compounds, giving the potential for the synthesis of precursors of methylglyoxal. We have done this with Lewis lung tumour in mice (R.B. Brandt & R.P. Gilbert, unpublished work 1974). We inhibited the growth of the tumour by reducing thiamine levels in the animal, thus presumably increasing methylglyoxal production. This approach is consistent with the clinical approach, when methylglyoxal derivatives are given (Regelson & Holland 1963), but bad side-effects have been obtained and other approaches may be needed, as indicated by Professor Slater.

Dianzani: Generally speaking, chemotherapeutic effects of this type are difficult to modulate. If you give a large amount of the toxic substance you will damage not only tumour cells but also the immunologically competent cells. The development of the tumour results not only from the multiplication of tumour cells but from the strength of the opposing activity of the immunological cells. If you use a substance blocking mitosis in tumour cells, you will also block mitosis in normal cells, including the immune cells. The problem is to find substances which inhibit tumour cells and not normal cells. The fact that we have preliminary evidence that rat hepatoma possesses lower aldehyde dehydrogenase activity than normal liver cells seems promising. We have, however, to explore this point further.

Fodor: Professor Slater mentioned that the methylglyoxal acetal of ascorbic acid is considerably less toxic than methylglyoxal. I wonder whether the malealdehyde acetal derivative of ascorbic acid that we have also made, containing one mole of ascorbic acid and one mole of malealdehyde, could be of use? It might also be worth re-investigating 4-hydroxypentenal in that form, because so far as we know the electronic structure required for the formation of an enediol acetal, that compound would also meet that requirement. It would perhaps be worth investigating the toxicity of the cyclic acetal compounds to see whether they are the right carriers for 4-hydroxy- and 4-oxopentenal.

Connors: As a general point, I do not see how the discussion so far relates to the bioelectronic theory of cancer. It is likely that, if one succeeds in concentrating a reactive chemical species inside a tumour cell one will get some selective antitumour action whether an aldehyde is used or an alkylating agent or some other electrophilic reactant. One does not have to invoke the proposed bioelectronic theory to explain these results.

Williams: As I see it, one of the things the ketoaldehydes are claimed to do is to affect cell division, but there are many other things that they do which have nothing to do with cell division, such as the steps mentioned by Dr Connors. Another effect they may have is on some specific protective devices, and these could be a point of deliberate attack. If you remove the protective device, the theory goes, you would be successfully combating a tumour. Dr Dianzani has looked at cell damage resulting from lipid peroxidation. Are the reactive aldehydes stopping the removal of peroxides that have been generated? If so, one would be looking for the systems which protect the cell by keeping levels of peroxides very low. Here I come back to the selenium system of glutathione peroxidase (see p. 244). This peroxidase system is present in the cell to protect it, through glutathione, from lipid peroxidation. If the selenium is removed the liability of that cell to damage from peroxide is increased, and I believe that one of the first things that aldehydes do is to remove selenium. Another such device which Dr Dianzani studied is *P*-450, which protects in a different way. It is an oxygen attack for protection or detoxification. Dr Dianzani said that perhaps *P*-450 could be damaged as well by aldehydes, and that would be another removal of a protection. This is all very different from anything to do with the restriction of mitosis. I wonder whether, as well as *P*-450, you have examined whether these cells are now protected to the same degree from lipid peroxides in themselves, and whether aldehydes in fact do not kill directly, but remove protection?

Dianzani: The block of mitosis as induced by aldehydes may be due to their inhibitory action on tubulin, which is involved in chromosomes movements.

As regards the action of the aldehydes on the smooth endoplasmic reticulum of the liver cell, I can add that we have tested the action of the aldehydes on the metabolism of CCl_4 by the smooth endoplasmic reticulum. Preliminary observations have shown that if the membranes from the smooth endoplasmic reticulum are preincubated with aldehydes before CCl_4 is added, the diene conjugation band in the microsomal lipids is much lower than in non-preincubated membranes.

Williams: As I see it, many people here have concentrated on the SH group as one of the main targets for aldehydes, but SeH is more reactive with aldehydes than SH.

Kornberg: I want to come back to the question of concentrations of aldehydes at the affected site. We have seen that for example the activity of β-galactosidase is the same with or without the aldehyde, which means that removal of any essential amino group from this enzyme can play no part in its activity. So one must assume that at the concentrations used, the mechanism we have been discussing, which depends on the interaction of methylglyoxal at a concentration sufficient to cause adduct formation, is not likely to play any role in at least lactose utilization. However, as Professor Dianzani was careful to point out, the concentrations he used were high enough to show major inhibition of all SH activities examined. The effects that Professor Williams is talking about would be found at a concentration which might be 1% of that used. I do not believe that millimolar amounts of aldehydes are found in the cell.

Dianzani: A further point is that the intracellular targets for aldehydes are probably very numerous, so it is questionable how much of the aldehyde produced can reach the different targets. There are probably many shields between the production sites of the aldehydes and a specific target situated in a remote part of the cell. This might explain why such high concentrations of aldehydes are required in our experiments.

Slater: If you freeze-clamp samples of normal rat tissues (e.g. liver) and measure their content of malonaldehyde-reacting substances (see Slater 1972) you find levels of approximately 0.5 μmol/l, much lower than those Professor Dianzani has referred to, and this is why he emphasized that the concentrations of methylglyoxal and other aldehydes that he used are high. The amounts found *in vivo* are about 1 μmol/l, rather than mmol/l, although within cells there might be varying concentrations of these aldehydes leading to localized higher concentrations. However, when we add them to highly complex systems, as with whole-animal experiments, where there are many 'shields' protecting sensitive sites, it may well be that we require millimolar rather than micromolar concentrations to obtain significant effects.

Bondy: I am beginning to be discouraged! Initially, being naive, I thought that the hypothesis we were considering was concerned with factors controlling the organization and differentiation of cells. Presumably there was some disturbance of these controlling factors associated with the development of malignancy; but if this disturbance could be corrected, malignancy could perhaps be reversed. This is, of course, a large order, but it appeared to be what was being aimed at. But we are now talking about killing cells, which is a different matter. It is easy to *kill* a cancer. There is the problem of killing the cancer cells without killing the host, of course, and that is what Tom Connors was talking about. But agents preventing mitosis or interfering with major enzyme systems are unlikely to be highly selective, because every organism contains normal cells which must continue to multiply rapidly and have many of the growth characteristics of cancer cells. The difference is mainly that they know when to stop growing, whereas cancer cells don't. But if you use an agent that kills cells in rapid mitosis, you will necessarily also attack bone marrow and intestinal cells. This gets us no further than we were. I was hoping we would be hearing about something with a higher level of specificity than the ability to poison enzymes or disrupt tubulin or interfere with mitosis.

Williams: I think you are right, but when one comes down from the general theory to look at the specifics, one has to ask, using whatever chemical or tissue one chooses to use, what is actually observed? My understanding is that the aldehydes were chosen to study the general hypothesis about compounds which might interfere specifically in the way in which the cancer cell reproduces itself, and that they might make it into a normal cell. This would have been extraordinarily interesting. However, Dr Dianzani's study of the effects these aldehydes have has simply demonstrated a number of chemical reactions in no way different from normal chemical reactions, or from the normal biological changes of the cell in response to these chemical reactions. So far, it seems from the observations made that the effects of aldehydes do not belong specifically in the area of Szent-Györgyi's approach to the problem of cancer. They belong more to the general approach to cancer chemotherapy that is already in use.

Neuberger: Let me take up the suggestion that a more selective approach to chemotherapy might be made by having aldehydes generated in the cancer cell. This means that there must be some metabolic specificity of the cancer cell distinguishing it from the normal cell from which it is derived, consisting of either a quantitative or a qualitative difference from the normal cell. This seems to be the only rational approach. Unless one finds something specific in the metabolism of the cancer cell, the suggested approach would not be likely to get anywhere.

Slater: I can't give a specific answer to this, but I would agree entirely with Professor Williams's reply to Professor Bondy about the way in which we have reached a seeming impasse with the aldehydes as toxic substances. We certainly began by hoping to find relatively *specific* effects produced by these aldehydes *in vivo* but we have discovered that, although they can have certain highly specific actions (e.g. on thiol-dependent enzymes), when given parenterally they act as rather unspecific cytotoxic agents. Whether the use of adducts will improve this is not yet known. What Professor Neuberger says is completely true. I was trying to draw attention to one possible way in which we might proceed in the future. There is one clue that may be of use. We have heard about the production of aldehydes from lipid peroxidation. Tumour cells generally have a severe disturbance of lipid peroxidation by comparison with their normal analogues. This may give us a lead into a difference between normal and tumour cells that is of practical importance. Lipid peroxidation, of course, is associated with the production of a variety of low molecular weight aldehydes.

References

ESTERBAUER, H., ERTL, A. & SCHOLZ, N. (1967) The reaction of cysteine with α,β-unsaturated aldehydes. *Tetrahedron 32,* 285-288

REGELSON, W. & HOLLAND, J.F. (1963) Clinical experience with methylglyoxal bis(guanylhydrazone) dihydrochloride: a new agent with clinical activity in acute myelocytic leukemia and the lymphomas. *Cancer Chemother. Rep. 27,* 15-26

SLATER, T.F. (1972) *Free Radical Mechanisms in Tissue Injury,* Pion Ltd., London

TILLIAN, H.M., SCHAUENSTEIN, E., ERTL, A. & ESTERBAUER, H. (1976) Therapeutic effects of cysteine adducts of α,β-unsaturated aldehydes on Ehrlich ascites tumour of mice. *Eur. J. Cancer 12,* 989-993

Carcinostatic activity of methylglyoxal and related substances in tumour-bearing mice

PETER J. CONROY

Cancer Center of the University of Rochester Medical Center, Rochester, New York

Abstract Methylglyoxal treatment of tumour cells *in vitro* primarily depresses protein synthesis, in contrast to *trans*-4-hydroxypent-2-enal (HPE) which preferentially inhibits DNA synthesis. Methylglyoxal and HPE are potent carcinostatic agents *in vitro* but relatively ineffective *in vivo*. Both aldehydes have a short half-life *in vivo* which may explain their poor carcinostatic properties when administered other than peritumorally. Several possibilities of increasing the effective half-life were investigated including *(i)* multiple intraperitoneal injections, *(ii)* concomitant administration of an inhibitor of glyoxalase I, *(iii)* administration of aldehyde–cysteine adducts, and *(iv)* continuous intravenous infusion.

Methylglyoxal (36 mg/kg i.p., twice daily) was slightly less effective in inhibiting the growth of the solid form of Ehrlich carcinoma than a dose of 72 mg/kg (inj. 1); 36 mg/kg (inj. 2) (46.2% compared to 51%). The aldehyde was more effective against the ascitic form of the tumour, with 99.76% inhibition of growth after giving 72 mg/kg twice daily for five days followed by 36 mg/kg for five days.

The glyoxalase I inhibitor *S*-(*p*-bromobenzyl)-glutathione did not significantly enhance the activity of methylglyoxal against the solid form of the tumour. Nicotinamide (1% w/v in the drink) was similarily inactive. Methylglyoxal in combination with nicotinamide was significantly more effective ($P < 0.05$) than methylglyoxal alone (36 mg/kg, twice daily) in inhibiting the growth of the ascitic tumour. Methylglyoxal–*N*-acetyl-L-cysteine was four times less toxic than methylglyoxal alone but was marginally less effective against the ascitic form of the tumour. Doses of these adducts equivalent to 144 mg/kg per day of methylglyoxal were more effective ($P < 0.05$) than the optimal regime of methylglyoxal in inhibiting the solid tumour (67.5% inhibition compared to 51%).

Treatment of mice bearing the ascitic form of Sarcoma 180 with five daily doses (i.p.) of an HPE–cysteine adduct equivalent to a dose of HPE alone of 32–256 mg/kg per day significantly increased survival time by comparison with controls. The adduct was 2–3 times more effective, dose-for-dose, than HPE alone in inhibiting tumour growth.

Purified buffered methylglyoxal has an LD_{50} on continuous infusion into the right lateral tail vein in mice of more than 3.0 mg/g per day (seven days at 2.8

ml/day). Local oedema followed by tail necrosis occurs at doses in excess of 0.25–0.5 mg/g per day in mice bearing the solid forms of the syngeneic tumours: squamous carcinoma D; lymphosarcoma 1 (WH/Ht mice); and spontaneous mammary D5056 (CBA/CA mice). A maximum tumour volume growth delay of 3.4 days at Day 17 ($P < 0.001$) after transplantation was observed after infusion of 0.5 mg/g per day methylglyoxal on Days 11–17 in the CBA/CA D40 syngeneic mammary tumour. Tumour regrowth after termination of therapy eliminated the significant difference between control and methylglyoxal-treated tumours by Day 27. Methylglyoxal infusion (0.12 mg/g per day for six days) did not inhibit the growth of the ascitic form of Ehrlich carcinoma or Gardner lymphosarcoma in CBA/CA mice. Results are discussed in relation to the use of aldehyde–carrier adducts in tumour therapy.

Methylglyoxal markedly inhibits the growth of Ehrlich ascites carcinoma, L1210 and L4946 leukaemia, TA_3 mammary carcinoma, 6C3HED lymphosarcoma and Sarcoma 180 in mice at daily dose levels of approximately 80 mg/kg body weight, injected intraperitoneally (Apple & Greenberg 1967). Single doses of about 225 mg/kg body weight significantly inhibited advanced leukaemia and produced apparently indefinite survivors among mice bearing either 6C3HED lymphosarcoma or TA_3 carcinoma. The *in vivo* antitumour activity of methylglyoxal was significantly enhanced by its combination with DL-glyceraldehyde (Apple & Greenberg 1968). Similar observations have been made by Jerzykowski *et al.* (1970) with the Ehrlich ascites tumour in mice and Yoshida ascites hepatoma in rats. The potentiation of the antitumour effects of methylglyoxal by DL-glyceraldehyde may be related to the conversion both enzymically (Bonsignore *et al.* 1968) and non-enzymically (Riddle & Lorenz 1968) of glyceraldehyde to methylglyoxal. Egyud & Szent-Györgyi (1968) obtained little therapeutic effect of methylglyoxal with solid tumours of Sarcoma 180 on either intraperitoneal or subcutaneous injection; however, significant increases in survival time were obtained in mice bearing the ascitic form of the tumour, compared to saline-treated control animals. Methylglyoxal is rapidly rendered ineffective *in vivo* primarily by its conversion into D-lactate via the glyoxalase enzyme system which requires reduced glutathione as a cofactor. Broad substrate specificity is shown for both aliphatic and aromatic α-ketoaldehydes, the rate-determining step being a hydride migration (Vanderjagt & Han 1973) with the final product being an α-hydroxy acid.

Vince & Wadd (1969) suggested that the selective inhibition of glutathione may potentiate the carcinostatic activity of methylglyoxal and related α-ketoaldehydes. Glyoxalase I has an apparent hydrophobic area in the region of its active centre which may be occupied by alkyl and aryl groups substituted by thioether linkage to glutathione. Vince & Daluge (1971) reported that the cytotoxic activity of methylglyoxal (2.3 μg/ml) against L1210 leukaemia cells *in vitro* was increased 14–18-fold by the addition of the glyoxalase inhibitor, *S*-(2-benzoylethyl)-glutathione. However, mouse liver glutathionase rapidly metabolized such *S*-substituted glutathione derivatives, rendering them inactive when tested *in vivo*. *S*-(*p*-Bromobenzyl)-glutathione, a metabolically stable inhibitor of glyoxalase I, was synthesized by Vince *et al.* (1973); such an inhibitor could be expected to enhance significantly the *in vivo* antitumour activity of methylglyoxal and related compounds. Jerzykowski *et al.* (1975) observed that glyoxalase II was absent or nearly absent in 16 types of tumour tissue while glyoxalase I levels were similar to those of five normal tissues. Where glyoxalase II was present (7/16), the level of activity was about 10% of that of the appropriate control values. Earlier investigations on the glyoxalase system showed values either lower or higher than those for normal tissues (Cohen 1945; Strzinek *et al.* 1970). These observations of Jerzykowski and co-workers appear to be of potential significance in interpreting the antitumour effects of ketoaldehydes.

The primary effect of methylglyoxal and related ketoaldehydes against *Escherichia coli* and several mammalian cell lines is the inhibition of protein synthesis (Egyud & Szent-Györgyi 1966; Otsuka & Egyud 1967); the polymerization step is inhibited at the ribosomal level. Gregg (1968) showed that methylglyoxal did not induce synchronous growth *in vitro* in mouse lymphoma cells (L-5178Y) or Chinese hamster ovary cells but prevented further cell division within 10–15 minutes of the addition of the aldehyde. Studies of the incorporation of labelled precursors into DNA, RNA and protein showed that only protein turnover was affected rapidly enough to be considered as the primary locus of inhibition by added methylglyoxal.

The inhibition of protein synthesis in a variety of bacterial and mammalian cells by methylglyoxal may be reversed by the addition of equimolar amounts of cysteine (Egyud 1965, 1967, 1968; Egyud & Szent-Györgyi 1966; Baylor & Egyud 1967), suggesting that the inhibitory activity of the compounds tested depended on the SH content of dividing cells. Methylglyoxal can bind to exposed guanine residues of tRNA molecules (Litt 1969) and the structure of the guanine–ketoaldehyde adduct is supported by the observations of Shapiro *et al.* (1969). The amino acid acceptor activity was restored after the tRNA was stripped of the ketoaldehyde. Vince & Daluge (1971) proposed that the observed protection by thiol compounds of methylglyoxal toxicity does not establish that the aldehyde results from its interaction with critically placed SH groups. Rather, methylglyoxal may inhibit protein synthesis by reversibly binding with tRNA, thus preventing its participation in protein synthesis.

Protection by thiol against the inhibitory effects of ketoaldehydes may then be due to displacement of the aldehydes from the guanine residues by thiol compounds.

Szent-Györgyi and co-workers have ascribed a role to a glyoxal derivative that functions as an electron-accepting growth inhibitor which regulates cell division in both bacterial and mammalian cell systems (Szent-Györgyi 1960, 1968*a*,*b*, 1974, 1976). Schauenstein (1967) in the first of several reports showed that the peroxidation of polyunsaturated fatty acids leads to the production of water-soluble products of low molecular weight, some of which have powerful inhibitory effects on cell division in mammalian cell lines in culture (reviewed by Schauenstein *et al.* 1977). Two of the most active compounds were *trans*-4-hydroxy-2-octenal (HOE) and *trans*-4-hydroxypent-2-enal (HPE), which have broadly similar biological properties.

Moderate to good inhibition of tumour growth has been observed with the following subcutaneously implanted lines: Ehrlich carcinoma (Schauenstein *et al.* 1968, 1969), Sarcoma 180 (Schauenstein *et al.* 1971*a*), Nemeth-Kellner lymphoma (Schauenstein *et al.* 1971*b*), Harding-Passey melanoma (Krings & Tritsch 1973) after treatment with HPE (30–40 mg/kg body weight by peri-tumoral injection). Zollner *et al.* (1975) have shown that the therapeutic effects of HPE, when injected locally, against the Ehrlich solid tumour in mice were related to the water solubility of the aldehyde and its relative reactivity towards SH groups. In general, local treatment with HPE or HOE of animals bearing subcutaneous tumours resulted in skin lesions (0.5–1.0 cm) at the injection site in 80% of cases. All these aldehyde-induced defects tended to heal promptly. Local treatment of ascitic tumours *in vivo* with HPE or HOE has not proved successful, since the aldehyde is cleared from the peritoneal cavity with remarkable rapidity (Conroy *et al.* 1975; Schauenstein *et al.* 1977). After an injection of 5mM-HOE, only 33% was detectable after 5 min and 12% after 10 min, giving a $t_{1/2}$ of about 3 min (Schauenstein 1967). Similar measurements of the persistence of HPE in the peritoneal cavity and subcutaneous tissue of mice revealed that the $t_{1/2}$ of the aldehyde was also comparatively short (2–3 min) (Schauenstein *et al.* 1968).

One approach to increasing the effective $t_{1/2}$ of HPE *in vivo* has been by the formation of reversible adducts of the aldehyde with cysteine (Esterbauer *et al.* 1976). In general, such adducts could serve as depot forms of the aldehydes. There also exists the possibility that the adduct(s) may penetrate cell membranes more effectively than the reactive aldehyde alone. Significant increases in the survival time of mice bearing the ascitic forms of Sarcoma 180 and Ehrlich carcinoma were observed after treatment with the adduct(s) of HPE and cysteine, injected intraperitoneally at 256 or 360 mg/kg respectively on Days 3–7 after implantation of the tumour (Conroy *et al.* 1977; Tillian *et al.* 1976).

Bickis *et al.* (1969) have reported that HPE strongly inhibits the oxidation of tricarboxylic acids, glycolysis and the incorporation of radioactive precursors into DNA, RNA, protein and lipids in ascites tumour cells. All these observed inhibitions could be prevented by adding glutathione to the incubation medium. Slater *et al.* (1973) and Conroy *et al.* (1975, 1977; Conroy 1976) have confirmed that the effect of HPE treatment *in vitro* of certain murine tumour cells (at concentrations in excess of 0.16 mM) was to depress primarily the incorporation of [^{3}H]thymidine into DNA; [^{3}H]uridine and [^{3}H]leucine incorporation into RNA and protein respectively were considerably less affected. In general, these effects are more pronounced with tumour cell lines, since the production of similar results with normal tissues requires up to four times the dose of HPE effective with the tumour cell lines. Schauenstein *et al.* (1970) have demonstrated that the respiratory inhibition induced by HPE in Ehrlich ascites cells *in vitro* is manifested by a drastic drop in the cytoplasmic NAD level and a partial blockage of the respiratory chain at aldehyde concentrations between 1 mM and 3 mM. HPE inhibits the activity of several SH-containing enzymes, including glyceraldehyde-3-phosphate dehydrogenase and lactate dehydrogenase (Schauenstein 1967; Bickis *et al.* 1969), and the inhibition may be reversed by adding cysteine or glutathione. The double bond of HPE reacts avidly with SH compounds, forming the corresponding 4-hydroxyalkanal, which rearranges to give a cyclic semiacetal (Esterbauer 1970; Esterbauer *et al.* 1975). The relative resistance of normal cells to HPE treatment may be the result of a higher content of low molecular weight SH compounds, providing a better protection of functional SH groups of essential enzymes (Zollner & Schauenstein 1973).

The data of Szent-Györgyi, Egyud and co-workers cited above have indicated that only protein turnover is affected rapidly enough to be considered as the primary locus of inhibition by methylglyoxal in bacterial and mammalian cells. The addition of exogenous thiols reversed the inhibitory effects of methylglyoxal. In general, hydroxy-saturated aldehydes and ketoaldehydes preferentially inhibit protein synthesis in neoplastic cells by one or more of the following mechanisms:

(a) Reduction of intracellular cysteine;

(b) Reversible inhibition of enzymes containing thiol groups essential for catalytic activity; or

(c) Reversible binding to guanine residues in tRNA or RNA.

In contrast, our preliminary studies with the unsaturated hydroxyalkenal, HPE, indicated that inhibition of the incorporation of [^{3}H]thymidine into

DNA of several mouse tumour lines was the primary effect of aldehyde treatment. Our investigations subsequently concentrated on determining *(i)* the concentration dependence, *(ii)* the time course and *(iii)* the effect of addition of exogenous thiols such as cysteine and glutathione on the type of inhibition induced by treatment of tumour cells *in vitro* with HPE. Section I will deal with the results achieved in this area.

Section II (p. 282) outlines and summarizes several approaches towards increasing the carcinostatic effectiveness of methylglyoxal and HPE against mouse tumours *in vivo*, including *(i)* multiple intraperitoneal injections, *(ii)* concomitant administration of a glyoxalase I inhibitor, *(iii)* administration of aldehyde-cysteine adducts, and *(iv)* continuous intravenous infusion.

SECTION I: IN VITRO STUDIES

Materials

All reagents were of analytical grade or the highest grade commercially available. Sources are as outlined in Conroy *et al.* (1977). Radiochemicals were purchased from the Radiochemical Centre, Bucks., England: [6-^{3}H]thymidine, 5.0 Ci/mmol; [5-^{3}H]uridine, 5.0 Ci/mmol; [4,5-^{3}H]DL-leucine, 1.0 Ci/mmol.

Radioactive precursor incorporation

The following quantities of labelled precursors were used per ml of tumour cell suspension: [^{3}H]thymidine, 0.25 μCi/ml; [^{3}H]uridine, 0.25 μCi/ml; [^{3}H]leucine, 2.5 μCi/ml. Details are as described in Conroy (1976) and Conroy *et al.* (1977).

Preparation of trans-4-hydroxypent-2-enal (HPE) and HPE–cysteine adduct

Methods for the synthesis, purification and testing of both compounds are fully described elsewhere (Conroy *et al.* 1975, 1977).

Tumour sources and maintenance

Details of tumour sources, passaging and harvesting are given in Conroy (1976) and Conroy *et al.* (1977).

RESULTS OF IN VITRO STUDIES

Effects of HPE on incorporation of DNA precursors into Sarcoma 180 cell suspensions in vitro

Sarcoma 180 cells were incubated at 37 °C with and without HPE for 30 min and ^{3}H-labelled precursors were added to obtain an incorporation profile, illustrated in Fig. 1. The incorporation of [^{3}H]thymidine was primarily affected at doses of aldehyde in excess of 0.16 mM. Total incorporation of labelled thymidine into DNA was reduced to about 10% of the control level at 0.16 mM-HPE, whereas the incorporation of [^{3}H]uridine into RNA and of [^{3}H]leucine into protein was reduced to approximately 50% of the control incorporation at the same aldehyde concentration.

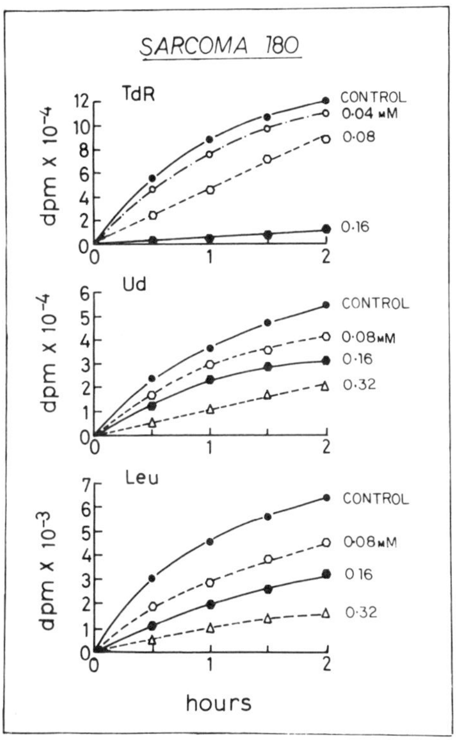

FIG. 1. Effect of 30-min preincubation of Sarcoma 180 cells (1 × 10^6/ml) with 0.04–0.32 mM-HPE per ml of cell suspension on the subsequent incorporation of isotopic precursors of cell components over 2 h at 37 °C. Samples were withdrawn every 30 min for radioactivity assay. [^{3}H]Thymidine (TdR) was incorporated into DNA; [^{3}H]uridine (Ud) principally in RNA; [^{3}H]-leucine (Leu) into protein.

Since the effect of HPE treatment was primarily to depress the incorporation of labelled thymidine into DNA of the tumour cells, this tritiated precursor was selected for investigation of the protective effects of cysteine against the inhibition of DNA protein synthesis by added aldehyde, and also to study the response of the tumour cells *in vitro* to pretreatment in various ways with the adduct formed between HPE and cysteine.

Effects of HPE and HPE–cysteine adduct on [^{3}H]thymidine incorporation into Sarcoma 180 tumour cells in vitro

We used several types of preincubation procedure at 37 °C of tumour cells with different drug regimes before adding [^{3}H]thymidine (see legend to Fig. 2). The effectiveness of the different preincubations was compared to two control procedures:

(i) 0.04–1.28 mM-HPE per ml cell suspension, added 30 min before [^{3}H]-thymidine addition (preincubation procedure (1) in Fig. 2).

(ii) 0.04–1.28 mM-cysteine per ml cell suspension, added 30 minutes before [^{3}H]thymidine addition (preincubation procedure (2) in Fig. 2).

The dose–response relationship of these two treatments is shown in Fig. 2 (upper plot) and repeated for purposes of comparison in the two lower dose–response plots. Treatment of tumour cells with HPE alone or cysteine alone depresses [^{3}H]thymidine incorporation by comparison with untreated controls and produces a sigmoidal dose–response curve.

In preincubation procedure (3) (Fig. 2), tumour cells received cysteine at 35 min followed by HPE 30 min before the addition of labelled precursor. The dose–response plot is shifted slightly to the right, indicating a small degree of protection by the cysteine against added aldehyde. The addition of cysteine alone to the cell suspension only reduces the rate of incorporation of precursor by 4% compared to solvent-treated controls and this is considered insignificant.

Premixing the HPE and cysteine at a molar ratio of 1:2, as in procedure (4), produced a much more pronounced shift of the dose–response plot towards that of cells treated with cysteine alone. A broadly similar effect was obtained by adding cysteine 5 min before the simultaneous addition of HPE and labelled thymidine (procedure 5). It is significant that the responses of tumour cells to preincubation procedures (4) and (5) lie between the plots indicated as controls (procedures 1 and 2). The result of procedure (5), compared to that obtained in procedure (3), suggests that the preincubation period of 30 min is sufficient to allow a proportion of the HPE–cysteine adduct formed *in situ* to decay, releasing HPE and cysteine.

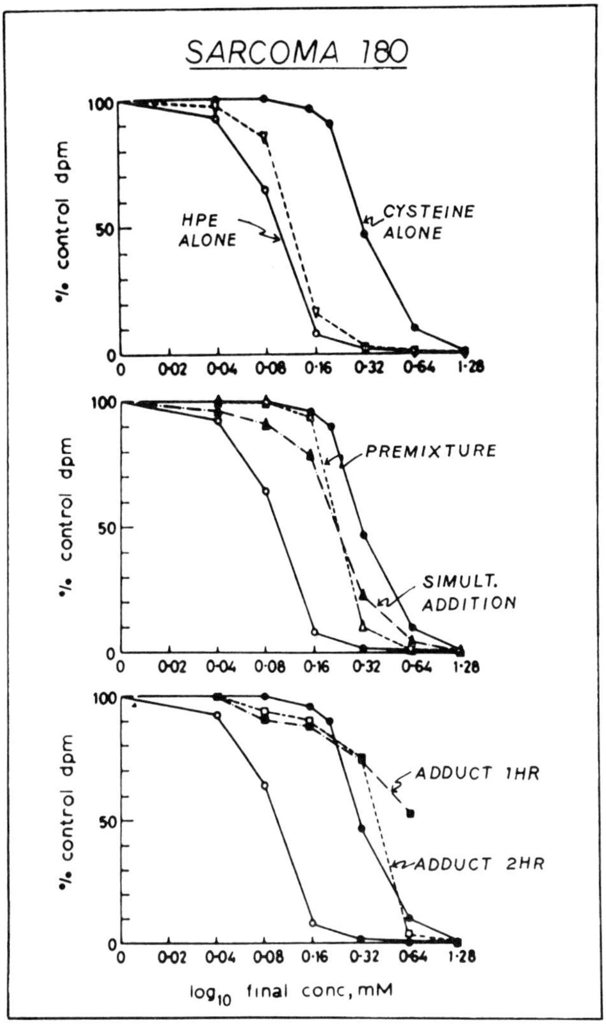

FIG. 2. Effect of preincubation at 37 °C with HPE, cysteine and the HPE–cysteine adduct on the subsequent incorporation of [^{3}H]thymidine (TdR) into DNA of Sarcoma 180 cells (1×10^6/ml). Preincubation procedures: (1) ○——○, HPE added alone, 30 min before TdR. (2) •——•, cysteine added alone, 30 min before TdR. (3) ∇-----∇, 0.16 mM-cysteine added at 35 min, followed by HPE at 30 min before TdR. (4) △-----△, HPE premixed with cysteine and added 60 min before TdR. (5) ▲—·—▲, 0.16 mM-cysteine added 5 min before the simultaneous addition of HPE and TdR. (6) ■—·—■, HPE-cysteine added 60 min before TdR. (7) □-----□, HPE-cysteine added 120 min before TdR. Each plotted point is the mean of 3–10 determinations.

This was confirmed when we compared the response of tumour cells to the HPE–cysteine adduct (purified by thin layer chromatography). Preincubation procedure (6) (Fig. 2) illustrates the response obtained from 60 min of incubation with the purified adduct. When the responses are compared, dose-for-dose, the adduct is less effective as an inhibitor than cysteine alone at concentrations over 0.2 mM. Increasing the preincubation with purified HPE–cysteine adduct to 120 min produced a similar response up to 0.32 mM. Thereafter it declined rapidly to 3% of the solvent-treated control at 0.64 mM. The adduct appears to be breaking down with time in the medium, releasing two moles of cysteine for every mole of HPE. It is possible that cysteine and HPE, at concentrations above 0.32 mM, act synergistically in depressing [^{3}H]thymidine incorporation into DNA. The effects of procedure (3) are comparatively slight. Preincubation procedures (4) and (5) are broadly equivalent. To obtain the same degree of depression of [^{3}H]thymidine incorporation as after treatment with HPE alone (procedure 1), 2–3 times the dose was required. Similarly, treatments (6) and (7) are 4–7 times less effective on a molar basis than treating Sarcoma 180 cells with HPE alone. The evidence on the effects of HPE–cysteine on the incorporation of [^{3}H]thymidine into Sarcoma 180 tumour cells is consistent with the view that the extended breakdown of such an adduct of HPE and cysteine, releasing the reactive aldehyde, may increase the half-life of HPE *in vitro*.

Effect of glutathione on the reduced incorporation of [^{3}H]thymidine into Sarcoma 180 tumour cells before or after treatment with HPE

The plan of various types of treatment of Sarcoma 180 cells at 37 °C after the addition of [^{3}H]thymidine is shown in Fig. 3. The effect of different times of treatment with glutathione (GSH) in reversing the inhibitory effects of HPE on [^{3}H]thymidine incorporation into tumour DNA was assessed by comparison with a control procedure (1) in which HPE was added 20 min after [^{3}H]thymidine. The depression of incorporation of labelled precursor into Sarcoma 180 tumour cells, produced by exposure to HPE, is well established within 5 min of the addition of more than 0.16 mM-aldehyde.

In treatments (2) and (3) tumour cells received GSH 5 min before or after the addition of either 0.1 or 0.16 mM-HPE. At both aldehyde concentrations, treatment (2) resulted in a pronounced protection by GSH against the aldehyde-induced depression of [^{3}H]thymidine incorporation into tumour DNA. The addition of GSH to the cell suspension reduced the rate of incorporation of precursor by not more than 10% compared to solvent-treated controls (Fig. 4). However, although treatment (3) resulted in broadly similar overall effects at both concentrations of HPE, interesting differences in response time were evident. Fig. 4 illustrates that tumour cells receiving

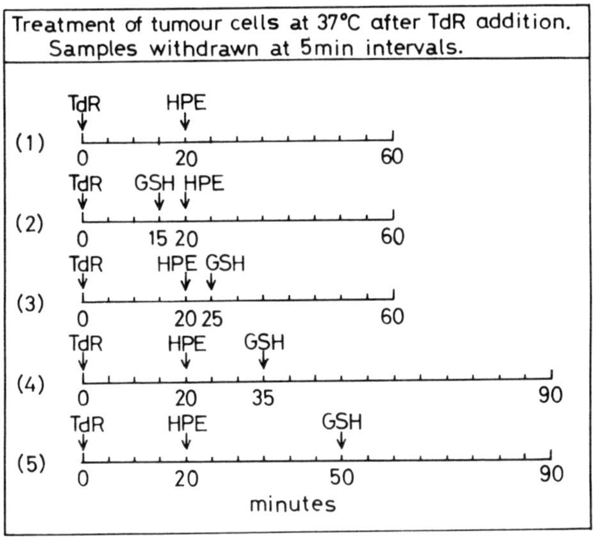

FIG. 3. Plan of treatment of Sarcoma 180 cells (1 × 10^6/ml) at 37 °C with HPE and glutathione (GSH) after the addition of [^{3}H]thymidine (TdR). Samples were removed at 5 min intervals over 60 or 90 min. 0.1 or 0.16 mM-HPE per ml of cell suspension was added at 20 min, preceded or succeeded by 0.32 mM-GSH added 5 min before or 5, 15 or 30 min after HPE.

GSH 5 min after addition of 0.1 mM-HPE do not respond for 10 min (lag period) and then progressively recover over a further 15 min to the level of cells treated with GSH 5 min before aldehyde addition. At 0.16 mM-HPE, treatment (3) resulted in an extended lag period of 20 min and then gradual recovery over the next 15 min.

The effects of treatments (4) and (5) are shown in Fig. 4. Treatment (4) was marginally more effective than (5) in reversing the lesion induced by the aldehyde. However, in both cases overall recovery was minimal. Treatment of Sarcoma 180 tumour cells with 0.16 mM-HPE depressed total [^{3}H]thymidine incorporation into DNA to 50% of the control level within 10 min and to approximately 10% by 120 min after exposure to the drug. GSH (0.32 mM) did not alleviate the aldehyde-induced depression of incorporation of labelled precursor, when added more than 15 min after HPE (treatments 4 and 5). In contrast, when GSH was given 5 min after HPE (treatment 3), a substantial degree of recovery occurred.

Therefore, in the period 5–15 min after the addition of HPE a biochemical lesion has been induced in tumour cells which is essentially irreversible by added thiols (at least at the concentration of thiol used). It is interesting that these results are in contrast to those obtained in mammalian cell lines treated

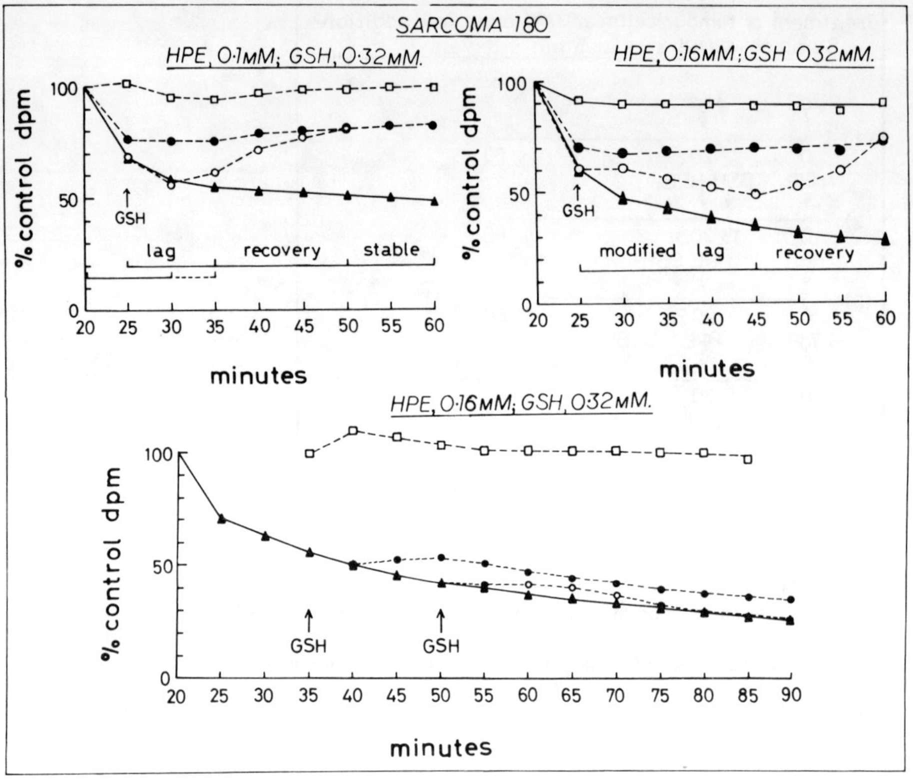

FIG. 4. Effect of the various treatments outlined in Fig. 3 on the incorporation of [^{3}H]thymidine into DNA of Sarcoma 180 cells. ▲——▲, procedure 1; •----•, procedure 2 or 4; ○----○, procedure 3 or 5; □----□, GSH alone.

with methylglyoxal. The inhibition of protein synthesis induced by methylglyoxal could be reversed by the addition of equimolar amounts of cysteine.

SECTION II: IN VIVO STUDIES

Materials

Methylglyoxal was obtained as a 40% w/v aqueous solution (lots 04089 and 07047) from Aldrich Chemical Co. *S*-(*p*-Bromobenzyl)-glutathione was the generous gift of Dr R. Vince, University of Minnesota.

Preparation of an aqueous solution of purified methylglyoxal

Routine steam distillation of 25 ml of a 40% w/v aqueous methylglyoxal solution on average yielded approx. 50 ml of a pale yellow solution of the aldehyde (0.25–0.4M, pH 3.5). Spectral analysis showed a single sharp peak (λ_{max} = 282 nm, λ_{min} = 255 nm) and the absence of the pronounced shoulder present in the unpurified, partly polymerized methylglyoxal stock solutions (λ_{max} = 292 nm, λ_{min} = 265 nm). Semicarbazide assay of the distillate gave a molar extinction coefficient of 183.4 mol^{-1} cm^{-1} at 282 nm. This value is considerably greater than those previously quoted (27.5 mol^{-1} cm^{-1} and 75 mol^{-1} cm^{-1}). The pH of the distillate (3.5) indicates that some low molecular weight acidic contaminants have codistilled with monomeric methylglyoxal. 0.2–2.3 ml of a 20% w/v aqueous solution of NaOH was generally sufficient to neutralize these contaminants.

Preparation of methylglyoxal and methylglyoxal–N-acetyl-L-cysteine solutions for toxicity testing and antitumour evaluation

All drug solutions were filter-sterilized through Millipore filters (Millex-GS, 0.22 μm) and stored at 4 °C until required. In all toxicity and drug therapy schedules, animals were weighed daily. Drugs were administered on the basis of 10 ml of drug solution equivalent to 1 kg of body weight for i.p. injections.

Methylglyoxal. The steam distillate of methylglyoxal was neutralized to pH 7.4 with 20% w/v aqueous NaOH. For toxicity or cancer therapy courses with methylglyoxal alone, NaCl was added to a final concentration of 0.9% w/v for i.p. injection or methylglyoxal was diluted with an appropriate concentration of phosphate-buffered saline for infusion.

Methylglyoxal–N-acetyl-L-cysteine (1:1: and 1:2). Crystalline *N*-acetyl-L-cysteine was added to a range of steam-distilled methylglyoxal solutions (3.6–28 mg/ml), then N_2-flushed for 10 min before incubation in stoppered glass tubes for 2 h at 37 °C. NaCl was added to these solutions of the methylglyoxal–*N*-acetyl-L-cysteine adduct to produce the same osmolality as physiological saline after neutralization to pH 7.4 with 20% w/v NaOH.

Tumour sources and maintenance

The D40 mammary adenocarcinoma used arose spontaneously in a retired breeder within the inbred CBA/CA mouse colony at Brunel University, Middlesex, England. It was successfully transplanted and maintained by biweekly implantation of 1 mm^3 fragments subcutaneously in the flanks of syngeneic female CBA/CA mice.

Lymphosarcoma 1 and Squamous carcinoma D were the generous gift of Dr H.B. Hewitt of the Gray Laboratory, Middlesex, England. These spontaneous syngeneic neoplasms arose in the colony of Wh/Ht mice maintained at the Gray Laboratory. The tumour cells were routinely passaged on the 7th or 10th day respectively after trocar implantation of 1 mm^3 fragments into the flanks of mice derived from an inbred colony maintained at Brunel, using a triad mating system. The original triads were obtained from the colony at the Gray Laboratory.

Ehrlich carcinoma ES3A[1]/582 was maintained by Dr A. Szent-Györgyi at the NFCR Laboratory at Woods Hole, Mass. by weekly serial transplantation of 5×10^6 cells i.p. in female CD1 mice obtained from Jackson Laboratories, USA.

Tumour dissociation procedure

The method was essentially as described by Hewitt (1966) with minor modifications. Briefly, the procedure consisted of four serial digests lasting 10 min each at 37 °C of a tumour mince in an enzyme 'cocktail' containing 0.4% trypsin, 0.08% pancreatin and 0.02% w/v DNAase. Erythrocytes present in the original tumour mince were removed by two washes in a lysing buffer (7.47 g NH_4Cl, 2.02 g Tris-HCl in 1 litre brought to pH 7.2 with concentrated HCl). The dissociated tumour cells were recombined and finally resuspended in cold (4 °C) mouse or fetal calf serum–Tyrode's solution for reinjection into syngeneic host mice. Tumour cells were counted and their viability was assessed under phase-contrast microscopy with a Fuch-Rosenthal haemocytometer. In general, the method typically resulted in a viable tumour cell yield of approx. 3×10^7 cells/g.

Lateral tail vein cannulation in mice

The method used is a modification of the system described by Paul & Dave (1975), and is fully described elsewhere (Conroy & Elvin 1979). The aim is to insert a polyethylene cannula into the lateral tail vein of the mouse so that with adequate protection and suspension the animal retains almost complete mobility but cannot dislodge the cannula. It is advisable to equilibrate the animal with an infusion of phosphate-buffered saline for one hour before infusing drug solutions.

Tumour volume evaluation

Tumour volumes were computed according to the formula $\frac{\pi}{6}(d_1 \times d_2 \times d_3)$

where d_1, d_2, d_3 are the measured diameters in three mutually opposed directions.

Method of evaluation for ascitic tumours

Tumour cells were removed from the peritoneal cavity on the 10th day after transplantation, with 3–4 washes of heparinized 0.9% w/v saline (5 ml) into 25 ml graduated cylinders, and the volume was noted. The total number of cells present per animal was calculated from: Haemocytometer count × dilution factor × total wash volume.

Cellular viability was determined by Trypan Blue exclusion.

RESULTS OF IN VIVO EXPERIMENTS

Carcinostatic activity of methylglyoxal in vivo against the solid form of Ehrlich ES3A^1/582 carcinoma in female CD1 mice

Table 1 summarizes the results of the various therapy courses investigated. In general, some variation was noted in the apparent growth rate of the solid tumour in the inguinal region and the range of control tumour weights obtained. In this respect, the control variation in expt. II (range 0.580–1.265 g) was rather more satisfactory than that of expt. I (range 0.405–1.340 g). These differences may be due to variation in the precise site of tumour cell inoculation and the subsequent development of a suitable vascular supply.

Experiments I and II both indicate that, in the conditions of this study, the glyoxalase I inhibitor *S*-(*p*-bromobenzyl)-glutathione was only marginally effective in potentiating the antitumour activity of methylglyoxal (expt. I, 57.5% inhibition of tumour growth compared to 46.2% with methylglyoxal alone; expt. II, 56.8% inhibition compared to 51%, respectively). The enzyme inhibitor was totally inactive ($P < 0.999$) when injected alone at 50 mg/kg, twice daily. These results indicate that the glyoxalase I inhibitor *(i)* may be inactivated rather rapidly by mouse liver glutathionase, so that insufficient material reached sites distant from the site of administration, or *(ii)* was incapable of penetrating the tumour cells, or *(iii)* was not injected in a therapeutically useful dose. Similar comments can be made on the effects of including 1% w/v nicotinamide in the drinking water of tumour-bearing mice treated with methylglyoxal (Table 1).

The results obtained with the methylglyoxal–*N*-acetyl-L-cysteine adduct (expt. III) show that, first, there is a dose-dependent inhibition of tumour

TABLE 1

Carcinostatic activity of methylglyoxal against the solid form of the Ehrlich ES3A[1]/582 carcinoma in female CD1 mice

Expt.	*Treatment*	*Mean tumour wt. ± S.E.M. (g)*	*Inhibition of tumour growth (%)*	*P value (2-tailed t-test)*
I	Saline-treated controls	0.530 ± 0.106 (10)	–	–
	Glyoxalase I inhibitor[a], 50mg/kg	0.532 ± 0.103 (5)	0	<0.999
	Methylglyoxal, 36mg/kg	0.285 ± 0.029 (5)	46.2	<0.10
	Methylglyoxal, 36mg/kg + glyoxalase I inhibitor[a], 50mg/kg	0.225 ± 0.033 (5)	57.5	<0.05
II	Saline-treated controls	0.769 ± 0.069 (10)	–	–
	Methylglyoxal, 72mg/kg (inj. 1); 36mg/kg (inj. 2)	0.377 ± 0.038 (5)	51.0	<0.005
	Methylglyoxal as above + glyoxalase I inhibitor	0.332 ± 0.046 (5)	56.8	<0.001
	Nicotinamide (1% w/v in drink)	0.626 ± 0.034 (5)	18.6	<0.20
	Methylglyoxal as before + nicotinamide (1% in drink)	0.314 ± 0.056 (5)	59.2	<0.001
III	Methylglyoxal–*N*-acetyl-L-cysteine, 36mg/kg	1:1 0.516 ± 0.063 (5)	32.9	<0.05
		1:2 0.566 ± 0.067 (5)	26.1	<0.10
	Methylglyoxal–*N*-acetyl-L-cysteine, 72mg/kg	1:1 0.512 ± 0.075 (5)	33.4	<0.05
		1:2 0.387 ± 0.055 (5)	49.7	<0.005
	Methylglyoxal–*N*-acetyl-L-cysteine, 144mg/kg	1:1 0.252 ± 0.047 (5)	67.2	<0.001
		1:2 0.248 ± 0.066 (5)	67.8	<0.001

Female CD1 mice (20–25 g) were injected (subcutaneous, inguinal) with 5×10^6 Ehrlich ES3A[1]/582 carcinoma cells. Treatment consisted of two i.p. injections daily, 6 h apart, beginning on Day 1 after tumour transplantation and continuing for 10 days.
[a]*S*-(*p*-Bromobenzyl)glutathione.

growth, secondly, administration of the adduct at doses equivalent to 144 mg/kg methylglyoxal alone was slightly more effective in inhibiting tumour growth (av. 67.5%, $P < 0.001$) than methylglyoxal alone either at 36 mg/kg, twice daily (46.2% $P < 0.02$) or at 72 mg/kg (inj. 1); 36 mg/kg (inj. 2) (51%, $P < 0.005$). However, since the LD_{50} for the adduct is 216 mg/kg (twice daily for 10 days, 432 mg/kg daily total), there is not much margin for improvement with the current dose regime. Only marginal differences in effectiveness were noted between the 1:1 and 1:2 adducts. The LD_{50} for a purified solution of methylglyoxal in female CDl mice (20–25 g) was 54 mg/kg (twice daily for 10 days, 108 mg/kg total daily dose). Thus the methylglyoxal–*N*-acetyl-L-cysteine adduct is four times less toxic than methylglyoxal alone, on an equivalent basis.

Tumour-bearing mice easily tolerate 36 mg/kg methylglyoxal, twice daily, but find difficulty with either of the following dose regimes:

(i) 72 mg/kg, twice daily for five days; 36 mg/kg, twice daily for five days.

(ii) 72 mg/kg (inj. 1); 36 mg/kg (inj. 2) for 10 days.

The toxicity tests showed that i.p. injection of methylglyoxal at 72 mg/kg, twice daily, proved lethal if continued much beyond five days. Disorientation and signs of generalized depression of the central nervous system became increasingly apparent.

Carcinostatic activity of methylglyoxal in vivo against the ascitic form of Ehrlich ES3A[1]/582 carcinoma in female CD1 mice

Table 2 summarizes the results obtained against the ascitic tumour and reflects the mean total cells per animal ($\times$ 10^6) after 10 days of therapy. Rather less control variation was noted in these experiments than in the analogous treatments of the solid tumour (Table 1). A typical growth curve is illustrated in Fig. 5. Although the treatment of ascitic tumours by intraperitoneal injection is considered 'local therapy', and therefore not therapeutically useful, drugs which do not require metabolic activation and which are found to be inactive when tested in this manner commonly do not exhibit much carcinostatic activity against solid tumours. The results of this study should be considered with this reservation in mind.

Toxicity problems became apparent within 3–4 days in mice receiving 72 mg/kg methylglyoxal plus 50 mg/kg glyoxalase I inhibitor: only one animal out of five survived the therapy course (expt. I). The remainder died 5–9 days after transplantation. The survivor proved to be tumour-free. This interesting result deserves further investigation. Methylglyoxal at 72 mg/kg (five days) followed by 36 mg/kg (five days) inhibited tumour growth by 99.97% (P <0.001), a very effective confirmation of the results of Apple & Greenberg (1967, 1968) since two of these treated animals had less than 1 $\times$ 10^5 observable tumour cells 10 days after the transplantation of 5 $\times$ 10^6 cells. Treatment of tumour-bearing mice with methylglyoxal at 36 mg/kg, twice daily, was rather less effective (97.8%) but still quite significant (P <0.001).

In contrast to the results obtained against the solid form of the tumour, nicotinamide (1% w/v in the drinking water) alone inhibited tumour growth by 47.2% (P <0.001). In combination with methylglyoxal (36 mg/kg, twice

TABLE 2

Carcinostatic activty of methylglyoxal against the ascites form of the Ehrlich ES3A[1]/582 carcinoma in female CD1 mice

Expt.	*Treatment*		*Mean total cells/ mouse × 10^6 ± S.E.M.*	*Inhibition of tumour growth (%)*	*P value (2-tailed t-test)*
	Saline-treated controls		1098 ± 89 (10)	—	—
I	Glyoxalase I inhibitor[a], 50mg/kg		1006 ± 144 (5)	8.4	<0.60
	Methylglyoxal, 72mg/kg (5 days); 36mg/kg (5 days)		0.365 ± 0.098 (5)	99.97	<0.001
	Methylglyoxal as above + glyoxalase I inhibitor[a]		Nil[a] (1)	100.00	<0.001
	Saline-treated controls		1248 ± 82 (6)	—	—
II	Nicotinamide (1% w/v in drink)		659 ± 83 (5)	47.2	<0.001
	Methylglyoxal, 36mg/kg		27.2 ± 12.2 (5)	97.8	<0.001
	Methylglyoxal, 36mg/kg + nicotinamide		4.52 ± 1.86 (5)	99.60	<0.001
	Saline-treated controls		1196 ± 87 (5)	—	—
III	Methylglyoxal–*N*-acetyl-L-cysteine, 36mg/kg	1:1	646 ± 73 (5)	46.0	<0.005
		1:2	586 ± 69 (5)	52.0	<0.001
	Methylglyoxal–*N*-acetyl-L-cysteine, 72mg/kg	1:1	155 ± 21 (5)	87.0	<0.001
		1:2	181 ± 15 (5)	84.8	<0.001
	Methylglyoxal–*N*-acetyl-L-cysteine, 144mg/kg	1:1	17.77 ± 0.55 (5)	98.5	<0.001
		1:2	10.80 ± 0.72 (5)	99.1	<0.001

Female CDl mice (20–25 g) were injected intraperitoneally with 5 × 10^6 Ehrlich ES3A[1]/582 carcinoma cells. Treatment consisted of two i.p. injections daily , 6 h apart, beginning on Day 1 after tumour transplantation and continuing for 10 days.

[a]Only one mouse out of five survived treatment.

[b]*S*-(*p*-Bromobenzyl)glutathione.

daily) it was significantly more effective than methylglyoxal alone (99.6% inhibition compared to 97.8%). Nicotinamide thus appears to have some cytostatic activity which may supplement the cytotoxic capabilities of methylglyoxal and its derivatives.

Treatment of tumour-bearing mice with methylglyoxal–*N*-acetyl-L-cysteine at a dose equivalent to 144 mg/kg methylglyoxal proved marginally more effective in inhibiting tumour growth than methylglyoxal alone at 36 mg/kg, but this is not statistically significant. In general, the effects of the adduct were dose-dependent and the results obtained do not support the view that the

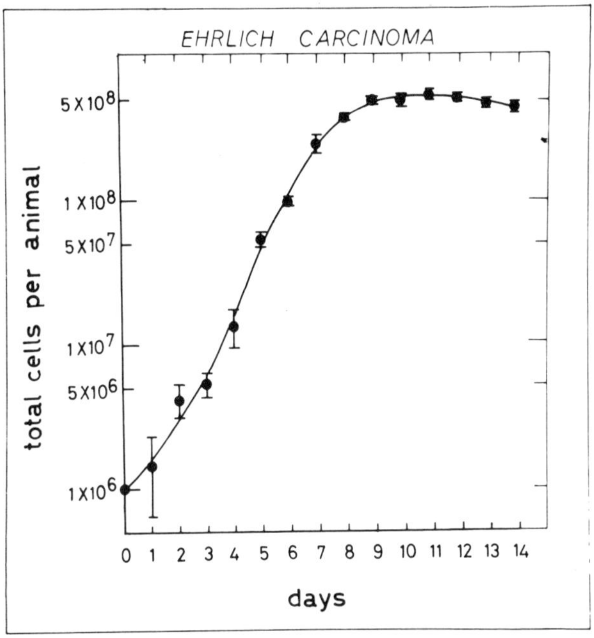

FIG. 5. Growth curve of Ehrlich carcinoma (ascitic form; 1×10^6 cells injected per mouse at Day 0). Plot of log total cells per animal against time in days. Points represent mean values obtained from at least three animals.

substance offers any significant advantage over methylglyoxal alone, at least in the conditions employed here. There was no significant difference between the patterns obtained with the 1:1 and 1:2 adducts.

Carcinostatic activity of an HPE–cysteine adduct against Sarcoma 180 in vivo

The LD_{50} in mice for a single i.p. injection of 4-hydroxypentenal is 55 mg/kg. HPE injected in four equal doses spaced at two-hourly intervals gives a LD_{50} in CBA/CA mice of 180 mg/kg per day. Injection of an HPE – cysteine adduct into mice produced an LD_{50} value of 362 mg/kg per day (expressed as mg HPE assumed to be released from the adduct). Thus the HPE–cysteine adduct is much less toxic than the free aldehyde injected intraperitoneally; in comparison to a single i.p. injection of HPE as the free aldehyde, HPE–cysteine is approximately seven times less toxic. The *in vivo* carcinostatic activity of the HPE–cysteine adduct, compared to HPE alone, against Sarcoma 180 CBA/CA mice is shown in Fig. 6. The dose of HPE

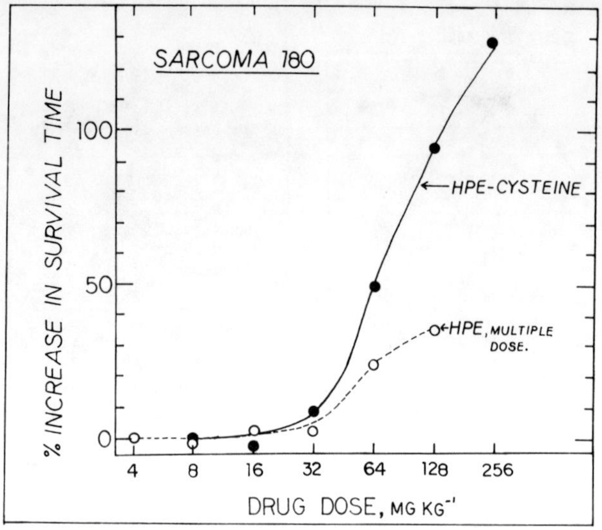

FIG. 6. Carcinostatic activity of an HPE–cysteine adduct *in vivo* compared to HPE alone against Sarcoma 180 in CBA/CA mice. o-----o, HPE alone, injected i.p. on Days 3 and 4 after transplantation in four divided doses; •——•, HPE–cysteine given on Days 3-7 after transplantation in single doses.

plotted for the adduct is the amount assumed to be released on the breakdown of the adduct with time. For any given dose above 32 mg/kg per day, treatment with HPE–cysteine i.p. is 2–3 times more effective than HPE given as the free aldehyde. The results obtained are consistent with the view that the extended breakdown of an adduct of HPE and cysteine, releasing the reactive aldehyde, increases the half-life of HPE *in vivo*.

Methylglyoxal toxicity after continuous intravenous infusion

Continuous intravenous infusion of methylglyoxal into mice in excess of 500 mg/kg per day (six days at 2.80 ml/day) produced local oedema followed by necrosis of the tail. Little or no normal tissue response was observed after an infusion of 120 mg/kg per day for six days in either CBA/CA or WH/Ht mice. The LD_{50} over a three-day period was found to be more than 3.0 mg/g per day.

Effect of continuous intravenous infusion of methylglyoxal on the growth of CBA/CA D40 mammary tumour

The response of the CBA/CA D40 mammary tumour is illustrated in Fig. 7.

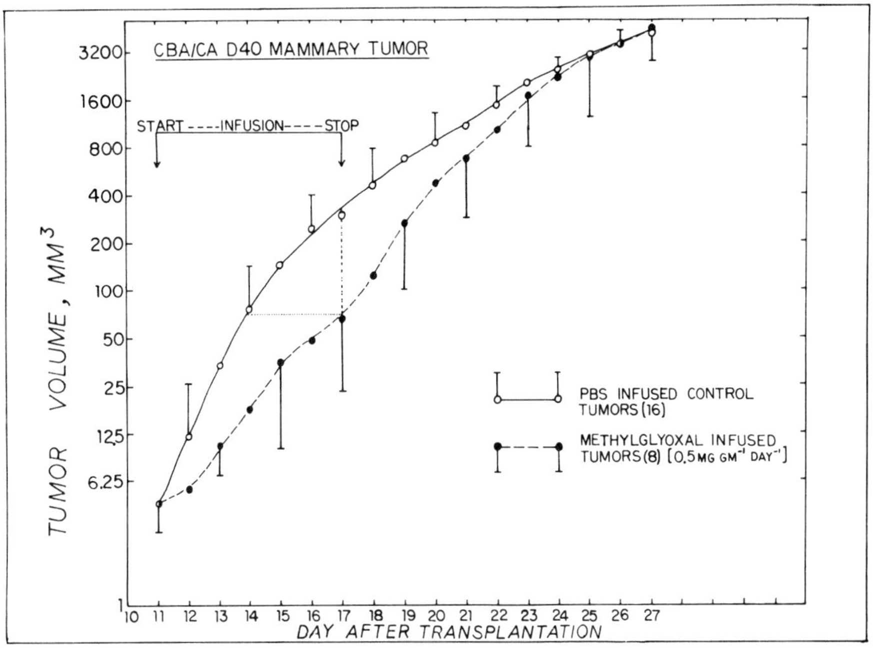

FIG. 7. Effect of continuous intravenous infusion of 500 mg/kg per day of methylglyoxal (Days 11–17) on the growth of CBA/CA D40 mammary tumour. Tumour volume (mm^3) is plotted against days after transplantation.

Continuous intravenous infusion at 500 mg/kg per day for six days, beginning at Day 11 after transplantation, produced a maximum tumour growth delay of 3.4 days at Day 17 ($P < 0.001$) (Fig. 8). This represents a reduction of tumour volume compared to control tumours of about 77% ($P < 0.001$) (Table 3). However, the tumour regrew after the termination of the therapy regime so that by Day 27 after transplantation no significant difference remained between control and methylglyoxal-treated tumours (Fig. 7). The overall result is significant; however, it may be the maximum possible tumour inhibition in this system since unacceptable local tissue toxicity occurs at drug doses greater than 500 mg/kg per day.

Effect of continuous intravenous infusion of methylglyoxal on the growth of Lymphosarcoma I and Squamous carcinoma D in WH/Ht mice

Methylglyoxal (500 mg/kg per day) was ineffective in inhibiting the growth of both Lymphosarcoma 1 and Squamous carcinoma D when the drug was infused on Days 3–9 after transplantation.

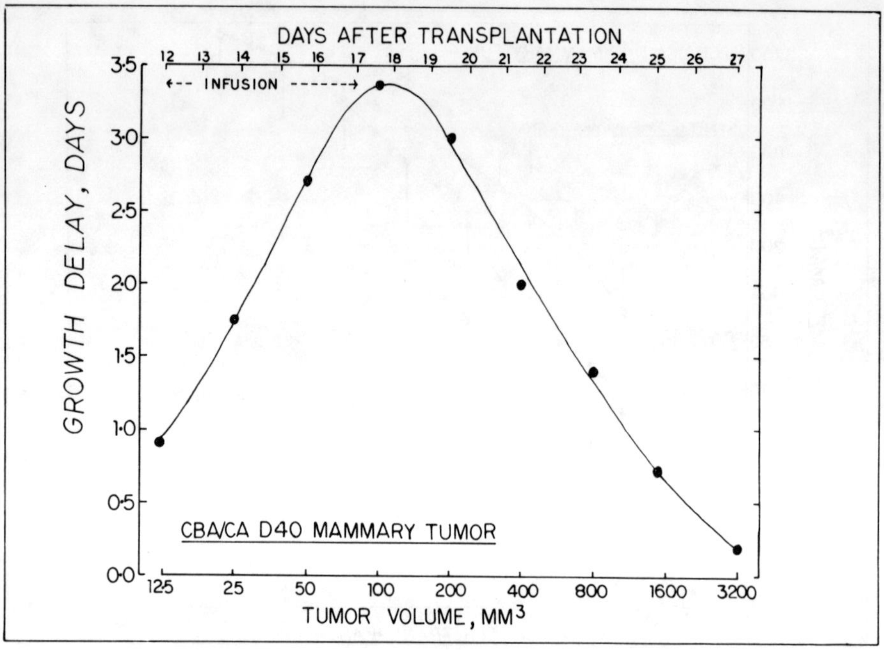

FIG. 8. Effect of continuous intravenous infusion of 500 mg/kg per day of methylglyoxal (Days 11–17) on growth delay (days) and tumour volume (mm^3) in CBA/CA D40 mammary tumour.

TABLE 3

The effect of continuous intravenous infusion of purified methylglyoxal on the growth of CBA/CA D40 mammary tumour in mice

Treatment	*Mean tumour volume (mm^3) ± 95% C.I. at Day 10*	*Mean tumour volume as % control ± C.I.*	*Mean tumour radius (mm) ± 95% C.I. at Day 17*	*Mean tumour radius as % control ± 95% C.I.*
Phosphate-buffered saline (control) (16)	290 ± 105.2	—	4.125 ± 0.493	—
Methylglyoxal (8)	66 ± 42.0[a]	22.8 ± 7.2[b]	2.535 ± 0.506	61.5 ± 5.0[c]

Methylglyoxal was infused at 500mg/kg per day for six days (2.7 ml per day) into mice bearing CBA/CA D40 mammary tumours (5×10^5 cells implanted subcutaneously at two sites per mouse).

[a]Growth delay to reach 100 mm^3 on Day 17 = 3.37 days.

[b]$P < 0.001$.

[c]$P < 0.05$ (two-tailed *t*-test).

Effect of continuous intravenous infusion of methylglyoxal on the growth of the ascitic form of Ehrlich carcinoma and Gardner lymphosarcoma in CBA/CA mice

The effect of an infusion of 120 mg/kg per day of methylglyoxal for five days after the transplantation of 1×10^6 Ehrlich carcinoma or Gardner lymphosarcoma cells in CBA/CA mice is shown in Fig. 9. No significant difference was noted in either the number or the viability of cells recoverable from the peritoneal cavity of mice receiving methylglyoxal compared to control animals receiving buffered saline.

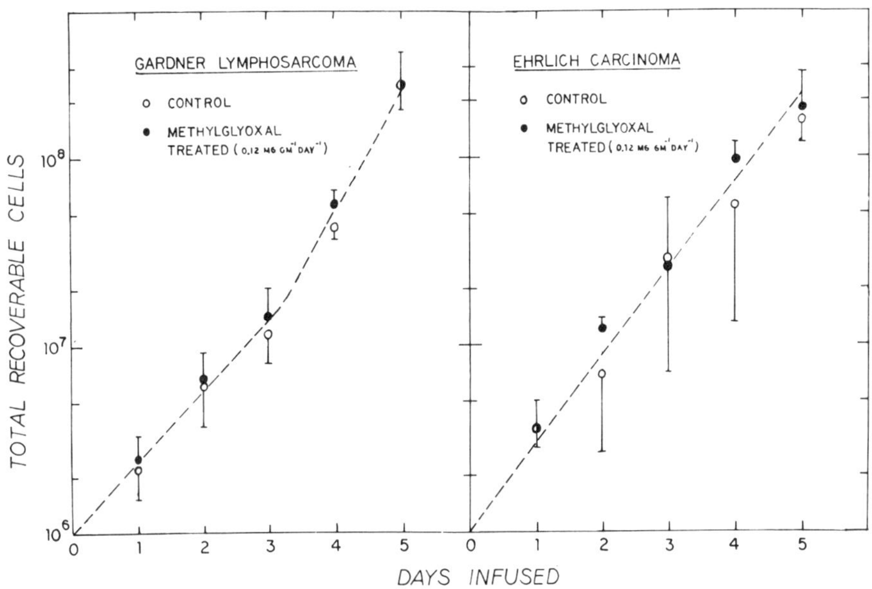

FIG. 9. Effect of continuous intravenous infusion of 120 mg/kg per day of methylglyoxal (Days 0–6) on the growth of the ascitic form of Ehrlich carcinoma and Gardner lymphosarcoma in CBA/CA mice.

DISCUSSION

Treatment of Sarcoma 180 cells *in vitro* with 4-hydroxypentenal resulted in a primary depression of the incorporation of [^{3}H]thymidine into DNA. The addition of cysteine or glutathione before the tumour cell suspensions were treated with HPE provided partial or substantial protection against the depression of [^{3}H]thymidine incorporation produced by the aldehyde.

However, glutathione was ineffective when given more than five minutes after HPE: an essentially irreversible lesion occurred in these tumour cells 5–15 minutes after the aldehyde treatment. These results indicate that the mode of action of methylglyoxal against mammalian tumour cells may differ from that of HPE in at least two respects, since methylglyoxal primarily affects protein synthesis and the resulting inhibition can be reversed by adding thiol.

Methylglyoxal injected intraperitoneally was only moderately effective in inhibiting the growth of the solid form of Ehrlich carcinoma. Marginal potentiation of its carcinostatic effectiveness was obtained by *(i)* optimizing the total daily dose of methylglyoxal, *(ii)* concomitant administration of the glyoxalase I inhibitor, *S*-(*p*-bromobenzyl)-glutathione, and *(ii)* the inclusion of 1% w/v nicotinamide in the drink. In general, potentiation of the carcinostatic effects of methylglyoxal was achieved with the glyoxalase I inhibitor or nicotinamide against the ascitic form of Ehrlich carcinoma when the aldehyde was given in a suboptimal drug fractionation scheme. None of the treatment schedules approached the effectiveness of the optimal dose regime of methylglyoxal alone.

Since the LD_{50} for twice-daily administration of methylglyoxal is approximately 108 mg/kg total dose there appears to be relatively little margin for improvement with an increase in dose fractionation. In general, there are at least three inherent limitations to multiple intraperitoneal injections of this type: *(i)* acute toxicity towards normal tissue, *(ii)* the low therapeutic index of the compound administered alone, and *(iii)* rapid detoxification of methylglyoxal by the glyoxalase enzyme system.

Methylglyoxal–*N*-acetyl-L-cysteine was four times less toxic than methylglyoxal on a dose-equivalent basis. Doses of the adduct equivalent to 144 mg/kg methylglyoxal were more effective against the solid tumour than treatment with the optimum regime of methylglyoxal alone. The increased carcinostatic effectiveness of an aldehyde administered as a thiol adduct was confirmed by the results obtained with the HPE–cysteine adduct. This adduct was 2–3 times more effective, dose-for-dose, than the optimal intraperitoneal schedule of HPE alone in inhibiting tumour growth *in vivo*. However, *in vitro* treatment with the HPE–cysteine adduct was 4–7 times *less* effective on a molar basis than treatment of Sarcoma 180 cells with HPE alone. The apparent paradox of the protection against the effect of HPE on tumour cells by thiols *in vitro* and the increased effectiveness of the HPE–cysteine adduct *in vivo* compared to HPE alone may be reconciled by assuming that an extended release of the reactive aldehyde has occurred *in vivo*, thereby prolonging the effective half-life.

Theoretically, continuous administration of methylglyoxal should obviate the problems of multiple daily injections of the compound and maintain adequate blood levels of the drug. Two methods are currently available: the Alzet minipump, and continuous intravenous infusion. The minipump has a total capacity of 170 μl and delivers aqueous or weak ethanolic solutions of drugs at the rate of 1 μl/h for seven days or 2.5 μl/h for three days. Surgical insertion of the pump containing a 2.5 M solution of methylglyoxal into tumour-bearing mice will result in a drug dose of 7.2 mg/kg per hour (173 mg/kg per day) at a release rate of 1 μl/h. The forseeable limitations are the unit cost of the product. The major advantage of the system is that it is very simple to insert the pump either intraperitoneally or subcutaneously in mice.

The major limitation of the effective use of continuous intravenous infusion of methylglyoxal as the free aldehyde is the severe local tissue toxicity elicited in the tail region of cannulated mice at infusion doses greater than 250–500 mg/kg per day. The moderate carcinostatic effectiveness of methylglyoxal against the relatively slow growing CBA/CA D40 mammary tumour (volume doubling time 1.35 days, 50–400 mm^3) should be compared to the lack of effect of the compound against Lymphosarcoma 1 (volume doubling time 0.8 days, 50–800 mm^3) and Squamous carcinoma D (volume doubling time 1.10 days, 50–800 mm^3). These tumours arose spontaneously and are syngeneic with the host, so the contribution made by the host's immune system in inhibiting tumour growth may be considered small (TD_{50}'s in the range of 1–10 cells).

There may be at least three ways of increasing the therapeutic effectiveness of methylglyoxal given by continuous intravenous infusion: (1) synthesis of methylglyoxal adducts such as methylglyoxal–cysteine, methylglyoxal–ascorbate or methylglyoxal–methylamine for testing in the continuous infusion system, on the hypothesis that an extended breakdown of the adduct with time will occur; (2) inhibition of the glyoxalase enzyme system; and (3) concomitant administration of cytostatic agents such as dibutyryl cyclic AMP or nicotinamide.

This paper has concentrated on methods of increasing the carcinostatic effectiveness of methylglyoxal by intraperitoneal injection compared to the effects obtained by continuous intravenous infusion. In general, continuous intravenous infusion of the free aldehyde at the dose levels used in this study was less effective than were the corresponding schedules for intraperitoneal administration in inhibiting tumour growth. The effectiveness of the approaches outlined above is now being investigated.

One of the assumptions implicit in the theoretical framework developed by Szent-Györgyi is that there exists some general growth regulatory system involving protein semiconductivity which requires the continuous intracellular production of biologically specific aldehydic materials which interact with structural proteins. The hypothesis implies that cancer cells may *(i)* be deficient in the output of such aldehydic materials, *(ii)* metabolize such substances differently from normal cells, or *(iii)* fail to respond to such aldehydes in an appropriate manner. Our approach towards testing this hypothesis has been to increase the exogenous supply of aldehydes such as methylglyoxal and HPE. In general, we obtained no significant improvements in the carcinostatic effectiveness of the aldehydes tested, either by increasing the total dose administered or by inhibiting aldehyde metabolism. This does not invalidate the theory, since our conventional methods of testing it may not be adequate. A more realistic approach could involve the stimulation of specific peroxidative products in neoplastic cells by manipulation of membrane fluidity and the initiation of the free-radical process leading to such an event. At this time, there exists neither the detailed information on membrane structure and response to external signals nor the testing methods to evaluate such a proposal.

ACKNOWLEDGEMENTS

Part of the work described in this paper was supported by the Cancer Research Campaign. Thanks are also due to Paul Elvin for permission to include results obtained jointly with him.

References

APPLE, M.A. & GREENBERG, D.M. (1967) Arrest of cancer in mice by therapy with normal metabolites. I. 2-oxo-propanal. *Cancer Chemother. Rep. 51*, 455-464

APPLE, M.A. & GREENBERG, D.M. (1968) Arrest of cancer in mice by therapy with normal metabolites. II. Indefinite survivors among mice treated with mixtures of 2-oxo-propanal and 2,3-dihydroxy-propanal. *Cancer Chemother. Rep. 52*, 687-696

BAYLOR, M.B. & EGYUD, L.G. (1967) The effect of methylglyoxal on phage infection of *Escherichia coli. Virology 31*, 380-382

BICKIS, I.J., SCHAUENSTEIN, E. & TAUFER, M. (1969) Effects of hydroxypentenal on the metabolism of normal and malignant cells. *Monatsh. Chem. 100*, 1077-1104

BONSIGNORE, A., CASTELLANI, A., FORNAINI, G., LEONCINI, G. & SEGNI, P. (1968) Pyruvic and L(+)-lactic acid formation from glyceraldehyde via methylglyoxal in the soluble fraction of liver and in reconstructed systems. *Ital. J. Biochem. (Engl. Ed.) 17*, 65-76

COHEN, P. (1945) Glyoxalase activity of liver from rats fed *p*-dimethylaminoazobenzene. *Cancer Res. 5*, 626-630

CONROY, P.J. (1976) *The Inhibitory Effects of Some Aldehydes on Tumour Cells*. Ph. D. Thesis, Brunel University, Middlesex

CONROY, P.J., NODES, J.T., SLATER, T.F. & WHITE, G.W. (1975) Carcinostatic activity of 4-hydroxy-2-pent-en-1-al against transplantable murine tumour lines. *Eur. J. Cancer 11*, 231-240

CONROY, P.J., NODES, J.T., SLATER, T.F. & WHITE, G.W. (1977) The inhibitory effects of a 4-hydroxy-pentenal:cysteine adduct against sarcoma 180 cells in mice. *Eur. J. Cancer 13*, 55-63

CONROY, P.J. & ELVIN, P. (1979) Continuous intravenous infusion of cyclophosphamide in tumor bearing mice. *J. Lab. Clin. Med.*, in press

EGYUD, L.G. (1965) Studies on autobiotics: chemical nature of retine. *Proc. Natl. Acad. Sci. U.S.A. 54*, 200-202

EGYUD, L.G. (1967) Studies on cell division: the effect of aldehydes, ketones and α-ketoaldehydes on the proliferation of *E. coli. Curr. Mod. Biol. 1*, 14-20

EGYUD, L.G. (1968) Studies on cell division: the antagonism of α-ketoaldehydes and thiols. *Curr. Mod. Biol. 2*, 128-132

EGYUD, L.G. & SZENT-GYÖRGYI, A. (1966) On the regulation of cell division. *Proc. Natl. Acad. Sci. U.S.A. 56*, 203-207

EGYUD, L.G. & SZENT-GYÖRGYI, A. (1968) Cancerostatic action of methylglyoxal. *Science (Wash. D.C.) 160*, 1140

ESTERBAUER, H. (1970) Kinetics of the reaction of sulphydryl compounds with α,β-unsaturated aldehydes in an aqueous system. *Monatsh. Chem. 101*, 782-810

ESTERBAUER, H., ZOLLNER, H. & ZOLLNER, N. (1975) Reaction of glutathione with conjugated carbonyls. *Z. Naturforsch. Sect. C Biosci. 30*, 466-473

ESTERBAUER, H., ERTL, A. & SCHOLZ, N. (1976) The reaction of cysteine with α,β-unsaturated aldehydes. *Tetrahedron 32*, 285-289

GREGG, C.H.T. (1968) Inhibition of mammalian cell division by glyoxals. *Exp. Cell Res. 50*, 65-72

HEWITT, H.B. (1966) The effect on cell survival of inhalation of oxygen under high pressure during irradiation *in vivo* of a solid mouse sarcoma. *Br. J. Radiol. 39*, 19-24

JERZYKOWSKI, T., MATUSZEWSKI, W., OTRZONSEK, N. & WINTER, T. (1970) Anti-neoplastic activity of methylglyoxal. *Neoplasma (Bratisl.) 17*, 25-35

JERZYKOWSKI, T., WINTER, T., MATUSZEWSKI, W. & SZCZUREK, Z. (1975) Glyoxalase II activity in tumours. *Experientia (Basel) 31*, 32-33

KRINGS, R. & TRITSCH, H. (1973) Wirkung von HPE auf das Harding-Passey Melanom der Maus. *Arch. Dermatol. Forsch. 247*, 108-113

LITT, M. (1969) Structural studies on transfer ribonucleic acid. I. Labelling of exposed guanine sites in yeast phenylalanine transfer ribonucleic acid with kethoxal. *Biochemistry 8*, 3249-3253

OTSUKA, H. & EGYUD, L.G. (1967) Nucleic acid and protein synthesis of malignant ascites cells in the presence of liver extracts and methylglyoxal. *Cancer Res. 27*, 1498-1499

PAUL, M.A. & DAVE, C. (1975) A simple method for long-term drug infusion in mice: evaluation of guanazole as a model. *Proc. Soc. Exp. Biol. Med. 148*, 118-122

RIDDLE, V. & LORENZ, F.W. (1968) Nonenzymic polyvalent anion-catalysed formation of methylglyoxal as an explanation of its presence in physiological systems. *J. Biol. Chem. 243*, 2718-2724

SCHAUENSTEIN, E. (1967) Autoxidation of polyunsaturated esters in water: chemical structure and biological activity of the products. *J. Lipid Res. 8*, 417-428

SCHAUENSTEIN, E., WÜNSCHMANN, B. & ESTERBAUER, H. (1968) The selective *in vivo* destruction of subcutaneously implanted Ehrlich ascites tumour cells by 4-hydroxy-enals and especially by 4-hydroxypentenal. *Z. Krebsforsch. 71*, 21-27

SCHAUENSTEIN, E., WÜNSCHMANN, B. & ESTERBAUER, H. (1969) Experiments on the therapeutic effect of hydroxypentenal on the Ehrlich ascites solid tumour of mice. *Z. Krebsforsch. 72*, 325-328

SCHAUENSTEIN, E., VERDINO, H. & TAUFER, M. (1970) The inhibitory mechanism induced in tumour cells by 4-hydroxy-2-pentenal (HPE). *Monatsh. Chem. 10*, 1189-1202

SCHAUENSTEIN, E., ERNET, M., ESTERBAUER, H. & ZOLLNER, H. (1971*a*) Experiments on the therapeutic effect of 4-hydroxypentenal. II. Inhibition of the growth of Sarcoma 180. *Z. Krebsforsch. 75*, 90-94

SCHAUENSTEIN, E., ZOLLNER, M., ERNET, M. & ESTERBAUER, H. (1971*b*) Experiments on the therapeutic effect of 4-hydroxypentenal. III. Inhibition of the growth of solid Nemeth-Kellner lymphoma. *Z. Krebsforsch. Klin. Onkol. 16*, 140-144

SCHAUENSTEIN, E., ESTERBAUER, H. & ZOLLNER, H. (1977) *Aldehydes in Biological Systems* (Gore, P.H., transl.), Pion/Academic, London

SHAPIRO, R., COHEN, B.I., SHIUEY, S.J. & MAURER, H. (1969) On the reaction of guanine with glyoxal, pyruvaldehyde and kethoxal, and the structure of the acylguanines. A new synthesis of N^2-alkylguanines. *Biochemistry 8*, 238-245

SLATER, T.F., CONROY, P.J., FRAVAL, H.N., JOSE, P.J., MCBRIEN, D.C.H., NODES, J.T., SAWYER, B.C. & WHITE, G.W. (1973) Biochemical and anti-tumour properties of hydroxy- and ketoaldehydes. *Abstracts 2nd Meeting Eur. Assoc. Cancer Res., Heidelberg*, p. 182

STRZINEK, R.A., VELA, G.R., SCHOLES, V.E. & NORTON, S.J. (1970) Further studies on liver glyoxalase activity in mice bearing lymphosarcoma. *Cancer Res. 30*, 334-337

SZENT-GYÖRGYI, A. (1960) *Introduction to a Submolecular Biology*, Academic Press, New York & London

SZENT-GYÖRGYI, A. (1968*a*) Bioelectronics. *Science (Wash. D.C.) 161*, 988-990

SZENT-GYÖRGYI, A. (1968*b*) *Bioelectronics: A Study in Cellular Regulations, Defense and Cancer*, Academic Press, New York & London

SZENT-GYÖRGYI, A. (1974) Electronic biology and its relation to cancer. *Life Sci. 15*, 863-875

SZENT-GYÖRGYI, A (1976) *Electronic Biology and Cancer. A New Theory of Cancer*, Dekker, New York & Basel

TILLIAN, H.M., SCHAUENSTEIN, E., ERTL, A. & ESTERBAUER, H. (1976) Therapeutic effects of cysteine adducts of α,β-unsaturated aldehydes on Ehrlich ascites tumour of mice. *Eur. J. Cancer 12*, 989-993

VANDERJAGT, D.L. & HAN, L.P. (1973) Rate determining hydride migration in the glyoxalase-I catalysed disproportionation of α-ketoaldehydes. *Fed. Proc. 32*, 668 (Abstr.2560)

VINCE, R. & DALUGE, S. (1971) Glyoxalase inhibitors. A possible approach to anticancer agents. *J. Med. Chem. 14*, 35-37

VINCE, R. & WADD, W.B. (1969) Glyoxalase inhibitors as potential anticancer agents. *Biochem. Biophys. Res. Commun. 35*, 593-598

VINCE, R., WOLF, M. & SANFORD, C. (1973) Glutaryl-*S*-*p*-brombenzyl-L-cysteinylglycine. A metabolically stable inhibitor of glyoxalase I. *J. Med. Chem. 16*, 951-953

ZOLLNER, H. & SCHAUENSTEIN, E. (1973) On the different response of succinate oxidation and succinate dehydrogenase to the action of 4-hydroxypentenal in Ehrlich ascites tumour and normal liver cells. *Z. Krebsforsch. 79*, 108-113

ZOLLNER, H., ESTERBAUER, H. & SCHAUENSTEIN, E. (1975) Relationship between chemical constitution and therapeutic activity of α,β-unsaturated aldehydes in the Ehrlich solid tumor of the mouse. *Z. Krebsforsch. 83*, 27-30

Discussion

Esterbauer: You showed an inhibition (lesion) of DNA synthesis *in vitro* which could be reversed when glutathione was added between 5 and 15 min after the addition of the aldehyde. At a later time, however, the inhibition could not be reversed by glutathione.

I would suggest that the target enzyme adds with its SH group to the aldehyde, giving an adduct in the open-chain form first. As we know from our kinetic measurements, at this stage the addition can easily be reversed by excess of a thiol compound. Later on, the cyclic hemiacetal is present and at this point the reaction is nearly irreversible. So probably what you are seeing is the open-chain form going to the cyclic form.

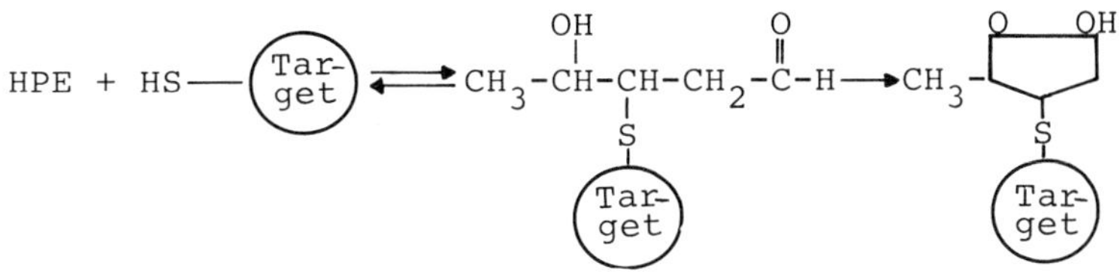

Conroy: That is an excellent suggestion. One thing that intrigued us was the fact that we knew, particularly with methylglyoxal, that it was possible to reverse the effect of the inhibition of protein synthesis by methylglyoxal with cysteine. In general methylglyoxal has a protein-inhibitory effect. Its primary effect is not directed towards DNA synthesis in mammalian cells. That may be something that indicates a degree of specificity for hydroxypentenal (HPE). A point that has been made in the discussions so far is the generally reactive nature of these aldehydes. This piece of evidence indicates that they can also have specific effects.

Esterbauer: Secondly, you see about a 100% increase in survival time when you give the HPE–cysteine adduct. Was it given intraperitoneally or intravenously? If the latter, there might be two ways in which such compounds can affect tumour growth, namely by local effects (see paper by Schauenstein & Esterbauer, this volume) and by effects via the circulation.

Conroy: We administered the HPE–cysteine adduct from Day 3 to Day 7, once a day, intraperitoneally. The tumour was produced by injecting cells into the peritoneal cavity. Animals treated with doses of the HPE–cysteine adduct of more than 128 mg/kg body weight survived considerably longer than untreated control mice. The treated animals in these dose ranges all died with a proliferated solid tumour and without ascites in the peritoneal cavity. This response could be classified as a 'local effect'. However, the results on the effect of continuous intravenous infusion of methylglyoxal on the growth of the solid form of the CBA/CA D40 mammary tumour cannot be classed as local effects: we did obtain a carcinostatic effect via the circulation in this case.

Esterbauer: Finally, you used a methylglyoxal–*N*-acetyl-L-cysteine adduct. Was it a hemiacetal or an adduct of the enol form?

Conroy: I am not certain of its structure. It is likely that it would be in an enol form. In the preparation we neutralized the starting materials and did the incubation in aqueous medium at pH 7.2 in the absence of oxygen.

Williams: Am I right in thinking that the concentrations of aldehyde that you were infusing into mice were roughly millimolar?

Conroy: Yes; 500 mg/kg per day, at an infusion rate of 2.8 ml/day of a solution of 4.5 mg/ml methylglyoxal (approximately 6 mM).

Williams: You have suggested that liposomes might be used to get these aldehydes into a cell in a form in which they would not be easily attacked. But will a liposome hold an aldehyde? It would surely diffuse straight out.

Conroy: It would depend on the lipid composition of the liposome. They are broadly divided into two groups, positive and negative, depending on the lipid composition (reviewed by Gregoriadis 1973). The molecular weight would need to be over 300. Some of the adducts that we want to make could have much higher molecular weights than that, particularly if we use a long-chain polymer. I agree that low molecular weight substances, particularly very water-soluble ones, would not remain in the liposome for long periods.

Brandt: You used as a glutathione inhibitor a bromobenzyl-substituted form of the molecule. Was this one of Vince's more effective inhibitors (Vince & Daluge 1971) that does cross cell membranes? Some of those he described worked in pure chemical systems but didn't function in cell systems because they were not entering the cells.

Conroy: We consulted Robert Vince about this (R. Vince, personal communication 1976). This was one of the few glyoxalase I inhibitors that he found to be metabolically stable and that did produce some therapeutic effect. I don't know why we could not demonstrate any additional therapeutic benefit over that of methylglyoxal alone. The glyoxalase inhibitor certainly made methylglyoxal more toxic, at least in terms of the overall tolerance of the tumour-bearing mice investigated. It may be that in the systems in which Vince tested them, such *S*-substituted glutathione derivatives were metabolically stable. In my mouse they were not, at least from the effects they produced.

Williams: Do sulphonamides inhibit glyoxalase? They do inhibit carbonic acid dehydratase! It may be unfair to make that parallel, but I was intrigued by the idea that if you made a molecule with a sulphonamide end-group and used glutathione to carry it, and as zinc is in the enzyme, glyoxalase might be blocked by such a drug.

Conroy: That is a useful suggestion and I must certainly try it.

References

GREGORIADIS, G. (1973) Molecular Trojan horses. *New Scientist 60*, 879-893

VINCE, R. & DALUGE, S. (1971) Glyoxalase inhibitors: a possible approach to anticancer agents. *J. Med. Chem. 14*, 35-37

Biochemical studies of transient intermediates in relation to chemical carcinogenesis

T.F. SLATER

Department of Biochemistry, and National Foundation for Cancer Research Laboratory, Brunel University, Uxbridge, Middlesex

Abstract Many chemical carcinogens must be metabolized to chemically reactive transient species before they can exert their full toxic action on mammalian cells. In general, this metabolic activation is performed by NADPH-dependent enzymes in the endoplasmic reticulum; the NADPH–cytochrome *P*-450 electron-transport chain is very important in this respect.

Biochemical studies on the chemical reactivities of such transient intermediates require the application of various fast-reaction and free-radical techniques: the use of such techniques is illustrated by reference to the metabolism of carbon tetrachloride.

CCl_4 is metabolized by liver endoplasmic reticulum in the presence of NADPH to a highly reactive product, probably $CCl_3^{\bullet}$; this activation of CCl_4 results in covalent binding of $CCl_3^{\bullet}$ and lipid peroxidation. The steady-state concentration of $CCl_3^{\bullet}$ is too low to be measured directly by e.s.r. spectroscopy but radical species can be accumulated with spin-trap techniques. The $CCl_3^{\bullet}$ radical can be generated by pulse radiolysis and the ensuing reactions with biologically important neighbouring species can be followed in the microsecond range by kinetic spectroscopy. The results point to the high reactivity of $CCl_3^{\bullet}$ and its restriction to a microenvironment within the endoplasmic reticulum.

Highly reactive electrophilic radicals (e.g. $CCl_3^{\bullet}$) can initiate lipid peroxidation in biomembranes and this is associated with changes in polyunsaturated fatty acids and in membrane fluidity.

The results are discussed in relation to carcinogen activation, to free-radical-mediated reactions in biomembranes, and to the general thesis that the production of reactive aldehydes by lipid peroxidation may act as a 'coarse control' of cell division.

Biochemical research proceeding in Brunel University with the support of the National Foundation for Cancer Research concerns three main projects:

(i) The testing for antitumour activity of novel compounds produced in other laboratories of the National Foundation for Cancer Research; this work has been done in collaboration with Dr P.J. Conroy and the main features are described elsewhere in this book (pp. 271-298).

(ii) Mechanisms of lipid peroxidation of biological membranes resulting in the production of unsaturated and keto-aldehydes, and their biological properties especially in relation to the inhibition of cell division. Our work on lipid peroxidation has been described recently at another Ciba Foundation Symposium (Slater 1979).

(iii) Free-radical mechanisms in chemical carcinogenesis and attempts to attenuate such free-radical reactions by the use of protective agents. In this paper I shall deal with studies that fall under this heading.

Cancer can be produced experimentally by many chemical and physical agents. Table 1 illustrates this diversity; some such materials are known to produce cancer in animals after a single administration, whereas other agents need repeated treatments to produce their effects. There is a wide variation among these different carcinogens in the total dose required to produce cancer as well, of course, as in the site and nature of the malignant lesion finally produced. Reviews that may be consulted for background information on these topics are by Süss *et al.* (1973), Ts'o & Di Paolo (1974) and Searle (1976).

Despite wide variations in chemical structure, in doses required, in the time for development of clinical signs, and in the nature of the malignant lesion produced, one important feature is common to many of these types of chemically produced cancer. In many cases the original substance is metabolized in the body to a chemically more reactive intermediate, the so-called ultimate carcinogen, which is the actual molecular species responsible for the biochemical disturbances that result in the malignant change or changes. This metabolism of the parent substance (or *activation* as it is often called), is generally enzymic and the products are mostly electrophiles such as carbonium ions, epoxides or free radicals. Examples of such activation

TABLE 1

A variety of agents that induce cancer in experimental animals (and in some instances are associated with human cancer) are listed to illustrate the diversity of chemical structures having this capability

General group	*Specific examples*
Aromatic amines	β-Naphthylamine
Azo dyes	*N,N*-Dimethylaminoazobenzene
Solvents	Carbon tetrachloride, benzene
Halogenated derivatives	Vinyl chloride
Nitrosamines	Dimethylnitrosamine
Fungal toxins	Aflatoxin B_1, sterigmatocystin
Alkaloids	Pyrrolizidine derivatives
Minerals	Asbestos
Metal salts	Chromates

processes are shown in Table 2. Miller & Miller (1974) have reviewed activation and chemical carcinogenesis. Since this symposium relates to the ideas of Albert Szent-Györgyi on electronic mechanisms, cell division and cancer, and specifically on the involvement of free radicals in such biological events, I shall concentrate on one class of activated carcinogen, that having a free-radical nature.

TABLE 2

The activation of various substances to electrophilic and free-radical products, including epoxidation of vinyl chloride and of benzo[*a*]pyrene, *N*-hydroxylation of aromatic amines followed by esterification or oxidation, and the formation of carbonium ions and free radicals

$$\text{Ionizing radiation} \xrightarrow{H_2O} OH\cdot$$

$$CCl_4 \xrightarrow{e^-} [CCl_4]^{-}\cdot \longrightarrow CCl_3\cdot$$

$$CH_2{=}CHCl \longrightarrow \overset{O}{CH_2 - CH_2} \longrightarrow R^+$$

$$RNH_2 \longrightarrow RNH-OH \longrightarrow RNH-OX,\ RNH-O\cdot$$

$$(CH_3)_2N-N{=}O \longrightarrow CH_3^+,\ CH_3-N(N{=}O)-O\cdot$$

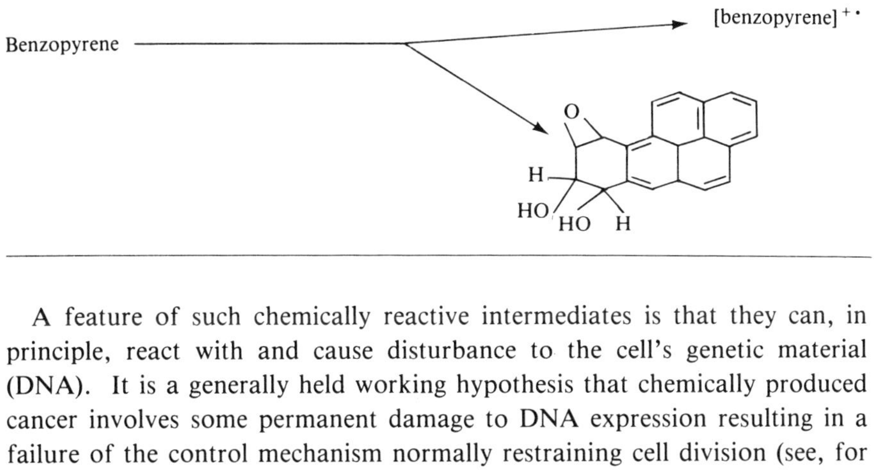

A feature of such chemically reactive intermediates is that they can, in principle, react with and cause disturbance to the cell's genetic material (DNA). It is a generally held working hypothesis that chemically produced cancer involves some permanent damage to DNA expression resulting in a failure of the control mechanism normally restraining cell division (see, for example, Barrett & Ts'o 1978 for a recent study of this basic assumption). However, since the metabolically produced electrophilic intermediates referred to above are chemically reactive, they can clearly interact with many cell components other than DNA and this can lead to serious cell dysfunction, even cell death. Appropriate doses of many chemical carcinogens—for example, dimethylnitrosamine (Barnes & Magee 1954; Magee *et al.* 1976), aflatoxins (Butler 1964, for other refs. see Wogan 1976), thioacetamide (Ambrose *et al.* 1949; Gupta 1956), and polycyclic hydrocarbons (Huggins *et*

al. 1963)—are well known to be associated with necrosis of various tissues. Fig. 1 illustrates some of the damaging reactions that may be expected to result from the production of highly reactive intermediates.

These types of reactive intermediate, the so-called ultimate carcinogens, are mainly produced by an enzyme sequence located in the lipoprotein membranes of the endoplasmic reticulum. The enzyme sequence is known as the NADPH–cytochrome *P*-450 electron transport chain (*or* the mixed function oxidase system, or the microsomal drug-metabolizing system). The components involved are shown in Fig. 2. Appreciable activities of the first enzyme of the sequence (NADPH–flavoprotein) can be found in many rat tissues, whereas cytochrome *P*-450 and the NADPH/ADP/Fe^{2+} lipid peroxidative sequence are much more restricted in location (see Table 3). As many carcinogens appear to require cytochrome *P*-450 (or *P*-448) for activation, the data shown in Table 3 indicate that the production of reactive intermediates by the *P*-450 route may be restricted in terms of the variety of available tissues; there is, however, another possibility, in relation to activation by the NADPH–cytochrome *P*-450 chain (see later).

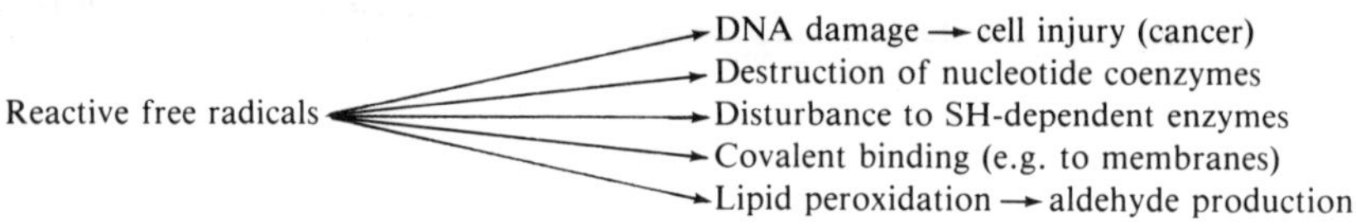

FIG. 1. Some damaging reactions that can occur inside a cell after the production of a reactive free-radical species (e.g. $CCl_3^{\bullet}$).

One enzyme activity referred to in Fig. 2 and Table 3 is the NADPH-dependent, ADP/Fe^{2+}-stimulated lipid peroxidation of the polyunsaturated fatty acids of the endoplasmic reticulum (for background refs. to this system see Hochstein & Ernster 1963; Slater 1968). Lipid peroxidation is associated with the production of low molecular weight aldehydes (see Schauenstein *et al.* 1977) that can inhibit cell division at low final concentration (Egyud & Szent-Györgyi 1966; see Schauenstein *et al.* 1977 for background refs.). In this respect it is interesting that tumour cells generally have a very low rate of lipid peroxidation in comparison to normal rat liver which has considerable peroxidizing activity. Table 4 illustrates this general observation with specific examples (for background refs. see Utsumi *et al.* 1965; Lash 1966; Burlakova 1975).

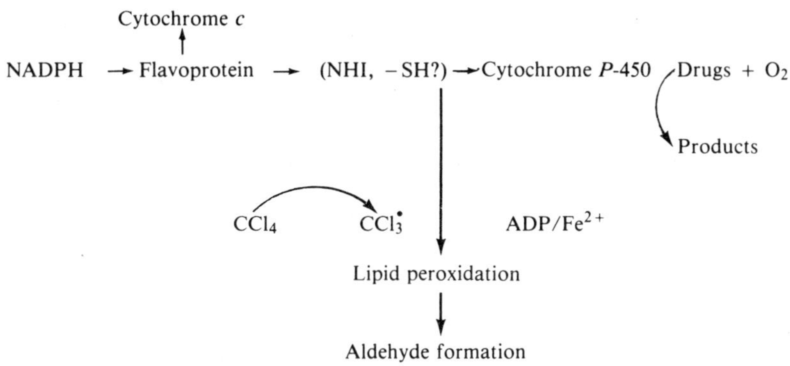

FIG. 2. The components of the NADPH–cytochrome *P*-450 electron transport chain. The locus of activation of CCl_4 to $CCl_3^{\bullet}$ is shown near to the NADPH–flavoprotein (see text); NHI and –SH stand for a component(s) having non-haem iron and thiol groups respectively. Evidence for non-haem iron is not yet conclusive but a component having non-haem iron characteristics has been detected in microsomes by Salerno & Ingledew (1975). The ADP/Fe^{2+}-stimulated lipid peroxidation is shown to involve electron diversion from the chain at the thiol-dependent site (Hochstein & Ernster 1963).

The studies outlined above on metabolic activation, on free-radical intermediates and on lipid peroxidation have led to a number of important concepts relating to biochemical pathology in general and carcinogenesis in particular. I shall use the enzymic activation of carbon tetrachloride as the main example to illustrate such concepts; the discussion that follows is

TABLE 3

The distribution of NADPH–cytochrome *c* reductase, cytochrome *P*-450 and NADPH–ADP/Fe^{2+}-dependent lipid peroxidation activity in microsomal fractions of several rat tissues. The values are related to adult liver as 100%

Tissue	*Cytochrome P-450*	*Cytochrome c reductase*	*Lipid peroxidation*
Adult liver	100	100	100
Newborn liver (2-4 days)	35	49	14
Kidney	14	67	12
Lung	Very low[a]	44	3
Brain	Very low[b]	16	2
Small intestine	2[c]	127	1
Placenta	7[d]	11[d]	8[d]

The data are from Benedetto *et al.* (1976) except for [a]Johannesen *et al.* (1977), [b]Cohn *et al.* (1977), [c]Stohs *et al.* (1976) and [d]Arvela *et al.* (1976).

TABLE 4

The lipid peroxidative activity of microsomal fractions prepared from rat liver samples and from two tumours: an aflatoxin-induced rat liver tumour propagated in nude mice and a mouse lymphosarcoma

Tissue	*Peroxidation (nmol O_2/min per g tissue)*
Adult rat liver	78 ± 3 (8)
1-day rat liver	2.1 (19)
2-day rat liver	4.4 (7)
Aflatoxin rat liver tumour in nude mice	Undetectable (4)
Mouse lymphosarcoma	Undetectable (4)

Peroxidative activity was measured at 25 °C as oxygen uptake in the presence of NADPH/ADP and $FeSO_4$ using the system described by Slater (1968). The aflatoxin-derived tumours were kindly provided by Professor T.A. Connors and the lymphosarcoma was passaged in WH/Ht mice and was obtained from Dr H. B. Hewitt.

relevant to the enzyme activation of many carcinogens. An understanding of how chemical carcinogens produce cell injury is basic to our general studies on the fundamental biology of cancer and relates in several ways to Albert Szent-Györgyi's views on the control of cell division.

A single dose of CCl_4 when administered to rats or to many other animal species produces centrilobular necrosis and fatty degeneration of the liver (for background refs. see Recknagel 1967; Slater 1972). Repeated doses of CCl_4 produce cirrhosis (Cameron & Karunaratne 1936; see also McLean *et al.* 1969) and may result in liver cancer (Reuber & Glover 1970). The liver is very sensitive to CCl_4 (for an illustration of the dose-response characteristics of the liver enzyme glucose-6-phosphatase see Reynolds & Yee 1967), and the sensitivity of the liver to injury by CCl_4 can be increased further by pre-dosing with agents that induce the NADPH–cytochrome *P*-450 chain, for example, phenobarbital (McLean & McLean 1966). The observation that necrosis is generally restricted to the centrilobular regions of the liver lobules whereas much higher concentrations of CCl_4 are present in other tissues, such as brain and adipose tissue (Van Oettingen 1955), points to a specific activation of CCl_4 to a reactive species in the centrilobular zones of the liver. Data consistent with an essential role of the NADPH–cytochrome *P*-450 system in such an activation process *in vivo* have been obtained by Gooding *et al.* (1978) who found that cytochrome *P*-450 is concentrated centrizonally, especially after phenobarbital treatment (Fig. 3).

The nature of the excited intermediates formed from CCl_4 in liver endoplasmic reticulum has been the subject of much debate but most people now accept that the trichloromethyl radical ($CCl_3^{\bullet}$) is the major contender as

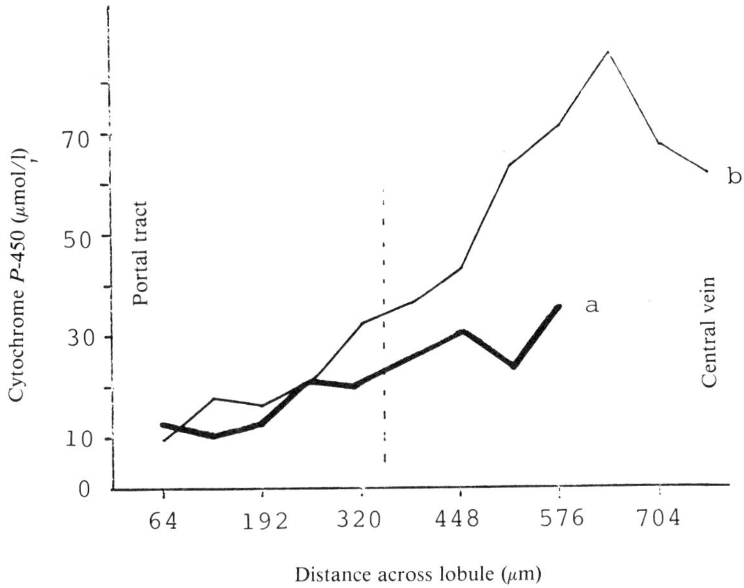

FIG. 3. The distribution of cytochrome *P*-450 in individual liver cells across the liver lobule from the periportal to the centrilobular zone. *P*-450 was measured by a microspectrophotometric method (Gooding *et al.* 1978). Trace *(a)* was obtained from normal control rat liver sections; trace *(b)* after induction with phenobarbital.

the primary excited species; this radical can be formed by dissociative electron capture (reaction 1).

$$CCl_4 \xrightarrow{e^-} [CCl_4]^{-\bullet} \longrightarrow CCl_3^{\bullet} + Cl^- \qquad (1)$$

When the *locus* of electron donation to CCl_4 from the NADPH–cytochrome *P*-450 chain was studied a few years ago (Slater & Sawyer 1971*a*) we found from metabolic inhibition studies of CCl_4-stimulated lipid peroxidation that the locus did not appear to be cytochrome *P*-450 as expected but to be in the region of the NADPH-flavoprotein (Fig. 4). Although this suggestion has remained controversial (see Glende 1972; Villarruel *et al.* 1977; Masuda & Murano 1978) recent work with highly purified microsomal NADPH–flavoprotein has shown that this component alone can stimulate NADPH-dependent lipid peroxidation *in vitro* (Table 5; see also Mishin *et al.* 1976 and Hirokata *et al.* 1978).

Since lipid hydroperoxides (produced by a flavoprotein-dependent lipid peroxidation) have been reported to 'activate' CCl_4 to $CCl_3^{\bullet}$ (O'Neal & Aust 1978) it is possible that some activation of CCl_4 occurs by this route *in vivo*. It

Agent	*Concentration (μmol/l)*	*Effect on drug metabolism (%)*	*Effect on CCl_4^- stimulated lipid peroxidation (%)*
Carbon monoxide	–	↓ 47	↑ 56
SKF 525A	100	↓ 59	N.S.
p-Chloromercuri-benzoate	100	↓ 58	↑ 52
Ageing	–	↓ 41	↑ 5
Cytochrome *c*	40	↓ 100	↓ 100
Propyl gallate (PG)	10	↑ 34	↓ 95

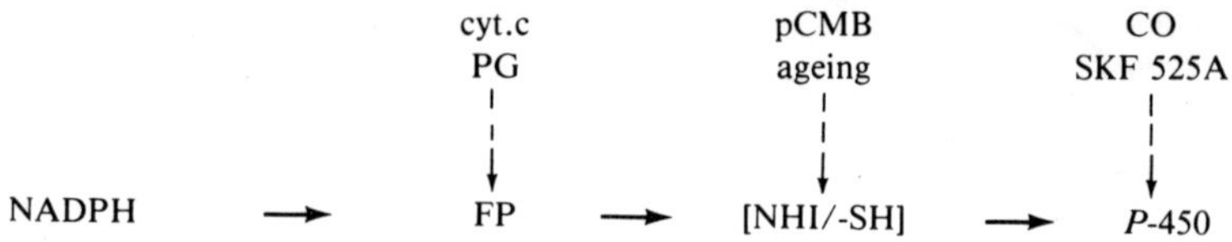

FIG. 4. The interactions of various inhibitors with the NADPH–cytochrome *P*-450 chain are shown below. The effects of the inhibitors on CCl_4-stimulated lipid peroxidation and on aminopyrene demethylation activity are shown above as increases (↑) or decreases (↓) relative to controls. Data from Slater & Sawyer 1971*a,b* and unpublished data of T.F. Slater, B. Sawyer & K. Cheeseman.

is probable, however, that electron flow out from the NADPH–cytochrome *P*-450 chain may occur at different sites and with different emphases, depending on the experimental conditions (see Slater 1975) which can affect the fluidity and interactions of the various membrane-linked components. In

TABLE 5

NADPH-dependent lipid peroxidation in phospholipid vesicles in the presence of highly purified rabbit liver NADPH–flavoprotein reductase, highly purified rabbit liver cytochrome *P*-450, or rabbit liver microsomes

	Additions	
	Endogenous	*+ NADPH*
Vesicles alone	<0.1	<0.1
plus reductase	<0.1	6.8 ± 0.8
plus *P*-450	1.2 ± 0.1	1.3 ± 0.5
plus reductase + *P*-450		3.5 ± 0.6
Microsomes	1.0 ± 0.1	3.8 ± 0.3

The peroxidation rate was measured by malonaldehyde production (nmol/ml). The additional components were incorporated into the vesicles by exposing the mixture of components to a short period of ultrasonic energy. Data from unpublished experiments of A. Blake, A. Stier & T.F. Slater.

conditions *in vitro*, the data of Slater & Sawyer (1971*a*) in combination with those of Benedetto *et al.* (1976) suggest that some additional component to the NADPH–flavoprotein (perhaps a non-haem iron thiol-dependent protein) is required for activation of CCl_4; cytochrome *P*-450 itself seems unlikely to be involved, in view of the results obtained with various inhibitory agents (Slater & Sawyer 1971*a*). However, whether such results can be strictly applied to endoplasmic reticulum *in vivo* is not clear at present, in view of the possible disturbances to the lipid–protein structure of the membranes during isolation of microsomal fractions.

The results described above indicate that electron flow out from the NADPH–cytochrome *P*-450 chain may occur at or near the flavoprotein locus. Other indications for this view are the production of the superoxide anion radical from the interaction of purified NADPH–flavoprotein and oxygen (Aust *et al.* 1972) and the oxidation of certain catechols by the purified NADPH–flavoprotein reductase (Kappus *et al.* 1977). As the NADPH–flavoprotein has a wide tissue distribution, at least in the rat (see Table 3), this offers the possibility for a broader tissue distribution of carcinogen activation than if cytochrome *P*-450 itself were essential, particularly of those carcinogens that require a direct electron capture as the initial event.

Returning now to the specific problem of CCl_4, there is much *indirect* evidence for the activation of CCl_4 to $CCl_3^{\bullet}$ (reviewed in Slater 1972) but direct evidence in terms of electron spin resonance spectroscopy has not been obtained despite many attempts (Calligaro *et al.* 1970; Keller *et al.* 1971; Slater 1972; Burdino *et al.* 1973).

The reasons for such lack of success appear to be that the concentration of $CCl_3^{\bullet}$ in liver cells after dosing rats with CCl_4 is below the level of detection of the spectrometers used, owing to a slow rate of metabolism of CCl_4 to the $CCl_3^{\bullet}$ free radical and to a high rate of reaction of the $CCl_3^{\bullet}$ radical with neighbouring molecular species. Some radical recombination, for instance, seems to occur, as Fowler (1969) has detected C_2Cl_6 in expired air from rabbits dosed with CCl_4. To attempt to clarify this difficult problem of the production of $CCl_3^{\bullet}$ (as a typical example of many excited transient carcinogen metabolites) and the subsequent rates of reaction with neighbouring species, we have recently been using two approaches: *(i)* spin-trap techniques and *(ii)* fast-reaction studies involving pulse radiolysis.

In spin trapping (see Perkins *et al.* 1970) a low concentration of a reactive radical (e.g. $CCl_3^{\bullet}$) interacts with a substituted nitroso compound to yield a relatively stable free-radical adduct, a nitroxy radical, which progressively accumulates as the initial reactive radical is formed and then trapped (e.g. by

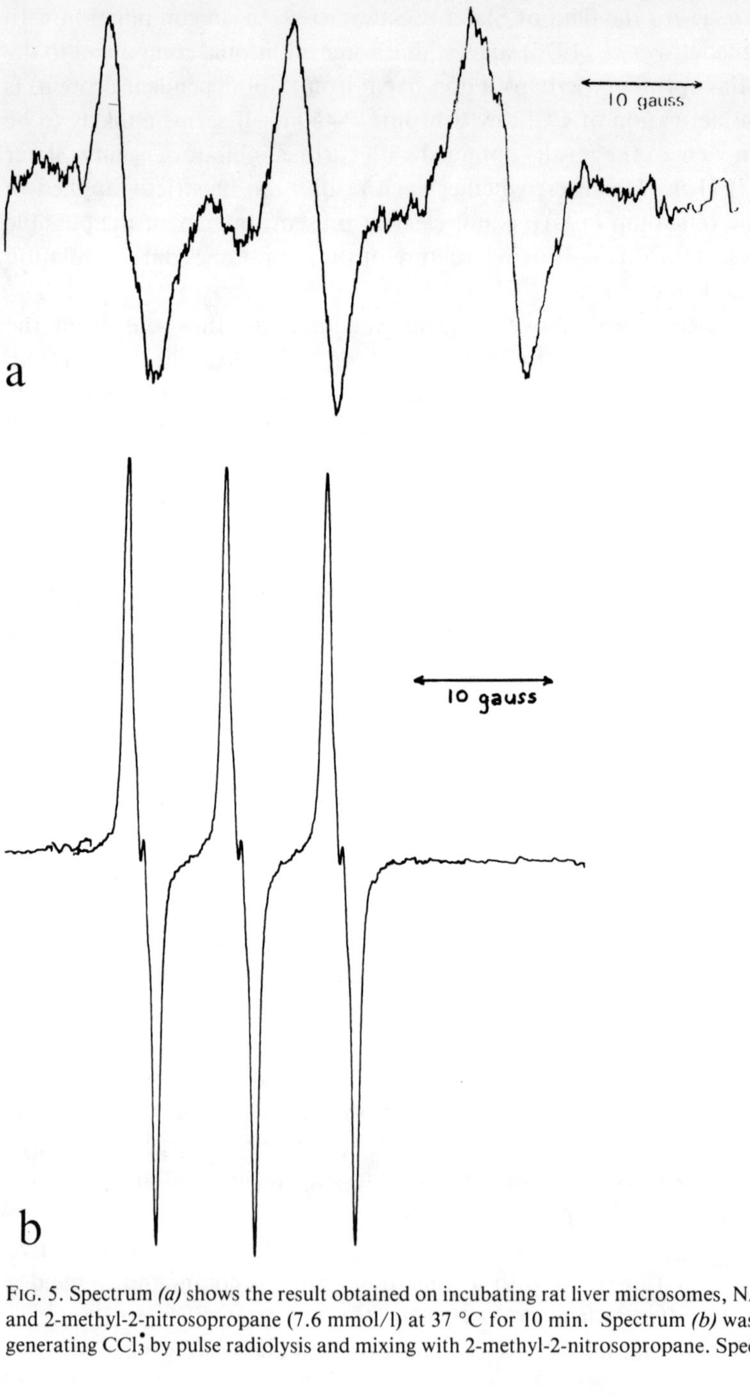

FIG. 5. Spectrum *(a)* shows the result obtained on incubating rat liver microsomes, NADPH, CCl_4 and 2-methyl-2-nitrosopropane (7.6 mmol/l) at 37 °C for 10 min. Spectrum *(b)* was obtained by generating $CCl_3^{\bullet}$ by pulse radiolysis and mixing with 2-methyl-2-nitrosopropane. Spectrum *(c)* was ↗

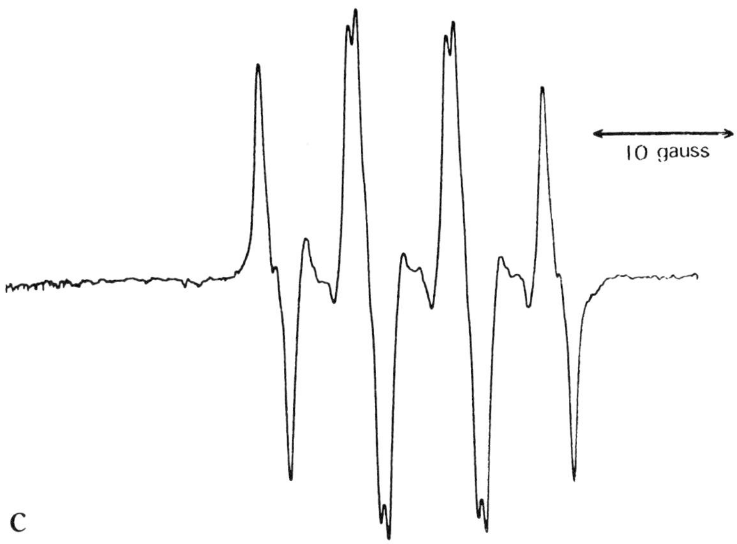

obtained as for *(b)* but using $^{13}CCl_4$ to generate $^{13}CCl_3^{\bullet}$. The e.s.r. spectra were obtained using a Varian E-3 spectrometer at room temperature; the nitroxide product was extracted into chloroform before e.s.r. analysis. Data from Ingall *et al.* 1978.

$$(CH_3)_3C-N=O + R^{\bullet} \rightarrow (CH_3)_3C-NR\overset{\bullet}{-}O \qquad (2)$$

2-methyl-2-nitrosopropane, reaction 2). Such nitroxy radicals yield a characteristic triplet e.s.r. spectrum (see Smith *et al.* 1976) and have been used extensively in other studies as spin labels.

When liver microsomes, NADPH and the spin trap 2-methyl-2-nitroso propane are incubated aerobically at 37 °C under appropriate conditions (Fig. 5) there is no significant production of a nitroxy radical product; however, addition of low concentrations of CCl_4 rapidly leads to the appearance of a product having a nitroxy-type behaviour when examined by e.s.r. spectroscopy (Fig. 5); the line separation was 15 G. The appearance of this material depends on the concentration of CCl_4 and on the time of incubation, between 0–5 min at 37 °C (Ingall *et al.* 1978). If this product were formed by the interaction of $CCl_3^{\bullet}$ with the spin trap, the e.s.r. spectrum should show additional hyperfine splitting when $^{12}CCl_4$ is substituted by $^{13}CCl_4$, owing to the nuclear spin effect (see Carrington & McLachlan 1967). However, no splitting of the e.s.r. signal was observed when $^{13}CCl_4$ was substituted for $^{12}CCl_4$ in these experiments.

To evaluate this point further we generated the trichloromethyl radical specifically using a 4 MeV linear accelerator adapted for pulse radiolysis. Details of the instrument used and the method for generating $CCl_3^{\bullet}$ radicals have been given elsewhere (Willson 1978; Willson & Slater 1975). With $CCl_3^{\bullet}$ generated by pulse radiolysis and the nitrosopropane spin trap in anaerobic conditions a triplet spectrum was observed (Ingall *et al.* 1978) that had a line separation of 9 G (Fig. 5). This time, using $CCl_3^{\bullet}$ generated by pulse radiolysis, the exchange of $^{12}CCl_4$ for $^{13}CCl_4$ produced hyperfine splitting (Fig. 5). These data are interpreted to indicate that when $CCl_3^{\bullet}$ is produced in *aerobic* conditions in liver microsomes, it rapidly interacts with neighbouring species (e.g. thiols, polyunsaturated fatty acids) to form secondary and tertiary radicals, and that these secondarily derived radicals are detected by the spin-trap technique. If this suggestion is valid, it appears that the $CCl_3^{\bullet}$ radical is really very reactive in its effects on biologically important substances; to get quantitative information on this point we have used pulse-radiolysis techniques coupled to kinetic spectroscopy (for background references see Willson 1978).

One of the first reactions studied in this way was that between CCl_4 and the phenothiazine drug promethazine. Phenothiazines are known to be good scavengers of electrophilic radicals (see Slater 1972) and promethazine is known to protect the liver *in vivo* against necrosis produced by CCl_4 (Rees & Spector 1961) and to be extremely effective *in vitro* in attenuating CCl_4-stimulated lipid peroxidation (Slater & Sawyer 1971*b*). General accounts of protection mechanisms in relation to CCl_4-induced liver injury have been given by Slater (1978, 1979).

The interaction of $CCl_3^{\bullet}$ and promethazine in aerobic conditions was studied by pulse radiolysis (Willson & Slater 1975) and a fast reaction rate of 1.1×10^9 l mol^{-1} s^{-1} was obtained. However, the reaction is slow in strict anaerobic conditions (Packer *et al.* 1978). These authors concluded that $CCl_3^{\bullet}$ is not very reactive as an electrophilic radical with promethazine but that it reacts rapidly with oxygen to yield a much more electrophilic agent. This suggestion has obvious implications for the role of oxygen in such radical-mediated reactions.

Table 6 gives some data for the *aerobic* interaction of $CCl_3^{\bullet}$ with several biologically-important compounds. In anaerobic conditions all the reactions listed were extremely slow.

The fast rates of reaction of $CCl_3^{\bullet}$ (in aerobic conditions) with the polyunsaturated fatty acids and with amino acids are especially noteworthy as $CCl_3^{\bullet}$ is produced *in vivo* in the lipoprotein environment of the endoplasmic reticulum. Free radicals of such high reactivity can be expected to have only a

TABLE 6

Reaction rate data for the aerobic interaction of $CCl_3^{\bullet}$ (generated by pulse radiolysis) with several biologically important materials

Substrate	*pH*	*K/(l mol⁻¹ s⁻¹)*
Tyrosine	12	1.6×10^8
	7	$<10^6$
Tryptophan	7	1.2×10^8
Cysteine	12	5.3×10^7
	4	$<10^6$
GMP	12	3.8×10^7
	8.5	$<10^6$
Arachidonic acid	7	Moderately fast
Ascorbic acid	7.5	2.7×10^8
Promethazine	7	$>5 \times 10^8$
Propyl gallate	7.5	2×10^8
$CCl_3^{\bullet} + O_2$	7	$>10^9$

Data from unpublished experiments of J. Packer, R.L. Willson & T.F. Slater. For details of the method used in pulse radiolysis see Willson (1978); for the procedure used to generate $CCl_3^{\bullet}$ see Willson & Slater (1975).

low localized concentration (assuming that their rate of production is not unusually high) and a restricted radius of diffusion (see Slater 1976). As the free radical is formed in a membrane surrounded by polyunsaturated fatty acids, amino acid residues, thiols, etc. with which it interacts (Table 6), the radius of diffusion can be represented as an inverse function of the chemical reactivity so that very reactive radicals such as $CCl_3^{\bullet}$ are essentially trapped in their microenvironment around the locus of formation (Slater 1976). They are restricted from interacting with DNA in the nucleus, for example, since this is in general outside the available sphere of diffusion for reactive radicals. This has an obvious implication for chemical carcinogenesis: chemically very reactive free radicals (or other similarly reactive excited species) formed in the membranes of the endoplasmic reticulum can be expected to be *unreactive as carcinogens* because of the diffusion barrier whereas chemical species of low intrinsic reactivity, although able to diffuse, will not be sufficiently reactive to bind covalently to DNA. The most active intermediates in terms of carcinogenicity can be expected to be intermediate in chemical reactivity, as indicated diagrammatically in Fig. 6.

This analysis presupposes *(i)* that an essential event in carcinogenesis is a chemical modification of DNA that allows transmission of the disturbance from generation to generation (as already mentioned, this is a generally accepted working hypothesis) and *(ii)* that the locus of the formation of the reactive species is far in molecular terms from DNA in the nucleus.

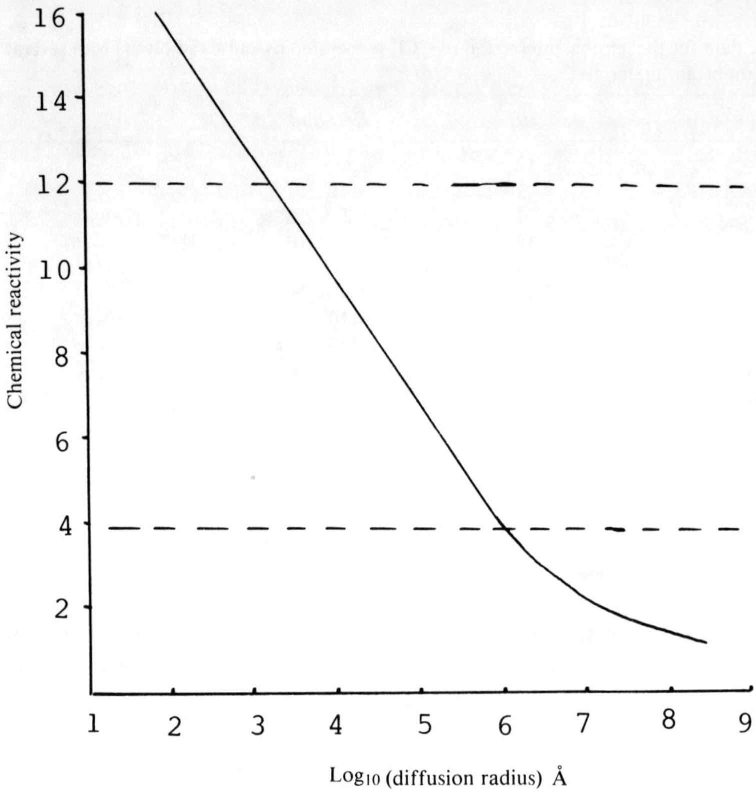

FIG. 6. A simple diagrammatic representation of the concept discussed in the text. Metabolic intermediates having high chemical reactivity (in arbitrary units) that are generated in the membranes of the endoplasmic reticulum are shown having very restricted diffusion from the site of formation; substances with low chemical reactivity can diffuse much greater distances. Effective carcinogenic agents *in vivo* are suggested to fall in the middle range of chemical reactivity (between the arbitrarily drawn dotted lines) to allow sufficient diffusion and yet maintaining relatively high reactivity (see Slater 1976).

In certain conditions, however, carcinogen activation may take place relatively close to susceptible DNA, for example as a result of the activation process itself being near to DNA; Jernstrom *et al.* (1976), Pezzuto *et al.* (1977) and Bresnick *et al.* (1977) have demonstrated the metabolism of 3,4-benz[*a*]-pyrene in the nuclear membrane of rat liver cells. Another example occurs in the artificial systems used for *in vitro* tests of carcinogenicity such as the Ames' test (Ames *et al.* 1975; see Serres *et al.* 1976). It is known that endoplasmic reticulum membranes adhere to the surface coat of the micro-organisms used in tests based on Ames' procedure, thereby greatly diminishing the distance that the activated materials have to diffuse in order to reach susceptible DNA. In these conditions it can be expected that some free radicals will show up positively as mutagens, whereas *in vivo* they may be too reactive to reach susceptible sites.

As already mentioned, the combination of $CCl_3^\bullet$ with oxygen increases the reactivity with respect to several biologically important materials (see Table 6); an increase in chemical reactivity implies (on the basis of the above analysis) a decrease in the probable radius of diffusion from the site of formation in the endoplasmic reticulum. If local depletion of oxygen can occur in the neighbourhood of $CCl_3^\bullet$ production, it may be envisaged that $CCl_3^\bullet$ will move further away from its site of formation before being 'activated' by oxygen, and hence will produce damage over a larger radius. An interesting aspect of oxygen in producing mutagenic effects has been recently described (Bruyninckx *et al.* 1972).

In appropriate conditions the formation of reactive electrophilic radicals in the endoplasmic reticulum can be expected to lead to an increased lipid peroxidation of membrane polyunsaturated fatty acids, and this can result in a wide range of toxic effects including some at a considerable distance from the initial site of peroxidation. This is because the secondarily-derived long-chain peroxy radicals and lipid hydroperoxides can diffuse in the plane of the membrane, whereas short-chain degradation products including aldehydes can escape from the membrane and cause general cellular disturbance, and even systemic effects (see Fig. 7). An increase in serum malonaldehyde in liver injury has been reported by Suematsu *et al.* (1977). The general pharma-

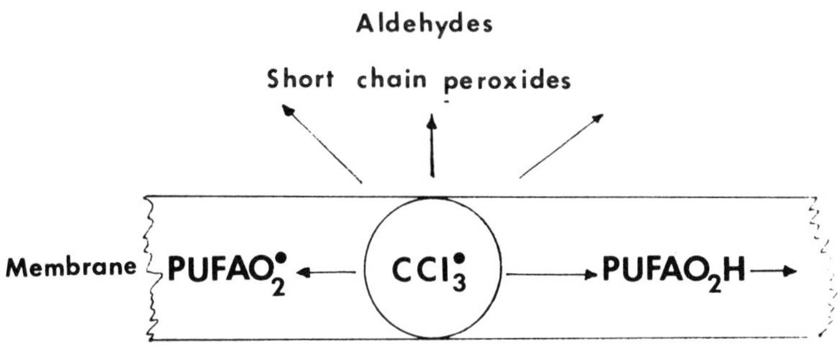

FIG. 7. The formation of a highly reactive species like $CCl_3^\bullet$ (or $CCl_3O_2^\bullet$) within a biomembrane is shown to be associated with only a small radius of diffusion (i.e. the intermediate is essentially trapped in its microenvironment). Secondarily formed lipophilic products ($PUFAO_2^\bullet$ and $PUFAO_2H$) can diffuse laterally in the plane of the membrane. Tertiary breakdown products (e.g. aldehydes, short-chain peroxides) are shown as capable of escaping from the membrane to produce effects at a distance (see Slater 1976).

cological and biochemical activities of the hydroperoxides and aldehydes are reviewed by Schauenstein *et al.* (1977); more specific aspects of immediate interest are given by Ugazio *et al.* (1976).

The various aldehydes produced by the peroxidation of unsaturated lipids have many interesting properties, some of which have been referred to elsewhere in this volume (see papers by Schauenstein & Esterbauer, Dianzani and Conroy). In addition, they readily form adducts with amino acids and polyamines (Chio & Tappel 1969; White & Slater 1974). With bifunctional aldehydes like malonaldehyde this means that cross-linking of protein chains is possible, as discussed by Chio & Tappel (1969). This reactivity of malonaldehyde (and related substances) suggests that such agents can modify the *fluidity* of membranes that are exposed to them. Using nitroxy-spin labels one

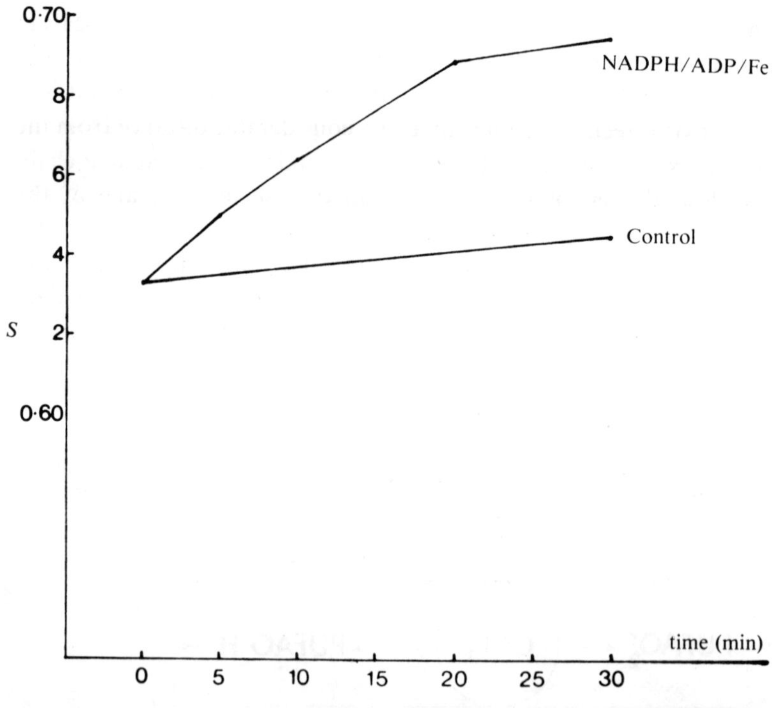

FIG. 8. The effects of peroxidizing rat liver microsomes at 37 °C with NADPH/ADP/Fe^{2+} on membrane fluidity are shown compared to the control situation (no stimulated lipid peroxidation). Fluidity was assessed by using a stearic acid-nitroxy spin label (Number 618, Synvar, Palo Alto, Cal., USA) and analysing its tumbling motion by e.s.r. spectroscopy. An order parameter '*S*' was calculated according to McConnell *et al.* (1972) and this is plotted on the ordinate against time; the order parameter is an inverse measure of fluidity. Unpublished data of A. Ingall, K.A.K. Lott, A. Stier & T.F. Slater.

can get an indication of membrane fluidity in relation to the tumbling motion of the particular spin label used (Smith *et al.* 1976).

When liver endoplasmic reticulum was peroxidized by NADPH/ADP/Fe^{2+} it was found that fluidity, as measured with a stearic acid nitroxy-spin label, decreased considerably (Fig. 8) (A. Ingall, K.A.K. Lott, T.F. Slater & A. Stier, unpublished work). Since cell division can be expected to be associated with an *increased* membrane fluidity (see, for example, Inbar *et al.* 1977; Campisi & Scandella 1978; but also Johnson & Edidin 1978) it can be seen that peroxidation and the associated production of aldehydes have complementary effects in decreasing cell division: both a decrease in fluidity, and the inhibitory action of aldehydes on DNA synthesis (see Schauenstein *et al.* 1977) and on microtubule function (see chapter by Dianzani in this volume, pp. 240-265). In this sense lipid peroxidation that results from the production of excited intermediates can be seen as a safety mechanism to inhibit cell division and even to result in death of the peroxidizing cell. Since many tumours do not peroxidize easily this could indicate that the safety mechanism has been lost, perhaps by an increased ability of tumour cell material to scavenge the peroxy radicals, or by a decreased rate of initiation, by a disturbance in the heterogeneous distribution of polyunsaturated fatty acids around the initiation sites (see Stier 1978), or by changes in the rates of metabolism of the aldehydic products. These possibilities are interesting features of membrane biochemistry that closely concern studies on cell division and cancer. It is through the studies on free radicals, lipid peroxidation and aldehydes outlined in this chapter that our work at Brunel has overlapped during many years the ideas of Albert Szent-Györgyi. Detailed examination of the points outlined above will be of importance, in my opinion, to our understanding of cell pathology in general, and will facilitate a critical evaluation of Albert Szent-Györgyi's views on the role of reactive aldehydes in cell division and cancer.

References

AMBROSE, A.M., DE EDS, F. & RATHER, L.J. (1949) Toxicity of thioacetamide in rats. *J. Ind. Hyg. Toxicol. 31*, 158-161

AMES, B.M., MCCANN, J. & YAMASAKI, E. (1975) Methods for detecting carcinogens and mutagens with the salmonella/mammalian microsome mutagenicity test. *Mutat. Res. 31*, 347-364

ARVELA, P., KÄRKI, N.T. & PELKONEN, R.O. (1976) Lipoperoxidation rates and drug-oxidising activities in the liver and placenta of some mammal species during the perinatal period. *Experientia (Basel) 32*, 1311-1313

AUST, S.D., ROERIG, D.L. & PEDERSON, T.C. (1972) Evidence for superoxide generation by NADPH-cytochrome c reductase of rat liver microsomes. *Biochem. Biophys. Res. Commun. 47*, 1133-1137

BARNES, J.M. & MAGEE, P.N. (1954) Some toxic properties of dimethylnitrosamine. *Br. J. Ind. Med. 11*, 167-174

BARRETT, J.C. & TS'O, P.O.P. (1978) Relationship between somatic mutation and neoplastic transformation. *Proc. Natl. Acad. Sci. U.S.A. 75*, 3297-3301

BENEDETTO, C., SLATER, T.F. & DIANZANI, M.U. (1976) NADPH-cytochrome c reductase, cytochrome P-450 and NADPH-linked lipid peroxidation in microsomal fractions obtained from rat tissues. *Biochem. Soc. Trans. 4*, 1094-1097

BRESNICK, E., STOMING, T.A., VAUGHT, J.B., THAKKER, D.R. & JERINA, D.M. (1977) Nuclear metabolism of benzo(a)pyrene and of trans-7,8-dihydroxy-7,8-dihydrobenzo(a) pyrene. *Arch. Biochem. Biophys. 183*, 31-37

BRUYNINCKX, W.J., MASON, H.S. & MORSE, S.A. (1978) Are physiological oxygen concentrations mutagenic? *Nature (Lond.) 274*, 606-607

BURDINO, E., GRAVELA, E., UGAZIO, G., VANNINI, V. & CALLIGARO, A. (1973) Initiation of free radical reactions and hepatotoxicity in rats poisoned with carbon tetrachloride or bromotrichloromethane. *Agents Actions 4*, 244-253

BURLAKOVA, E.B. (1975) Bio-antioxidants and synthetic inhibitors of radical processes. *Russian Chem. Rev. 44*, 871-880

BUTLER, W.H. (1964) Acute liver injury in ducklings as a result of aflatoxin poisoning. *J. Pathol. Bacteriol. 88*, 189-196

CALLIGARO, A., CONGIU, L. & VANNINI, V. (1970) Comparative study of biochemical and morphological lesions of the liver of rats treated with tetrachloromethane and bromotrichloromethane. *Rass. Med. Sarda 73*, 365-377

CAMERON, G.R. & KARUNARATNE, W.A.E. (1936) Carbon tetrachloride cirrhosis in relation to liver regeneration. *J. Pathol. Bacteriol. 42*, 1-21

CAMPISI, J. & SCANDELLA, C.J. (1978) Fertilisation induced changes in membrane fluidity of sea urchin embryos. *Science (Wash. D.C.) 199*, 1336-1337

CARRINGTON, A. & MCLACHLAN, A.D. (1967) *Introduction to Magnetic Resonance*, Harper & Row, London

CHIO, K.S. & TAPPEL, A.L. (1969) Synthesis and characterisation of the fluorescent products derived from malonaldehyde and amino acids. *Biochemistry 8*, 2821-2827

COHN, J.A., ALVARES, A.P. & KAPPAS, A. (1977) On the occurrence of cytochrome P-450 and aryl hydrocarbon hydroxylase activity in rat brain. *J. Exp. Med. 145*, 1607-1611

EGYUD, L.G. & SZENT-GYÖRGYI, A. (1966) Cell division, SH, keto-aldehydes and cancer. *Proc. Natl. Acad. Sci. U.S.A. 55*, 388-393

FOWLER, J.S.L. (1969) Carbon tetrachloride metabolism in the rabbit. *Br. J. Pharmacol. 37*, 733-737

GLENDE, E.A. (1972) Carbon tetrachloride-induced protection against carbon tetrachloride toxicity. *Biochem. Pharmacol. 21*, 1697-1702

GOODING, P.E., CHAYEN, J., SAWYER, B. & SLATER, T.F. (1978) Cytochrome P-450 distribution in rat liver and the effect of sodium phenobarbitone administration. *Chem.-Biol. Interact. 20*, 299-310

GUPTA, D.N. (1956) Acute changes in the liver after administration of thioacetamide. *J. Pathol. Bacteriol. 72*, 183-192

HIROKATA, Y., SHIGEMATSU, A. & OMURA, T. (1978) Immunochemical study on the pathway of electron flow in reduced nicotinamide adenine dinucleotide-dependent microsomal lipid peroxidation. *J. Biochem. (Tokyo) 83*, 431-440

HOCHSTEIN, P. & ERNSTER, L. (1963) ADP-activated lipid peroxidation coupled to the TPNH oxidase system of microsomes. *Biochem. Biophys. Res. Commun. 12*, 388-394

HUGGINS, C., DEUEL, T.F. & FUKUNISHI, R. (1963) Protection of adrenal cortex by hydrocarbons against injury by 7,12-dimethylbenz(a)anthracene. *Biochem. Z. 338*, 106-113

INBAR, M., GOLDMAN, R., INBAR, L., BURSUKER, I., GOLDMAN, B., AKSTEIN, E., SEGAL, P., IPP, E. & BEN-BASSAT, I. (1977) Fluidity difference of membrane lipids in human normal and leukemic lymphocytes as controlled by serum components. *Cancer Res. 37*, 3037-3041

INGALL, A., LOTT, K.A.K., SLATER, T.F., FINCH, S. & STIER, A. (1978) Metabolic activation of carbon tetrachloride to a free radical product: studies using a spin trap. *Biochem. Soc. Trans. 6*, 962-964

JERNSTRÖM, B., VADI, H. & ORRENIUS, S. (1976) Formation in isolated rat liver microsomes and nuclei of benzo(a) pyrene metabolites that bind to DNA. *Cancer Res. 36*, 4107-4113

JOHANNESEN, K., DEPIERRE, J.W., BERGSTRAND, A., DALLNER, G. & ERNSTER, L. (1977) Preparation and characterisation of total rough and smooth microsomes from the lung of control and methyl cholanthrene-treated rats. *Biochim. Biophys. Acta 496*, 115-135

JOHNSON, M. & EDIDIN, M. (1978) Lateral diffusion in plasma membrane of mouse egg is restricted after fertilisation. *Nature (Lond.) 272*, 448-450

KAPPUS, H., KIECZKA, H., SCHEULEN, M. & REMMER, H. (1977) Molecular aspects of catechol and pyrogallol inhibition of liver microsomal lipid peroxidation stimulated by ferrous ion-ADP-complexes or by carbon tetrachloride. *Naunyn-Schmiedebergs Arch. Exp. Pathol. Pharmakol. 300*, 179-187

KELLER, F., SNYDER, A.B., PETRACEK, F.J. & SAUCIER, K.M. (1971) Hepatic free radicals in ethanol-treated and carbon tetrachloride-treated rats. *Biochem. Pharmacol. 20*, 2507-2511

LASH, E.D. (1966) The antioxidant and pro-oxidant activity in ascites tumours. *Arch. Biochem. Biophys. 115*, 332-336

MAGEE, P.N., MONTESANO, R. & PREUSSMAN, R. (1976) in *Chemical Carcinogens* (Searle, C.E., ed.) *(ACS Monograph 173)*, pp. 491-625, American Chemical Society, Washington, D.C.

MASUDA, Y. & MURANO, T. (1978) Role of cytochrome P-450 in CCl_4-induced microsomal lipid peroxidation. *Biochem. Pharmacol. 27*, 1983-1985

McCONNELL, H.M., WRIGHT, K.L. & McFARLAND, B.G. (1972) The fraction of the lipid in a biological membrane that is in a fluid state: a spin label assay. *Biochem. Biophys. Res. Commun. 47*, 273-281

McLEAN, A.E.M. & McLEAN, E.K. (1966) The effect of diet and 1,1,1-trichloro-2,2-bis-(p-chlorophenyl) ethane (DDT) on microsomal hydroxylating enzymes and on sensitivity of rats to carbon tetrachloride poisoning. *Biochem. J. 100*, 564-571

McLEAN, E.K., McLEAN, A.E.M. & SUTTON, P.M. (1969) Instant cirrhosis: an improved method for producing cirrhosis of the liver in rats by simultaneous administration of carbon tetrachloride and phenobarbitone. *Br. J. Exp. Pathol. 50*, 502-506

MISHIN, V., POKROVSKY, A. & LYAKHOVICH, V.V. (1976) Interactions of some acceptors with superoxide anion radicals formed by the NADPH-specific flavoprotein in rat liver microsomal fractions. *Biochem. J. 154*, 307-310

MILLER, J.A. & MILLER, E.C. (1974) in *Chemical Carcinogenesis*, Part A (Ts'o, P.O.P. & Di Paolo, J.A., eds.), pp. 61-85, Dekker, New York

O'NEAL, F.O. & AUST, S.D. (1978) Carbon tetrachloride activation: the role of microsomal mixed-function oxidase components. *Fed. Proc. 37*, 561

PACKER, J.E., SLATER, T.F. & WILLSON, R.L. (1978) Reactions of the carbon tetrachloride-related peroxy free radical ($CCl_3O_2\cdot$) with amino acids: pulse radiolysis evidence. *Life Sci. 23*, 2617-2620

PERKINS, M.J., WARD, P. & HORSFIELD, A. (1970) A probe for homolytic reactions in solution. Part III. Radicals by hydrogen abstraction. *J. Chem. Soc. B 2*, 395-400

PEZZUTO, J.M., LEA, M.A. & YANG, C.S. (1977) The role of microsome and nuclear envelope in the metabolic activation of benzo(a)pyrene leading to binding with nuclear macromolecules. *Cancer Res. 37*, 3427-3433

RECKNAGEL, R.O. (1967) Carbon tetrachloride hepatotoxicity. *Pharmacol. Rev. 49*, 145-208

REES, K.R. & SPECTOR, W.E. (1961) Reversible nature of liver cell damage due to carbon tetrachloride as demonstrated by the use of phenergan. *Nature (Lond.) 190*, 821-822

REUBER, M.D. & GLOVER E.L. (1970) Cirrhosis and carcinoma of the liver in male rats given subcutaneous carbon tetrachloride. *J. Natl. Cancer Inst. 44*, 419-427

REYNOLDS, E.S. & YEE, A.G. (1967) Liver parenchymal cell injury. V. Relationships between patterns of chloromethane-C^{14} incorporation into constituents of liver *in vivo* and cellular injury. *Lab. Invest. 16*, 591-603

SALERNO, J.C. & INGLEDEW, W.J. (1975) A microsomal iron-sulfur centre in rat liver. *Biochem. Biophys. Res. Commun. 65*, 618-624

SCHAUENSTEIN, E., ESTERBAUER, H. & ZOLLNER, H. (1977) *Aldehydes in Biological Systems* (Gore, P.H., transl.), Pion/Academic, London

SEARLE, C.E. (ed.) (1976) *Chemical Carcinogens (ACS Monograph 173),* American Chemical Society, Washington, D.C.

SERRES, F.J. DE, FOUTS, J.R., BEND, J.R. & PHILPOT, R.M. (eds.) (1976) *In Vitro Metabolic Activation in Mutagenesis Testing,* North-Holland, Amsterdam

SLATER, T.F. (1968) The inhibitory effects *in vitro* of phenothiazines and other drugs on lipid-peroxidation systems in rat liver microsomes, and their relationship to the liver necrosis produced by carbon tetrachloride. *Biochem. J. 106*, 155-160

SLATER, T.F. (1972) *Free Radical Mechanisms in Tissue Injury*, Pion Limited, London

SLATER, T.F. (1975) The role of lipid peroxidation in liver injury, in *Pathogenesis and Mechanisms of Liver Cell Necrosis* (Keppler, D., ed.), pp. 209-223, MTP, Lancaster

SLATER, T.F. (1976) Biochemical pathology in microtime, in *Recent Advances in Biochemical Pathology: Toxic Liver Injury* (Dianzani, M.U., Ugazio, G. & Sena, L.M., eds.), pp. 99-108, Minerva Medica, Torino, Italy

SLATER, T.F. (1978) Mechanisms of protection, in *Biochemical Mechanisms of Liver Injury* (Slater, T.F., ed.), pp. 745-801, Academic Press, London

SLATER, T.F. (1979) Mechanisms of protection against the damage produced in biological systems by oxygen-derived free radicals, in *Oxygen Free Radicals and Tissue Damage (Ciba Found. Symp. 65)*, Excerpta Medica, Amsterdam

SLATER, T.F. & SAWYER, B.C. (1971*a*) The stimulatory effects of carbon tetrachloride on peroxidative reactions in rat liver fractions *in vitro:* interaction sites in the endoplasmic reticulum. *Biochem. J. 123*, 815-821

SLATER, T.F. & SAWYER, B.C. (1971*b*) The stimulatory effects of carbon tetrachloride on peroxidative reactions in rat liver fractions *in vitro*: inhibitory effects of free radical scavengers and other agents. *Biochem. J. 123*, 823-828

SMITH, I.C.P., SCHREIER-MUCCILLO, S. & MARSH, D. (1976) Spin labelling, in *Free Radicals in Biology*, vol. 1 (Pryor, W.A., ed.), pp. 149-197, Academic Press, New York

STIER, A. (1978) Membrane fluidity, in *Biochemical Mechanisms of Liver Injury* (Slater, T.F., ed.), pp. 319-364, Academic Press, London

STOHS, S.J., GRAFSTRÖM, R.C., BURKE, M.D., MOLDEUS, P.H. & ORRENIUS, S.G. (1976) The isolation of rat intestinal microsomes with stable cytochrome P-450 and their metabolism of benzo(a)pyrene. *Arch. Biochem. Biophys. 177*, 105-116

SUEMATSU, T., KAMADA, T., ABE, H., KIKUCHI, S. & YAGI, K. (1977) Serum lipoperoxide level in patients suffering from liver diseases. *Clin. Chim. Acta 79*, 267-270

SÜSS, R., KINZEL, V. & SCRIBNER, J.D. (1973) in *Cancer: Experiments and Concepts*, Springer-Verlag, Heidelberg

TS'O, P.O.P. & DI PAOLO, J.A. (eds.) (1974) in *Chemical Carcinogenesis*, parts A and B, Dekker, New York

UGAZIO, G., TORRIELLI, M.V., BURDINO, E., SAWYER, B.C. & SLATER, T.F. (1976) Long-range effects of products of carbon tetrachloride-stimulated lipid peroxidation. *Biochem. Soc. Trans. 4*, 353-356

UTSUMI, K., YAMAMOTO, G. & INABA, K. (1965) Failure of Fe^{2+}-induced lipid peroxidation and swelling in the mitochondria isolated from ascites tumour cells. *Biochim. Biophys. Acta 105*, 368-371

VAN OETTINGEN W.F. (1955) in *The Halogenated Hydrocarbons, Toxicity and Potential Dangers (Public Health Service Publication 414),* U.S. Dept of Health, Education and Welfare, Washington D.C.

VILLARRUEL, M. DEL C., DE TORANZO, E.G.D. & CASTRO, J.A. (1977) Carbon tetrachloride activation, lipid peroxidation and the mixed function oxygenase activity of various rat tissues. *Toxicol. Appl. Pharmacol. 41*, 337-344

WHITE, G.W. & SLATER, T.F. (1974) The reaction of polyamines with aldehydic products of lipid peroxidation, in *Abstracts of 9th FEBS Meeting (Budapest)*, abstr. no. 367

WILLSON, R.L. (1978) Free radicals and tissue damage: mechanistic evidence from radiation studies, in *Biochemical Mechanisms of Liver Injury* (Slater, T.F., ed.), pp. 123-224, Academic Press, London

WILLSON, R.L. & SLATER, T.F. (1975) Carbon tetrachloride and biological damage: pulse radiolysis studies of associated free radical reactions, in *Fast Processes in Radiation Chemistry and Biology* (Adams, G.E., Fielden, E.M. & Michael, B.D., eds.), pp. 147-161, Wiley, London

WOGAN, G.N. (1976) The reduction of liver cell cancer by chemicals, in *Liver Cell Cancer* (Cameron, H.M., Linsell, D.A. & Warwick, G.P., eds.), pp. 121-152, Elsevier, Amsterdam

Discussion

Kornberg: In your experiments in which you quantified the effect of $CCl_3O_2^{\bullet}$ in aqueous solution on tryptophan, tyrosine and so on, what was actually being measured?

Slater: In these experiments the $CCl_3^{\bullet}$ radical is generated specifically by the action of the isopropanol radical

$$CH_3\dot{C}(OH)CH_3 + CCl_4 \rightarrow CH_3COCH_3 + CCl_3^{\bullet} + Cl^- + H^+$$

and its reactivity with compounds such as promethazine and tyrosine is studied by following the appearance of the characteristic absorption spectrum of the radical product. For example, promethazine reacts with $CCl_3^{\bullet}$ by a process of electron donation, to yield the highly coloured promethazine radical cation ($Pr^{+\bullet}$):

$$CCl_3^{\bullet} + Pr \rightarrow CCl_3^- + Pr^{+\bullet}$$

This radical has an intense optical absorption, so one can follow the sequence of the reaction by looking at the change in absorption. A sharp peak in the visible region appears 10–20 μs after $CCl_3^{\bullet}$ is formed in the solution. In other cases, with a substance X which does not have a characteristic optical absorption either initially or in the product we introduce competition between X and another species like promethazine where there is a highly coloured product and where we know the reaction kinetics. So this is a less direct measurement. Some of the reactions I have described for $CCl_3^{\bullet}$ have also been measured by conductivity procedures by Professor Dieter Asmus in Berlin; the results agree with those found by optical absorption. The use of pulse radiolysis has allowed us to determine the rates of reaction of $CCl_3^{\bullet}$ with various biologically important molecules for the first time. With our present data we believe that $CCl_3^{\bullet}$ by itself is not very reactive, but in the presence of oxygen it becomes much more reactive. We suggest that in the form of $CCl_3O_2^{\bullet}$ it gets trapped close to its site of formation by reacting with arachidonic acid, cysteine and so on. In the absence of oxygen, however, if our interpretations are correct it could diffuse further and get nearer to the nucleic acid sites before interacting first with oxygen and then possibly with nucleic acids. By this reasoning we could expect CCl_4 to be more reactive under partially anaerobic conditions than under fully aerated conditions when tested for mutagenicity with an Ames' test procedure. This may be, in a sense, a qualitative difference between anaerobic and aerobic systems: with aerobic conditions the $CCl_3^{\bullet}$ may be essentially trapped in its microenvironment due to the ready formation of $CCl_3O_2^{\bullet}$, whereas under largely anaerobic conditions interactions with species further away can occur.

Eley: I should like to connect these free-radical studies with the earlier discussion on conductivity in proteins. It has been shown by e.s.r. that unpaired electronic spins arising from X-ray ionization of solid proteins will diffuse at room temperature until eventually they localize on disulphide links, or failing these on oxygen atoms in $>C = O\,..\,HN<$ bridges (Gordy *et al.* 1955). We may assume the diffusing free radical species are positive holes, and there is evidence for their diffusion across molecular boundaries (Henricksen *et al.* 1963). We suggested some time ago (Cardew & Eley 1959) that an H atom (or any reactive free radical) might react with a protein to inject a positive hole into the valence band. We have a little evidence that such a process is possible, using a gas phase system (Eley & Willis 1961), but we have still to test whether, say, OH radicals in solution act in this way (we are planning to do this). I think it would be interesting to study whether the generation of OH or CCl_3 radicals in the presence of a protein in an aqueous or lipid (i.e. as in mitochondria) environment can lead to the diffusion of free spins in the protein, as studied by e.s.r.

Slater: This is a very interesting experimental idea. One of the problems, and something that we are looking at now, is to link up e.s.r. with our pulse radiolysis procedure. It can be done although there are considerable experimental difficulties. One problem is that $CCl_3^{\bullet}$ has a life-time in the μs range, which enormously complicates the collection of e.s.r. spectra. We hope, however, to add this technique to our present facility and then will certainly look at the interesting approach which you suggest.

Swartz: That experiment has been done in a way, by irradiating proteins at 96K(-196 °C). One sees initially a mixture of radicals. As the temperature is raised one sees a transfer of the unpaired spin preferentially to glycine–glycine linkages and to SH linkages. Henricksen did it too, but in a dry powder. Two interpretations have been offered. One is that one is getting migration of the unpaired spins through the protein, which is another way of saying conduction; alternatively, perhaps one is getting diffusion of small fragments. So far as I know, no one has been able to differentiate experimentally between the diffusion of something like a hydrogen atom around the protein, as opposed to something going down the chain of the protein.

Williams: A variety of experiments almost on these lines have been described. In the UK, E.M. Fielden in Manchester uses pulse radiolysis and R.C. Bray at Sussex University has also used this technique. There are groups too in Israel and in the USA. The experiment is the pulse radiolysis of water, looking for subsequent radical attack, particularly on cytochrome *c*. They don't attempt to interpret these results in terms of the conduction bands of the protein. They describe them strictly as chemical reactions between the haem iron centre and the radical diffusion.

Eley: In this connection I should like to draw attention to the work of Pecht (Pecht & Faraggi 1971), who followed the reaction of cytochrome *c* with hydrated electrons generated by pulse radiolysis in the surrounding aqueous solution, using optical methods. There is clearly a case for extending studies of this kind to solid and membrane systems.

Pethig: Professor R.B. Cundall's group are beginning to consider the possible relevance of protein and DNA conductivity in their work. They now have an electrical materials scientist in their group for this purpose.

Thomson: Wouldn't there be problems in aqueous solution, because even if you took an electron out, the solvent would surely make a considerable difference to the final effect?

Eley: Reacting free radicals with proteins in a lipid medium might be closer to the mitochondrial situation, and avoid problems arising in aqueous systems. Since I seem to be continually suggesting at symposia that free radicals will inject holes into proteins (cf. Eley 1971), it is certainly time I made a proper experimental test!

Neuberger: Professor Slater, I gather that the evidence for the existence of the complex radical $CCl_3O_2^{\bullet}$ depends on an e.s.r. spectrum. Is there any other clear evidence for the existence of this free radical? And if it does exist, has it some resemblance to the superoxide radical formed from oxygen? Does superoxide dismutase act on it?

Slater: Superoxide dismutase does not prevent the CCl_4 toxicity seen in liver microsomes (T.F. Slater & K. Cheeseman, unpublished data). Superoxide anion radicals can be formed as a secondary radical species through peroxidation pathways, but in the primary sequence of reactions, superoxide dismutase has no effect. The main evidence for the occurrence of the $CCl_3O_2^{\bullet}$ species is that if we generate $CCl_3^{\bullet}$ anaerobically by pulse radiolysis and study its reaction with arachidonic acid, promethazine, amino acids or propyl gallate, we can't detect any reaction. It may be going on, of course, but it is too slow for us to detect. If we introduce small amounts of oxygen the reaction goes rapidly. Analysis of the detailed kinetics of these interactions led Dr Packer and Dr Willson to the view that what is being formed is this peroxy radical (Packer *et al.* 1978).

Neuberger: Have you any idea from the conventional valency pattern how to write the formula of this radical?

Slater: I would write it simply as $Cl_3C\text{-}O\text{-}\dot{O}$; I presume it is behaving rather like a simple peroxy radical.

Thomson: There is independent e.s.r. evidence for the the analogous $CF_3O_2^{\bullet}$ radical, although not from irradiation of CF_4, so far as I am aware (VanderKooi & Fox 1967).

Slater: That is very interesting, although CF_4 is not, of course, toxic to cells in the same way as is CCl_4. For us, this whole notion of a $CCl_3O_2^{\bullet}$ radical is a very interesting one, because for years we have been saying how reactive $CCl_3^{\bullet}$ is, and now it seems that it is not $CCl_3^{\bullet}$ but an oxygen adduct which is the reactive species.

Williams: Could I make a suggestion about how an inorganic chemist would go about investigating such a species? We would do a matrix isolation experiment, in which you take CCl_4 at very low temperature ($<100K$) in a matrix of argon, with oxygen present. You then pulse the system with u.v., make your free radical, $CCl_3^{\bullet}$, and warm up. The oxygen already there diffuses through the argon and will make your species if it exists. Professor J.J. Turner of Nottingham University does this type of experiment.

Crumpton: I am interested in the flavoprotein that you think is responsible for the peroxidation of CCl_4. Can you say anything more about its subcellular distribution? You suggested that the peroxide radical may be formed in the nuclear membrane. Do you know whether a cell with little or no endoplasmic reticulum, such as a quiescent, unstimulated small lymphocyte, has the capacity to peroxidize CCl_4 and, if it does, whether this activity is located in the nuclear membrane?

Slater: We haven't looked at that particular cell system. We have studied (Benedetto *et al.* 1976) the distribution of the NADPH-flavoprotein in various rat tissues (not in lymphocytes) and tried to see how it reacts with substances like CCl_4. The reaction is complicated. It probably involves an SH and a non-haem iron protein component in addition to the flavoprotein. We looked for this system in the plasma membrane of rat liver cells (R. Le Page, K.

Cheeseman & T.F. Slater, unpublished data) but we have not found any activation of CCl_4 in the plasma membrane as opposed to the endoplasmic reticulum.

Thomson: In your initial step which generates $CCl_3^{\bullet}$ you described it as an electron capture, whereas Professor Dianzani suggested a mechanism which involved a chlorine atom. What exactly is the initiation step when the CCl_4 molecule changes into this radical?

Slater: I believe the initial step is dissociative electron capture, as I outlined in my paper. The important point is that one generates $CCl_3^{\bullet}$; I described the reaction as I did because it is a direct analogy to pulse radiolysis, where you form a transient $CCl_4^{-\bullet}$ radical anion.

Williams: As I understood it, you said that normal liver cells have high peroxidation activity against lipids, unlike other normal tissues, but that tumour cells (I presume from liver) had a low lipid peroxidation rate. You also said that this difference could affect cell division rate, which would be low in the normal liver cell, and high in the tumour cell. So the status of the tumour cell is different from the status of a normal cell in a particular way. This is now not the cancer initiation step, but the prolonged life of the cell and the control of its rate of division. (Of course, one way of curing cancers would be to stop their division altogether.) So, if cancer cells do have a special mechanism for division, that could be a point of attack. But is there in fact any relationship between the division rates of cells and their lipid peroxidative activity?

Slater: In the normal liver cell you do find a relatively high peroxidation rate dependent on NADPH in the endoplasmic reticulum; we know that this produces aldehydes. We also know that such aldehydes inhibit cell division when added to a variety of experimental systems *in vitro* (for references see Schauenstein *et al.* 1977); there is no controversy there. But we don't know whether the *production* of aldehydes in the normal liver cell by this peroxidative process affects cell division rate in this tissue. In tumour cells *in general*, including hepatoma cells, there is a very much depressed level of lipid peroxidation by comparison with normal rat liver endoplasmic reticulum as a standard. The study of lipid peroxidation and cell division in various types of *liver* tumour cells is something we haven't got involved in yet. It has been done by Burlakova's group in the USSR, in a large series of experiments. Burlakova (1975) has reported that cell division rate in certain tumours is correlated inversely with lipid peroxidation. We have tested the hypothesis that a high peroxidation rate correlates with a low rate of cell division (at least in a number of tumour cell lines) in *normal* rat liver cells by changing the rate of cell division, either by studying it after partial hepatectomy–taking part of the liver away—or by following it in rats over a 24-hour cycle, because during a normal day the rate of cell division in the liver changes. When we studied this with regenerating liver after partial hepatectomy or during the normal circadian rhythm, we found no correlation between the rate of peroxidation as measured by the malonaldehyde content *in vivo* and cell division (T.F. Slater, K. Cheeseman, C. Benedetto & B. Sawyer, unpublished data).

Neuberger: Do you measure potential peroxidation or actual peroxidation, because there is a glutathione peroxidase which removes peroxides?

Slater: We tried to measure actual peroxidation, by measuring *in vivo* levels of aldehyde products and of lipid peroxide (P.H. Beswick, K. Cheeseman & T.F. Slater, unpublished data). We have also done it *in vitro* by stimulating with either NADPH/ADP/Fe^{2+} or with ascorbate/Fe^{2+} and getting the maximum rate of peroxidation. In neither case do we find a correlation with mitosis in normal liver. In tissues like lung or brain, peroxidation levels are much lower than those in normal liver (Benedetto *et al.* 1976).

Connors: If there is no correlation between the mitotic rate of the various tissues studied and their rate of lipid peroxidation, this cannot be considered to be a general mechanism regulating cell division.

Slater: No. I wasn't putting this forward as a general mechanism. I was trying to stress that the distributions of lipid peroxidation and of cytochrome *P*-450 are much more restricted than the distribution of the flavoprotein.

Williams: What happens, then, if we take brain, say, where the level of lipid peroxidation in normal tissue is very low; would I still find that in a brain tumour there was an even further reduced peroxidation rate?

Slater: Perhaps yes! The general finding, mainly from the Russian literature (see Burlakova 1975) is that tumour cells peroxidize less readily than their corresponding normal cells. But the only correlation that I know between cell division *rate* and peroxidation *rate* has been reported by the Russian workers with tumour cell lines, and not in normal tissue.

Dianzani: My group in Turin studied three hepatomas, namely 5123 and 3924 A Morris hepatomas and AH-130 Yoshida hepatoma. In all three cases the rate of lipid peroxidation in homogenates incubated at 37 °C, as measured by malonaldehyde production, was extremely low by comparison with that observed in normal liver homogenates. The problem is, however, whether these hepatoma cells can be still considered as liver cells. In fact, after several transplantations in the peritoneal cavity, the hepatoma cells undergo profound and progressive changes. So one cannot be sure that the difference in lipid peroxidation between these cells and the normal ones is related to their tumour nature.

Another interesting point is that hepatoma cells are particularly insensitive to CCl_4, so their metabolizing chain must deal with this substance inefficiently. Furthermore, hepatoma cells contain substances able to inhibit lipid peroxidation in normal liver homogenates. Both a decreased efficiency of the smooth endoplasmic reticulum chain and the presence in the cell of large amounts of antioxidants may therefore explain the decreased extent of lipid peroxidation in liver tumours.

Slater: There is at least one further possibility. We know that when peroxidation occurs in rat liver after giving CCl_4, this lipid peroxidation is highly specific. It peroxidizes only the arachidonic acid of phosphatidylethanolamine (Benedetti *et al.* 1977). One explanation for this can be based on the asymmetric distribution of phospholipid within the endoplasmic reticulum (for references see Stier 1978). It is possible that around the activation site of CCl_4 there is a concentration of phosphatidylethanolamine. Peroxidation is highly specific for one type of fatty acid and for one type of phospholipid. This suggests that perhaps in a tumour cell there is some change in the heterogeneous distribution of phospholipid, and therefore one doesn't have the right lipid, so to speak, close to the locus of peroxidative radical initiation.

Dianzani: I agree. In 1968 Ruggieri and Fallani, working in Florence, showed a decrease in the arachidonic acid content of AH-130 Yoshida hepatoma cells, by comparison with normal liver cells. The decrease was especially relevant in microsomal phospholipids. Also in Turin, F. Feo *et al.* (unpublished results) confirmed this point.

References

BENEDETTI, A., CASINI, A.F., FERRALI, M. & COMPORTI, M. (1977) Early alterations induced by carbon tetrachloride in the lipids of the endoplasmic reticulum of the liver cell. II. Distribution of the alterations in the various lipid fractions. *Chem.-Biol. Interact. 17*, 167-183

BENEDETTO, C., SLATER, T.F. & DIANZANI, M.U. (1976) NADPH-cytochrome *c* reductase, cytochrome *P*-450 and NADPH-linked lipid peroxidation in microsomal fractions obtained from rat tissue. *Biochem. Soc. Trans. 4*, 1094-1097

BURLAKOVA, E.B. (1975) *Russian Chem. Rev. 44*, 871-880

CARDEW, M.H. & ELEY, D.D. (1959) The semiconductivity of organic substances. Part 3. Haemoglobin and some amino acids. *Disc. Faraday Soc. 27*, 115-128

ELEY, D.D. (1971) *Disc. Faraday Soc. 51*, 70

ELEY, D.D. & WILLIS, M.R. (1961) *The Electrical Conductivity of Solid Free Radicals and the Electron Tunnelling Mechanism (Symp. Electrical Conductivity Organic Solids)* (Kallman, H. & Silver, M., eds.), pp. 257-276, Interscience, New York

GORDY, W., ARD, W.B. & SHIELDS, H. (1955) Microwave spectroscopy of biological substances. I. Paramagnetic resonance in X-irradiated amino acids and proteins. *Proc. Natl. Acad. Sci. U.S.A. 41*, 983-996

HENRICKSEN, T., SANNER, T. & PIHL, A. (1963) Transfer of radiation-induced unpaired spins from proteins to sulphur compounds. *Radiat. Res. 18*, 163-176

PACKER, J.E., WILLSON, R.L. & SLATER, T.F. (1978) Reactions of the carbon tetrachloride-related peroxy free radical ($CCl3O_2^{\bullet}$) with amino acids: pulse radiolysis evidence. *Life Sci. 23,* 2617-2620

PECHT, I. & FARAGGI, M. (1971) The reduction of cytochrome c by hydrated electrons. *FEBS (Fed. Eur. Biochem. Soc.) Lett. 13*, 221-224

RUGGIERI, S. & FALLANI, A. (1968) Composizione lipidica delle frazioni citoplasmatiche dell' epatoma ascite di Yoshida e del fegato. *Sperimentale 118,* 313-335

SCHAUENSTEIN, E., ESTERBAUER, H. & ZOLLNER, H. (1977) *Aldehydes in Biological Systems* (Gore, P.H., transl.), Pion/Academic, London

STIER, A. (1978) in *Biochemical Mechanisms of Liver Injury* (Slater, T.F., ed.), pp. 291-364, Academic Press, London

VANDERKOOI, N. & FOX, W.B. (1967) ESR study of the photolysis of CF_3OF and OCF_3 at low temperatures: the CF_3O_2 radical in solution. *J. Chem. Phys. 47*, 3634-3639

General discussion

SZENT-GYÖRGYI'S HYPOTHESIS

Chairman's summary

Williams: I think we should start the final discussion from Szent-Györgyi's hypothesis. Dr Szent-Györgyi has considered 'life' and 'living organisms' in an overall way, looking for some organizing principle. In this symposium we have been showing how his idea of electronic controls within proteins is related to experimental tests and discussing how the idea has stood up to those tests. Many points remain untouched.

One of the general features of his idea is that organization in biology is in some way connected with electronic mobility. As a first point, I should add that this is not the only form of mobility in biological systems; nor is it the only way in which the very organization that he wants to think of could come about. It could be that electrolytic mobility of ionic systems is the important mobility. There are also other means of message transference which could operate through atomic movement, by which I mean coherent vibrational and rotational states of large molecules. All messages can be used to generate organization. Thus, although electronic mobility is one possibility, it may not be the only way in which organization can result. In this meeting, however, we are restricting ourselves to the ideas proposed by Dr Szent-Györgyi.

So, accepting his thesis, we first looked for the possibility of electronic conduction in biopolymers. Several papers dealt with the theory and experiment of conductivity in biopolymers. Theory clearly indicates a large band gap (about 10 eV) in most of these polymers—by themselves they are thermal insulators. Experiments gave band gaps of at least 3–6 eV. Those are large values and therefore the polymers in themselves are not electronic conductors, in the normal sense, but are insulators. That does not mean to say that one cannot introduce charges, in one way or another, either into the valence band or into the conduction band. If this is so, we next asked how this is to be done. At the same time we tried to determine whether the introduction of these charges brought about the type of electronic conductivity that we were led to look for. As I see it, theory gave no reason why charges should not be so introduced or why they should not be mobile. But biology is a living system as opposed to a man-made construction and the question is whether nature has assembled such an apparatus. It is important to distinguish between the theoretical support for its presence and its actual presence as determined by experiment.

The idea of introducing charges into a biopolymer involved the use of a small molecule as a donor or acceptor. Many small molecules can and do combine with large molecules in biological systems. Various small groups introduced into proteins do introduce the holes or charged states necessary for conduction: metal ions (many types of which occur in biological systems); flavin radicals; sulphur radicals and many other radicals. This does not give conduction in a protein band, however. We then asked: can these holes or other holes perhaps made by aldehydes interact with the valence band? At present, the experimental results do not establish whether they interact or not, even in isolated systems.

Eley: If I may interrupt, polyvinylcarbazole (pVK) as I said earlier is a vinyl polymer which behaves electrically not too differently from a protein. Its 1:1 complex with the acceptor trinitrofluorenone forms the basis of a commercial photocopier and there have been numerous photoconduction investigations in this area (cf. Mort & Pai 1976). For example, by studying the photoemission of holes from various electrode metals into pVK, Lakatos & Mort (1968) concluded that the highest valence band was less than 0.1 eV in width. It is believed that both electrons and holes in pVK: TNF are transported by a hopping mechanism (Gill 1972), and the drift mobilities are very low for both carriers, 10^{-5} to 10^{-8} $cm^2 V^{-1} s^{-1}$. Possibly some of the methods used in the pVK studies can be applied to proteins. At any rate, one can say that the concepts of semiconduction and photoconduction are well established for at least one synthetic polymer.

Williams: I accept that, but what I am trying to get across is the question of whether there is a biological polymer which has this property, not any polymer we care to devise. Of course we can design and make polymers that will conduct and have useful applications. So far, no experimental results reported here have convinced me that such mobility has been seen in proteins.

Eley: There is certainly one fairly clear-cut case of a semiconduction mechanism in biology. Digby (1965, 1967) has shown that the outer skin of certain crustacea is a highly conducting quinone–protein complex. His results show that a current passing from inside to outside of this membrane can cause a pH change on the surface which leads to the deposition of calcium carbonate. Digby (1966) has suggested a similar mechanism for the calcification of mammalian bone.

Williams: The results can be interpreted in terms of transfer between the quinone molecules; that does not invoke transfer along the backbone of a protein or of a polynucleic acid, which is the essence of what we are looking for—namely, the insertion of holes into the protein band. Your system could work perfectly well too on a 'hop' mechanism between the organic side-chains. Furthermore, the amount of quinones that lobsters and crabs have is not a common feature of all cells and, therefore, this cannot constitute a common, organizing, principle of cells.

To return to my theme: I am not refuting Dr Szent-Györgyi's ideas; they stand on good theoretical grounds. We are enquiring whether they have been substantiated in biology. As yet, no firm evidence has been presented here for such charge-transfer interactions with the backbone of proteins or DNA.

Another aspect stressed in the discussion is that *particular* small molecules have to be chosen as of special importance. Those selected, particularly by Dr Szent-Györgyi with his marvellous intuitive sense of where to look, were aldehydes and oxygen. We have not said much about oxygen, considering it more as a nuisance in most of the experiments because it generates radicals indiscriminately. However, oxygen is able to diffuse; it is one of the few radicals that can diffuse completely through a protein and certainly could lie right up against a polypeptide chain. Proteins are sufficiently mobile and sufficiently open for oxygen to gain access to any part of any protein. Oxygen is an excellent candidate for the introduction of a hole into the valence band. Molecules related to it, such as $OH^{\bullet}$, must also be exceptionally good candidates for this function. We chose to neglect such entities mainly because they are far too reactive and far too difficult to study. We were restricted, therefore, to the other possibility, which was suggested by Dr Szent-Györgyi as being present in life before the advent of oxygen. The basic idea is that the hole systems and the charge-transfer interactions with the protein came from special aldehydes, particularly the glyoxals. Consequently, we examined whether methylglyoxal interacted with proteins by means of a charge transfer. In certain circumstances various pieces of spectroscopic evidence (absorption or electron magnetic resonance spectra) were taken to support this view. The problem resided in relating the observations to the possibility of electron mobility within a valence band of a protein or of DNA. Again, as far as I understood the results, the experimental observations are insufficient to establish that point. They establish only that there are such e.s.r. and absorption spectra, but whether they are connected with the energy states of the backbone of DNA, RNA or proteins is unproven.

In chemical studies, general effects of aldehydes reacting with proteins were observed and several people here have tried to resolve the nature of the aldehyde interactions with proteins. This work led to a discussion of what these aldehyde compounds were. In some cases their identity has been established; in other cases, that seems to be a difficult task. The presence of ascorbic acid often affected whatever result was obtained—and that brings in another of Szent-Györgyi's ideas.

So far, I have dealt with the first part of the conjunction that Dr Szent-Györgyi has made between submolecular events and life. In his hypothesis cancer cells have lost the organizing principle common to all other cells, which we have looked for. So in some way one would expect cancer cells to be different physically. Some possibilities that we considered are that they might lack conducting proteins; or they might lack the charge-transfer and e.s.r. signals suggestive of these conducting systems. Therefore, we were right to look for those details within the cancer cells and compare them with normal cells. One could argue that the trouble with the procedure is that at the level at which this might be done there would be, say, only 10^{-20} species to be detected—so low that one could hardly detect anything. That, however, is always a false argument: a hypothesis must always be tested by experiment. If the hypothesis is untestable, one needs another hypothesis. So the question remains: did we find something different about the organization of cancer cells? It emerged from our discussions that in some ways cancer cells did in fact differ— in *bulk* properties. Damadian's results suggested that cancer cells had lost some of their organization. Their water molecules appeared to be more 'free'. There is something in these n.m.r. results of great interest, for it has not been explained here why a cancer cell should be like this. Has it anything to do with Szent-Györgyi's hypothesis?

Let us accept for the moment that the cancer cell is different and that Dr Szent-Györgyi has the right principle about the way it differs from the normal cell. Then, obviously, the answer to cancer (as given by Szent-Györgyi) is that we should cure the cancer cell by normalizing it with added compounds. Such added compounds, by the nature of this argument, could be aldehydes. We should then introduce aldehydes as drugs in the hope of generating charge-transfer interactions, thereby reintroducing the specific organizing principles which come from the conductivity or mobility of the electrons within proteins. But would any compound ever do that? On the basis of our discussion we can only say that, although aldehydes have certain effects on tumour cells, many other chemicals affect them as well and there is no direct connection between these results and the earlier ideas which we discussed.

Finally we have what I call, with all due respect to Professor Slater, a side-issue in our discussions: what free radicals do in biology, particularly with respect to cancer. Free radicals are extremely reactive species. They came into the picture because a hole in a protein (or anywhere else) might be detected as a free radical, and thus it was clearly right to think about the properties of free radicals. It was pointed out that they are electrophiles, which are known to be a major cause of cancer—how, we don't know. I called this a side-issue for us because it does not impinge on the argument that we came here to talk about. Cancer may be induced in various ways (see pp. 19-31). As I explained, a normal cell is then transformed. The transformed cell then enters a state of disorganized growth. It is the disorganization which our discussion has tackled and not the transformation step.

I hope that I have shown the lines of reasoning, as I see them, which have run through the symposium. Against each idea I have tried to set the major pieces of experimental evidence. I hope I have been fair.

Lewis: May I take up the point about untestable theories? It seems to me that the way in which we are likely to make progress in testing some of these electronic concepts is to move away from the already established techniques and ideas of biochemistry towards those which, in spite of the title of this symposium, we have discussed but superficially at this meeting. To establish a conduction model we must do conduction experiments, but these involve difficult practical problems and we need to devise better methods and procedures than are available at present. This is especially so if we want to confirm specifically that electron transfer occurs. Fortunately, experimental technique usually improves out of all recognition once it is resolved that a hypothesis is worthy of test.

At the same time we must begin to use some of the theoretical concepts which are not yet commonly used in biology but are already firmly established in solid-state physics. I refer particularly to two areas: the electronic states of disordered structures and one-dimensional solids. I realize that although biopolymers might be disordered in the crystalline sense they do not form strictly one-dimensional systems. They do, however, have a backbone and along such a backbone the latest theoretical calculations predict an easy movement of electrons or holes. We must bring such ideas into biology.

Another important consideration when weighing the validity of the electron transfer concept in biopolymers can best be illustrated by introducing its inorganic counterpart. The essential core of any modern electron device capable of at least artificial intelligence is a minute silicon crystal. Analysed by acceptable chemical and spectroscopic procedures, the crystal would be revealed as extremely pure silicon with only traces, perhaps almost undetectable traces, of elements close to silicon in the periodic table. The analysis would lead to the conclusion that the crystal did not have any obvious property likely to approach that of a human brain. The hidden properties of this crystal can be discovered only by the right electronic measurements based on an appropriate supporting theory. By analogy, is it not possible that the agents which determine whether a cell should be normal or malignant may be not only below the concentration levels we are accustomed to expect in biological systems but also as a consequence act by mechanisms not yet detected?

On the question of detectable concentrations of the active agent, whatever that turns out to be, there is one feature of Szent-Györgyi's idea of an acceptor causing desaturation of the valence band of a protein which is very appealing. It is that, once a mobile hole is created in the valence band, it could move away from the acceptor site along a pathway which extends over relatively long distances. Along such a pathway, holes could pass in a continuous stream for as long as the pathway is kept open by the acceptor. The acceptor then becomes a 'switch' in the pathway, controlling the flow of holes but not in itself being involved. The electronic influence of this 'switch' could then be out of all proportion to its chemical concentration in the system. At present we lack the tools to test this principle conclusively, but it is a well-established principle in solid-state electronics.

March: With regard to the possible importance of disorder: in one dimension, in the presence of disorder, all the electron states are localized. As Professor Lewis stressed, biological systems are not one-dimensional but, in three dimensions, beyond a certain degree of disorder the hole will be localized. It is important therefore that experiments are done with varying degrees of disorder.

Ladik: Professor Szent-Györgyi often says that we can control only what we understand. This explains how I came to physics after being trained as a chemist. In my opinion physics is not a 'field'; it is a method. I would put theory neither above nor below experiment. There seems to be a belief amongst experimentalists that only what is measured is true. That is only so if the experiment is good and the results are interpreted according to a good theory. Interplay between theory and experiment is vital for understanding. During our discussions I have been saddened by the lack of interplay between the theoretical work and the biochemical work, which is excellent in its own right, or between the physical measurement of conductivity and biochemical experiment. Neither theory nor experiment by itself can explain this very complex system. Work on inorganic semiconductors in solid-state physics has led to the discovery of new types of collective states. These are much simpler systems than biopolymers. Therefore one can expect still more new types of collective states in biopolymers. What is now untestable or inaccessible to experiment may be commonplace in a few years' time.

Laki: As Professor Williams has pointed out, much research is aimed at killing cancer cells but that is not the approach that Dr Szent-Györgyi proposes; he proposes to help the cancer cells to return to normal. Occasionally, as we know, some cancer cells recover spontaneously, so it should be possible to help cancer cells to recover. According to Dr Szent-Györgyi, a conducting protein is missing from cancer cells and if we replaced it, the cells might return to normal.

Phillips: At the beginning of the symposium Dr Szent-Györgyi pointed out the danger, of which I believe all biochemists are very much aware, of isolating from living systems the component molecules and studying them by themselves. Nevertheless, that sort of work has some achievements to its credit and that partly results, I believe, from the fact that, aware of these dangers, biochemists take great trouble to try to ensure that the isolated molecules or molecular systems that they study remain in something like a native state. Then they try to relate the properties of these isolated molecules back to those properties which can be studied in the native state: for example, the catalytic activity of an enzyme in an isolated pure state in a very dilute solution as compared with the known catalytic properties of that enzyme in cells. Let me say, perhaps somewhat provocatively, that none of the specimens we have been told about in the experiments on the conductivity of proteins seems to me to be conceivably relevant to biology. We have heard about proteins which can only be extremely denatured; we have heard them described as being, for example, 'dry', when they cannot conceivably contain less than 8 or 10% of liquid if they are to be in anything like a native biological condition; we have heard of a protein described as 'collagen', which can only really be analogous to raspberry jelly. I cannot believe that any of these studies have any relevance to biology.

On the other hand, I agree with Dr Ladik's view that one wants to achieve understanding through the interplay of theory and experiment. Conversely, we have heard nothing which suggests to me that the workers in this field pay attention to the understanding of the nature of protein structure which is coming from other types of work. I wonder whether they realize how much is now understood about the control of biological processes through the complicated nature of protein molecules and whether they have established the need, at the present level of understanding, for additional complicated properties taken over from solid-state physics? What is now known about biological structures needs to be taken much more into account, perhaps even before one begins to worry about these additional properties.

Pethig: I stressed in our paper that as a *first* step we wanted to investigate the possibility that methylglyoxal could react with proteins to make them more conductive. I believe we have obtained some experimental support for this concept. I also raised the problem (p. 84) of the possible biological relevance of the test samples. Apart from the points mentioned there, I should add that our starting dry lysozyme material possessed a high degree of enzymic activity (as specified by the Sigma Chemical Co.) and that we have made measurements on proteins with water contents ranging up to 30% by weight. To describe our collagen samples as being analogous to 'raspberry jelly' is not correct, since the collagen still retained some of its native structure. Our measurements at microwave frequencies are designed specifically to investigate those portions of the protein molecules that have remained 'intact' through our preparation procedures, and our present studies have gone beyond compressed proteins; we are studying the dielectrophoretic behaviour of cell suspensions in non-uniform electric fields in an attempt to study the conductivity of membrane-bound proteins in their natural state.

As Professor Eley indicated, a host of conductivity measurements have been made for dry proteins which, I agree, will have structures that differ from their biologically active ones, but despite that they all exhibit remarkably consistent conduction properties. This 'baseline' conductivity can be modified by the presence of methylglyoxal, and the basic underlying conductivity and associated energy-band structures should still be present in the natural protein forms. On the effects of protein structure, Nakahara *et al.* (1977) have concluded that there is a reproducible 100-fold difference in conductivity between the oxidized and reduced form of cytochrome *c*. Finally, Professor Phillips, we do try to keep abreast of developments in protein work—for example, we are aware of the work in Dickerson's laboratory (Takano *et al.* 1973; Mandel *et al.* 1977) on the conformational changes in cytochrome *c*, such as the movement of the phenylalanine-82 side-chain, when the protein is reduced.

Phillips: But that illustrates my point. Dickerson's idea of a moving phenylalanine group (Takano *et al.* 1973) was abandoned at least two years ago (Mandel *et al.* 1977). To keep abreast of the literature in such a field needs a great deal of effort.

Lewis: We are acutely aware of the fact that work on dry compressed proteins is not the same thing as work on proteins in cells and that the next generation of experiments must establish conditions closer to those within cells. For example, an attempt will be made to determine whether in certain circumstances an electronic pathway can be established via protein in artificial membranes.

Eley: I agree that the drying of proteins *in vacuo,* at say 100 °C, is a fierce procedure, which in most cases will denature the protein. However, the electrical properties of such dried proteins provides a baseline for further work, as Professor Lewis has also indicated. At a fairly early stage we established that the conductivity of haemoglobin did not change a great deal as it was progressively denatured, as judged by its absorption spectrum when taken into solution (Eley & Spivey 1960). This is perhaps surprising, but of course haemoglobin is not active in biological electron transfer. Another possible approach is to investigate the effects of inhibitors on the enzymic and conductivity behaviour of electron-transfer enzymes. We earlier had some success in this direction (Eley *et al.* 1972). In view of our present uncertainties about the microwave Hall technique used at the time, there is need for further studies in this direction. I repeat that all proteins seem to have a definite, very low conductivity in the dry state, which is increased either by electron donors such as water, or acceptors such as quinones and methylglyoxal. We are still a long way from demonstrating any biological significance of such conductivity behaviour. But I believe that a case had already been made as long ago as my last review (Eley 1968) for the need for increased activity in this field, and the recent work of Professor Lewis and Dr Pethig at Bangor seems to bear this out. When it becomes possible to study single crystals of biopolymers we shall be able to measure anisotropic effects and get more information on the importance or otherwise of hydrogen-bonding systems in the observed conduction. Thermoelectric power looks a fairly hopeful technique. In general terms, we need methods which will independently assess electron and hole mobilities in protein systems, with the detail we now possess for polyvinylcarbazole. The ability to study biological systems using electrodeless techniques is another area where advances are necessary. Monomers and prosthetic groups, biopolymers, and organized membranes and organelles, all require further investigation. Really, in relation to the magnitude of the task, as I see it, the present level of activity is very small, and I sincerely hope this conference will help to bring further workers into the field.

Williams: It is clear now what the problems are in the discussion of protein conductivity. I should like to move on to the way in which particular types of molecules could be used to introduce conduction into polymers.

Thomson: Both you and Professor Phillips have raised a point which I touched on in my paper. In the interactions of various amines with various ketoaldehydes, including methylglyoxal, although the spectral and e.s.r. changes are well established, we still are not certain of the precise chemical structure of the species that gives rise to the e.s.r. signals that have been described. Dr Szent-Györgyi has investigated several simple systems (Kimura & Szent-Györgyi 1969; Kon & Szent-Györgyi 1973) which show charge-transfer behaviour. I feel that reactions of some of these simple amine or diamine systems with methylglyoxal could bear more careful chemical characterization, to throw more light on the details of charge-transfer processes and other possible reactions, which still remain unclear.

Williams: I have done experiments on the conductivity of simple inorganic solids which were deliberately designed to contain charge-transfer characteristics (Williams 1965) and I doubt that this type of attack will be successful. I never found a good electronic conductor by this research.

Eley: What do you mean by good?

Williams: The conductivity was always below about 10^{-12} ohm^{-1} cm^{-1}. The activation energies were always of the order of 1 eV, and rarely lower.

Lewis: One has to ask how much conductivity is needed in a biological situation. How many electrons or holes per second might need to travel a given route to maintain a given biological process or communicate a biological signal? We do not know these numbers and until we are sure of our ground it is dangerous to assume that the levels of conductivity must be those to which we are accustomed in, say, electrolytes or conventional semiconductors. It should also be remembered that, in addition, restricted or localized conduction can give rise to long-range polarization, the ordering influence of which can greatly control cell structure and organization.

Williams: We don't know the answer to this question. With regard to the chemistry and the biological effects of the aldehydes, we have had a lot of discussion. Can we try to clear up any outstanding points?

Laki: I am wondering whether we can derive any useful information about the drugs which we assume cure cancer by putting both the cancer and the drug in the peritoneal cavity. (That would demonstrate that the drug kills the cancer cell.) However, if one implants intraperitoneally a tumour which rapidly makes metastases, addition of the drug intraperitoneally is all right, because the tumour has already entered the liver, spleen, lung and so on. If the drug works then, it really is curing the cancer and not just killing cancer cells. Almost anything kills cancer cells, but will not cure the cancer.

Connors: There are many different ways of evaluating anticancer drugs, ranging from simple tests where the drug is added to tumour cells *in vitro* to more 'relevant' tests where the drug is administered to animals with disseminated cancer. If one were to obtain positive results by the intraperitoneal injection of a drug, then, since this is a very sensitive test, further tests would be made where the tumour was transplanted in a different site or where the drug was administered by a different route.

Kornberg: May I clarify one point about the work with aldehydes? Although I am not an expert on cancer or epidemiology, I think I am right in saying that most human cancers are cancers of epithelial cells. Therefore, one would expect that whatever system is seen to be disorganized would be one which would have to apply certainly to epithelial cells. Although Dr Brandt presented some evidence that D-lactic acid could be formed in blood, nevertheless there is little, if any, reliable evidence for an enzymic path for the formation of methylglyoxal in epithelial cells from any breakdown product of carbohydrates. However, these aldehydes can be formed, as Professor Slater described, through the lipoperoxidase system, but (as we heard) that system is largely absent from epithelial cells. It seems to me that, if it is a crucial feature of the theory that the disorganization should be rectified by the biological provision of aldehydes, there must be a way of forming these aldehydes. And what is more, there must be a way of forming these aldehydes at a rate which is adequate to balance the known action of the systems that remove aldehydes—the glyoxalase system. Is there any evidence for this?

Slater: There is not much information on the endogenous concentrations of aldehydes in different types of tissues. That is something that we shall be investigating in the near future. With regard to lipid peroxidation, I reported results on the peroxidation that depends on NADPH. In tissues such as muscle and brain that type of peroxidation is hardly detectable but there are other routes for peroxidation. One such, which we are now studying, depends on ascorbic acid. With ascorbic acid we observe relatively high rates of peroxidation in brain and muscle (M. Ahmed, P. Brunswick & T.F. Slater, unpublished data).

Professor Williams referred to my results, lightheartedly, as a side-issue to Dr Szent-Györgyi's main theme. However, before the symposium I deliberately chose to try to draw attention to another aspect of Dr Szent-Györgyi's work rather than to discuss our results on the anti-tumour testing of aldehydes (see, for example, Conroy *et al.* 1975, 1977; Elvin *et al.* 1979) or on the effects of various inhibitory agents on glyoxalase activity (M. Ahmed, T.F. Slater & K. Cheeseman, unpublished data). Dr Szent-Györgyi has concentrated over the years some attention on the role of free radicals and of aldehydes. I tried, deliberately as a 'side-issue', to point out other aspects of the activity of free radicals in relation to cancer, to peroxidation and to aldehydes.

Williams: I should not, of course, have used the term 'side-issue'! Your work has great interest but it was not quite central to the general discussion. We could continue this discussion by asking what the nature of cancer is. But the answer is, really, that we do not know. It may be a multitude of different things and we have only many leads in our search for an answer. Again, with regard to the initiation of cancer, one could start a discussion from many different points—free radicals included. But we still don't understand, generally, what initiates tumours. I think that at this moment we should rather turn to Dr Szent-Györgyi for his overall, concluding impressions.

Szent-Györgyi: I am embarrassed, finding myself suddenly turned into a defendant by our chairman's criticism, and my brain works very slowly. One fact that is clear to me is that, after about 100 years and 3000 million dollars, we understand cancer no better than Rudolph Virchow did a century ago. This fact is proof to me that the direction now followed by the thousands of research workers is misguided and not fruitful. (It is said that cancer keeps alive more people than it kills.) Extrapolation leads nowhere. I see no chance for us to conquer cancer so long as we fail to understand what a cell is. How can I differentiate between a normal cell and a cancer cell when I don't know what a cell is? So, for a long time, in my group, we have tried to forget about cures. Research has been led prematurely in its search by the enormous amount of suffering caused by cancer: in the USA alone, 800 people die daily from cancer. This has derailed research and has made workers look for a cure before understanding. But how can we hope to cure cancer when we don't know what the essential difference is between a pebble and a cell? Cancer research should start with asking what life is, and what are the main parameters of the living state. One has to start building at the foundation and not at the roof. The problems that Professor Phillips has outlined are not new to us—we currently have several people working on them. Electronics have changed human life. A whole issue of *Science* this year was devoted to the 'electronic revolution' of human life. We are actually living through a profound electronic revolution. Human history has been changed by electronics. That nature should have neglected the wonderful possibilities of electronics when creating life seems impossible to me. Yet electronic aspects of life have remained entirely disregarded, neglected by biology. Biologists have taken no notice of electrons.

As I said, our aim was, and is, to be able to differentiate (in terms of the electron) between animate and inanimate, and then to go to a lower level of organization and hope to understand what a normal cell is, in terms of electronics.

References

CONROY, P.J., NODES, J.T., SLATER, T.F. & WHITE, G.W. (1975) Carcinostatic activity of 4-hydroxy-2-pent-en-1-al against transplantable murine tumour lines. *Eur. J. Cancer 11,* 231-240

CONROY, P.J., NODES, J.T., SLATER, T.F. & WHITE, G.W. (1977) The inhibitory effects of a 4-hydroxypentenal:cysteine adduct against sarcoma 180 cells in mice. *Eur. J. Cancer 13,* 55-63

DIGBY, P.J.B. (1965) Semiconductivity and electrode processes in biological material. I. Crustacea and certain soft-bodied forms. *Proc. R. Soc. Lond. B Biol. Sci. 161,* 504-525

DIGBY, P.J.B. (1966) Mechanism of calcification in mammalian bone. *Nature (Lond.) 212,* 1250-1252

DIGBY, P.J.B. (1967) Calcification and its mechanism in the shore-crab, *Carcinus maenas* (L.) *Proc. Linn. Soc. Lond. 178,* 129-146

ELEY, D.D. (1968) Semiconducting biological polymers, *in Organic Semiconducting Polymers* (Katon, J.E., ed.) p. 259-294, Dekker, New York

ELEY, D.D. & SPIVEY, D.I. (1960) Semiconductivity of organic substances. Part 6. A range of proteins. *Trans. Faraday Soc. 56,* 1432-1442

ELEY, D.D., MAYER, R.J. & PETHIG, R. (1972) Microwave Hall mobility measurements on heavy beef heart mitochondria. *Bioenergetics 4,* 389-402

ELVIN, P., CONROY, P.J. & SLATER, T.F. (1979) *Eur. J. Cancer,* in press

GILL, W.D. (1972) Drift mobilities in amorphous charge-transfer complexes of trinitrofluorenone and poly-n-vinylcarbazole. *J. Appl. Phys. 43,* 5033-5040

KIMURA, J.E. & SZENT-GYÖRGYI, A. (1969) Donor-acceptor interactions of nitrogen. *Proc. Natl. Acad. Sci U.S.A. 62,* 286-288

KON, H. & SZENT-GYÖRGYI, A. (1973) Charge transfer between amine and carbonyl. *Proc. Natl. Acad. Sci U.S.A. 70,* 3139-3140

LAKATOS, A.I. & MORT, T. (1968) Photoemission of holes from metals into the organic polymer poly-*N*-vinyl-carbazole. *Phys. Rev. Lett.* 1444-1447

MANDEL, N., MANDEL, G., TRUS, B.L., ROSENBERG, J., CARLSON, G. & DICKERSON, R.E. (1977) Tuna cytochrome *c* at 2.0 Å resolution. *J. Biol. Chem. 252,* 4619-4636

MORT, J. & PAI, D.M. (1976) *Photoconductivity and Related Phenomena,* Elsevier, Amsterdam

NAKAHARA, Y., KIMURA, K. & INOKUCHI, H. (1977) Electrical conductivity of cytochrome *c* anhydrous film. *Chem. Phys. Lett. 47,* 251-254

TAKANO, T., KALLAI, O.B., SWANSON, R. & DICKERSON, R.E. (1973) The structure of ferrocytochrome *c* at 2.45 Å resolution. *J. Biol. Chem. 248,* 5234-5255

WILLIAMS, R.J.P. (1965) in *Non-Heme Iron Proteins* (San Pietro, A., ed.), pp. 7-21, Antioch Press, Chicago

Index of contributors

Entries in **bold** *type indicate papers; other entries refer to discussion contributions*

Abdulnur, S. F. 48, **195**, 205, 206

Ball, J. R. **143**
Bondy, P. K. 126, 221, 242, 269
Bone, S. **83**
Brandt, R. B. 98, 129, 138, 139, 192, **211**, 220, 221, 222, 223, 266, 300
Butterich, J. **165**

Connors, T. A. 47, 267, 326, 338
Conroy, P. J. 191, 208, **271**, 299, 300
Crumpton, M. J. 100, 104, 187, 188, 324

Damadian, R. 79, **131**, 138, 139, 140, 141
Dianzani, M. U. 125, 221, 241, 242, **245**, 266, 267, 268, 326, 327

Eley, D. D. 43, 44, 45, 59, 61, 78, 79, 81, 82, 102, 103, 140, 160, 161, 189, 207, 322, 323, 330, 336, 338
Esterbauer, H. 170, 205, 206, **225**, 241, 242, 243, 244, 266, 298, 299

Fodor, G. 47, 49, **165**, 170, 171, 172, 173, 207, 243, 267
Fox, S. W. **175**, 188, 189, 190

Goldsmith, M. **131**

Kertesz, J. C. **33**
Kornberg, H. L. 48, 127, 141, 171, 172, 173, 268, 321, 338

Ladik, J. J. 44, 48, **51**, 58, 59, 60, 61, 62, 78, 80, 81, 98, 101, 102, 125, 126, 128, 162, 205, 206, 207, 334
Laki, K. **33**, 43, 45, 46, 47, 48, 49, 61, 104, 242, 335, 338
Lewis, T. J. 48, **65**, 79, 80, 81, 105, 333, 336, 338

March, N. H. 60, 162, 207, 334
Minkoff, L. **131**
Mujumdar, R. **165**

Neuberger, A. 43, 47, 189, 190, 191, 192, 205, 207, 221, 222, 244, 269, 323, 324, 326

Pethig, R. 46, 48, 62, 80, **83**, 98, 99, 100, 101, 104, 127, 139, 206, 207, 323, 335
Phillips, D. C. 61, 80, 101, 102, 104, 128, 206, 335, 336
Purcell, E. M. 164, 241

Reslova-Vasilukova, S. **19**

Schauenstein, E. **225**
Siegel, S. A. **211**
Slater, T. F. 128, 140, 162, 163, 169, 170, 171, 172, 174, 190, 208, 221, 242, 265, 268, 270, **301**, 321, 322, 323, 324, 325, 326, 327, 339
Suhai, S. **33**
Swartz, H. M. 98, 100, **107**, 129, 163, 164, 322

Syren, R. M. **175**
Szent-Györgyi, A. **3**, 127, 340

Thomson, C. **143**, 161, 162, 163, 164, 323, 324, 325, 337

Williams, R. J. P. **1**, **19**, 42, 44, 45, 47, 48, 49, 58, 62, 63, 81, 82, 98, 99, 101, 103, 104, 105, 124, 126, 127, 129, 138, 140, 170, 171, 172, 173, 174, 189, 190, 191, 192, 205, 206, 207, 208, 220, 222, 243, 244, 267, 268, 269, 299, 300, 323, 324, 325, 326, 329, 330, 331, 337, 338, 339
Windsor, C. R. **175**

Indexes compiled by M. R. Hardeman

Subject index

ab initio **SCF-MO calculation**
  quantum chemistry 149, 160
acetylacrylaldehyde
  see 4-oxopent-2-enal
aldehyde
  cancer chemotherapy 332
  capillary wall permeability 261
  carbon tetrachloride 246, 261
  glucose-6-phosphatase 257
  ligandin 260
  lysosome 259
  microsome enzyme 256
  mitochondrion respiratory rate 259
  mitochondrion swelling 260
  mitotic index 255
  protein synthesis inhibition 250
  tubulin activity 253
aldehyde–cysteine adduct
  dissociation 235
aldehyde–RSH adduct
  thymidine incorporation, inhibition 238
animal tumour
  carcinogen exposure time, relationship 25
animate and inanimate
  difference 4, 340
ascorbic acid
  charge transfer 154, 165
  evolutionary development 15
  glyoxalase I inhibitor 180, 190
  methylglyoxal carrier 165
  model system 154, 163
  reaction with 4-oxopent-2-enal 167
  tumour tissue, e.s.r. spectrum 121
ascorbic acid–methylglyoxal acetal
  biological reactivity 169
  ^{13}C n.m.r. data 166
  ^{1}H n.m.r. data 166

bioelectronic theory
  cancer 3, 340
biopolymers
  aperiodicity 56
  band structure calculation methods 51
  coherent potential approximation 56
  crystal orbital 51
  dark conductivity 59
  electron correlation 55
  linear combination of atomic orbitals 51
  local perturbation 56
  quantum mechanical treatment 51
  self-consistent-field method 51
bovine serum albumin
  e.s.r. spectrum 84
bovine serum albumin–methylglyoxal complex
  conductivity and dielectric relaxation time, temperature and 92
  conductivity, frequency variation 90

cancer
  bioelectronic theory 3, 340
  carcinogenesis, maintained state 124
  epithelial cell 338
  e.s.r. study 107, 118
  free radicals and 107
  n.m.r., prolonged relaxation 138
  n.m.r. scan, resolution 140
  peculiarities 20
  quantum chemical investigation 143
  submolecular biology 19
  theory 16, 30
  tissue water, dielectric constant 138
cancer disposition
  inheritable factors 29
  internal 'horizontal' changes 29

cancer initiation
change in DNA 24
mutation 23
cancer treatment
Haddow's Paradox 28, 171
carbon tetrachloride
aldehyde 246
effect on mammalian cell 245
e.s.r. spectrum 311
lipid peroxidation 246
malonaldehyde production, hepatocyte 247
'remote' damage 262
carbon tetrafluoride
324
carcinogenic substance
charge transfer interference 8
carcinostatic agent
lysogenic bacterium 173
new products 163
poly(amino acids) 185
charge transfer
ascorbic acid 154, 165
dimerization 8
dimethylglyoxal–formamide 153
ethylglyoxal–formamide 153
formamide–ketoaldehyde 149
α-hydroxytetronic acid–formamide 156
ketoaldehyde 148
methylglyoxal–formamide 150
permittivity 6, 7
perylene–chloranil 84
protein 33
protein–methylglyoxal 84
solvent dielectric constant 7
charge-transfer interactions
current views 144
quantum chemistry 143
quantum mechanical study 146
self-consistent-field molecular orbital method 146
'supermolecule' approach 146
chemical carcinogenesis
agents causing 302
free-radical mechanism 301, 303
transient intermediates 301
chemical reactivity
diffusion radius 314
crotonal–cysteine adduct
therapeutic action, mouse tumour 236
cytochrome *P*-450
liver cell distribution 307

defence mechanisms
evolution 26
dielectric loss factor
frequency and temperature variation 91
4,5-dihydroxypentenal
formation from 2-deoxyribose 239
dimerization
charge transfer 8
7,12-dimethylbenzanthracene
e.s.r. spectra of breast tumours 114–117, 121
dimethylglyoxal
charge transfer, formamide 153
2,4-dinitrophenylhydrazine
protein-bound aldehyde detection 228
DNA
electronic structure 51
methylglyoxal binding 208
solid-state aspects 51
doping
intramolecular charge transfer 7

Ehrlich ascites tumour cell
4-hydroxypentenal 230
Ehrlich carcinoma
ascitic form 287
methylglyoxal, continuous infusion 293
solid form 285
electrical conductivity
protein 4, 35, 44
electron mobility
6
protein 330
electron spin resonance spectroscopy
bovine serum albumin 84
oxygen effect 116
sample lyophilization 115
Emanuel universal kinetic curve
reproducibility 118
ethylglyoxal
charge transfer, formamide 153

FONAR (field-focusing n.m.r. spectroscopy)
131
FONAR image
human chest phantom 135
live human chest 136
formamide
charge transfer, ketoaldehyde 149
free radicals
biological role 332
cancer, historical review 107
damaging reactions 304
pulse radiolysis study 309
spin-trap technique 309

Gardner lymphosarcoma
methylglyoxal, continuous infusion 293
glutathione
glyoxalase 212
reaction with 4-hydroxypentenal 227
thymidine incorporation 280, 298
glyoxal
α-hydroxytetronic acid, charge transfer 157
glyoxalase
glutathione 212
D-lactic acid formation 10
zinc content 190
glyoxalase enzyme system
scheme 212
glyoxalase I
interaction with inhibitor 184
glyoxalase I inhibitors
amino acid ratios 178

ascorbic acid 180, 190
interaction with glyoxalase I 184
light absorption 189
(poly)tryptophan 177
protobiogenesis and 185
reaction kinetics 183
structure versus activity 184
tetronic, pentatonic polymers 179
thermal copoly(amino acids) 175
tritonic copoly(amino acids) 177

growth
controlling factors 22

Haddow's Paradox
cancer, 28, 171

haemoglobin
hole transition 78

hepatoma
lipid peroxidation 326

homopolynucleotide
excitonic effect 54
valence and conduction bands 54

homopolypeptide
electronic state, density 38

hopping electron
protein–methylglyoxal complex, model 94

HPE
see trans-*4-hydroxypent-2-enal*

4-hydroxyalkenal
antitumour activity 225

trans-**4-hydroxypent-2-enal**
carcinostatic agent 271
DNA precursor incorporation 277
effect on Ehrlich ascites cell 230
effect on normal rat organs 231
effect on transplanted tumour 231
effect on tubulin 254
formula 226
reaction with cysteine 227
reaction with glutathione 227
reaction with glyceraldehyde-phosphate dehydrogenase 229
reaction with SH group 226, 228
therapeutic action, mouse tumour 233
thymidine incorporation inhibition 237
tumour doubling time 232

4-hydroxypentenal–cysteine adduct
carcinostatic activity, mouse 233, 289
thymidine incorporation 278

α-hydroxytetronic acid
ascorbic acid model 155, 162
charge transfer, glyoxal 157
formamide, charge transfer 156

idiosyncratic cellular balance
concept 30

in vitro **study**
biological relevance 335

ketoaldehyde
charge transfer 148

D-lactic acid
determination 214
human plasma, mean concentration 215
human whole blood, mean concentration 215

D-lactic acid formation
influence of fluoride 214
influence of glycolytic substrates 217
influence of heparin 215
whole blood incubation 216

S-**lactoylglutathione methylglyoxal-lyase**
see glyoxalase I

ligandin
effect of aldehydes 260

lipid peroxidase activity
cell division rate 325

lipid peroxidation
carbon tetrachloride 246
hepatoma 326
membrane fluidity 316
NADPH-flavoprotein reductase 308
tumour microsomal fraction 306

liposomes
cancer chemotherapy 300

lymphosarcoma
methylglyoxal, continuous infusion 291

lysine side-chain
p-type conductor 11

lysozyme–methylglyoxal complex
Cole-Cole plot 92
Poole-Frenkel plot 88
Schottky plot 88

macrophage
free-radical release 125

malealdehyde
protein synthesis 250

manganous ion
e.s.r. spectrum 121

methylamine
methylglyoxal interaction *see methylglyoxal–methylamine interaction*
Schiff base formation 10

methylglyoxal
aqueous solution 283
ascorbic acid carrier 165
amino acid incorporation 252
biological activity 245
bromosulphophthalein binding 261
carcinostatic activity, mouse 271, 285
charge transfer, formamide 150
charge-transfer interaction 331
computer-optimized geometry 150
concentration in tissues 49
continuous intravenous infusion 290
DNA binding 208

methylglyoxal,
continued
Ehrlich carcinoma 293
electron acceptor 9
formamide, 'super-molecule' configuration 151
formation and catabolism 172, 192, 218
formation in micro-organisms 172
Gardner lymphosarcoma 293
glycolysis intermediate 218, 222
human plasma, concentration 220
intraperitoneal injection 285
keto and enol form, stability 196
lymphosarcoma 291
mitotic index, hepatoma cell 256
reaction with polylysine 38, 45
reaction with proteins 11
squamous carcinoma 291
toxicity 290
triplet state 198, 205
tumour growth inhibition 272
methylglyoxal–methylamine interaction
195
hydrogen bonding 198
solvent proton transfer 201
stacking interaction 199
vacuum 198
methylglyoxal–*N*-acetyl-L-cysteine
preparation 283
methylglyoxal production
human blood 211
molecules
electronic desaturation 5

NADPH–cytochrome *P*-450 chain
components 305
inhibitors 308
rat tissue distribution 305
NADPH–flavoprotein reductase
lipid peroxidation 308
non-crystalline solid
energy band and density-of-states scheme 68
nuclear magnetic resonance scanning
large sample scan 132
tumour detection 136, 141
whole body scan 131
4-oxopent-2-enal
DNA synthesis inhibition 170
reaction with ascorbic acid 167

partial charge transfer
oscillating electron 6
periodic table
electron donors and acceptors 7
permittivity
charge transfer 6, 7, 87
perylene–chloranil complex
charge-transfer interaction 84
conductivity data 87
poly(amino acids)
carcinostatic agents 185
poly(Glu, Trp)
fractionation 182
polylysine
reaction with methylglyoxal 38, 45
polylysine–methylglyoxal complex
^{13}C n.m.r. spectra 39
e.s.r. spectra 40
polypeptide
excitonic effect 54
valence and conduction bands 53
promethazine
trichloromethyl radical interaction 312
protein
charge transfer 33
conduction mechanism 65
conductivity–frequency plot 67
D^+–D hole transition 71
electrical conductivity 4, 35, 44
electron mobility 330
electronic energy states 68
electronic structure 51
energy band 33
hole transition, energy states 74
salt effect 42
solid-state approach 33, 51
protein function
integrated system 4
protein isoelectric point
salt effect 33
protein melting temperature
salt effect 33
protein–methylglyoxal complex
a.c. conductivity 90
bovine serum albumin 84
casein 84
charge-transfer interaction 84
collagen 84
conductivity, standard equation 85
conductivity, temperature variation 86
dielectric data 87
dispersion spread 72
electrical measurements, biological relevance 84
electronic and dielectric properties 83
e.s.r. spectrum 46
hopping electron model 94
lysozyme 85
mean hop distance 95
thiol group 98
protein salt effect
electronegativity difference 34
protein SH group
reaction with 4-hydroxy-pentenal 226, 238
pulse radiolysis
free radical 309

quantum chemistry
ab initio SCF-MO calculation 149, 160

rat liver microsomes
e.s.r. spectrum 121

sample lyophilization
free-radical content 116
manganous ion content 116
sarcoma 180
thymidine incorporation 280
Slater orbital
147
spin-trap technique
free radical 309
squamous carcinoma
methylglyoxal, continuous infusion 291
submolecular biology
electron 1, 340
Szent-Györgyi hypothesis
general discussion 329
thermal copoly(amino acids)
molecular structure 188
thymidine incorporation
aldehyde–RSH adduct 238
glutathione 280, 298
4-hydroxypentenal 237
4-hydroxypentenal–cysteine 278
sarcoma 180 cell 280
$\alpha \leftrightarrow \beta$ transformation
history of life 13
trichloromethyl radical
biomembrane 315
generation from isopropanol radical 321
interaction with promethazine 312, 321
peroxidation 324
tritonic copoly(amino acids)
light absorption 183
tubulin activity
effect of aldehydes 253
tumour cell
antioxidant concentration 125
n.m.r. study 131
tumour growth
cell multiplication 27, 317
tumour microsomal fraction
lipid peroxidative activity 306, 325
tumour tissue
growth versus free-radical content 109
α, β-unsaturated aldehydes
in mammalian tissue 238

Van der Waals force
5

Walker 256 tumour
e.s.r. spectrum 114